Universitext

Universitext

Universitext is a series of textbooks that presents material from a wide variety of mathematical disciplines at master's level and beyond. The books, often well class-tested by their author, may have an informal, personal, even experimental approach to their subject matter. Some of the most successful and established books in the series have evolved through several editions, always following the evolution of teaching curricula, into very polished texts.

Thus as research topics trickle down into graduate-level teaching, first textbooks written for new, cutting-edge courses may make their way into *Universitext*.

More information about this series at http://www.springer.com/series/223

Roger Godement

Analysis III

Analytic and Differential Functions, Manifolds and Riemann Surfaces

Translated by Urmie Ray

 Springer

Roger Godement
Paris, France

Translated by Urmie Ray

Translation from the French language edition: Analyse mathematique III by Roger Godement,
Copyright © Springer-Verlag GmbH Berlin Heidelberg 2002.
Springer-Verlag GmbH Berlin Heidelberg is part of Springer Science+Business Media

ISSN 0172-5939   ISSN 2191-6675   (electronic)
Universitext
ISBN 978-3-319-16052-8   ISBN 978-3-319-16053-5   (eBook)
DOI 10.1007/978-3-319-16053-5

Library of Congress Control Number: 2015935234

Springer Cham Heidelberg New York Dordrecht London
© Springer International Publishing Switzerland 2015

Printed on acid-free paper

Springer International Publishing AG Switzerland is part of Springer Science+Business Media
(www.springer.com)

Table of Contents of Volume III

VIII – Cauchy Theory

§ 1. Integrals of holomorphic functions – § 2. Cauchy's Integral Formulas – § 3. Some Applications of Cauchy's Method

In Chapter VII, § 4 we showed how a significant part of the classical theory of holomorphic or analytic functions on ℂ can be obtained from Fourier series. In fact, our universal method for constructing them – a fundamental idea of Cauchy – is to integrate holomorphic functions along curves drawn in their domains of definition and thereby obtain a version of the "fundamental theorem of differential and integral calculus" (FT) for holomorphic functions, and then to deduce countless consequences.

I will present only very few of them. The general theory of analytic functions is of unlimited scope[1] and the results needed in mathematical fields where holomorphic functions are encountered are, on the other hand, very limited in most cases. For instance, a famous result like Riemann's theorem on conformal mapping of simply connected domains is rarely used, although it is recommended to be familiar with it for the sake of "general knowledge"; as for classifying simply connected Riemann surfaces, which would be far more useful, it would need far too complicated developments. The very basic results and methods that we will present in this chapter are, for example, quite sufficient for the chapter devoted to the theory of Riemann surfaces or for the one on elliptic and modular functions.

---

[1] The two volumes by Reinhold Remmert, *Funktionentheorie* (Springer, 1995, also available in English edition), more than 700 very compact pages, can give some idea of the general theory of analytic functions, but do not cover Riemann surfaces, elliptic and automorphic functions, differential equations in the complex domain, special functions, etc., areas that would require thousands of additional pages and that, at any rate, have been the subject of specialized presentations. Other numerous available presentations include Walter Rudin, *Real and Complex Analysis* (McGraw-Hill, 1966, also available in French), Jean Dieudonné, *Calcul Infinitésimal* (Hermann, 1968), in particular useful for its many exercises, Eberhard Freitag & Rolf Busam, *Funktionentheorie* (Springer-Verlag, 1995), which lists several other books, Serge Lang, *Complex Analysis* (Springer, many editions), John B. Conway, *Functions of One Complex Variable* (2 vol., Springer, 1978–95), Carlos A. Berenstein & Roger Gray, *Complex Variables. An Introduction* (Springer, 1991).

1

Learning to use basic ideas is, therefore, much better than learning a multitude of general theorems, albeit ingenious and deep, unless, of course, the aim is to specialize in general theory.

Cauchy theory (it would be much better to say: Cauchy and Weierstrass) has been and continues to be the subject of many accounts merely differing from each other on detailed points of presentation or style; not seeing the need to reproduce them an umpteenth time, I have tried, whenever possible, not to follow them, in particular regarding homotopy. As will be seen in the next chapter, apart from the residue theorem, another easy consequence, Cauchy's method falls within the much more general framework of multivariate differential forms.

§1. Integrals of Holomorphic Functions

1 – Preliminary Results

(i) *The fundamental theorem (FT) of differential and integral calculus* (Chap. V, §3). In its simplest form, take a continuous function f on an interval $I \subset \mathbb{R}$ and, choosing an $a \in I$, set

$$F(x) = \int_a^x f(t)dt$$

a differentiable function such that $F'(x) = f(x)$ for all $x \in I$ is thus obtained. Conversely, any primitive for f is given up to an additive constant by this formula.

If we start with a *regulated* function[2] f – a less simple case –, the previous formula defines a continuous function F admitting right-hand and left-hand derivatives at every $x \in I$, given by

$$F'_d(x) = \lim_{\substack{h=0 \\ h>0}} \frac{F(x+h) - F(x)}{h} = f(x+0) = \lim_{\substack{h=0 \\ h>0}} f(x+h)$$

and by a similar left-hand formula. In particular, the derivative $F'(x)$ exists outside some countable set D of discontinuous points of f. Conversely, if there is regulated function f and a *continuous* function F in I which, outside some *countable* subset of I, admits a derivative equal to $f(x)$, then, up to a constant, F is again given by the standard formula (Chap. V, §3, n° 13). F is then said to be a *primitive* for f.

For simplicity's sake, we will say that a function F is *of class* $C^{1/2}$ in I if it is a primitive for a regulated function which we will always write F'; it exists except possibly for a countable number of values of the variable: an unimportant ambiguity that can be removed by setting $F'(x) = F'_d(x)$ for all x. For this it is not sufficient that F be differentiable outside some countable set. We adopt the notation $C^{1/2}$ because C^0 means that F is continuous, a less restrictive condition, whereas C^1 means that F' exists everywhere and is continuous, a more restrictive one.

---

[2] Recall that a function f defined on an interval $I \subset \mathbb{R}$ is said to be regulated if it satisfies the following three equivalent conditions: (a) it has both right and left limits at all points of I, ; (b) for any compact interval $K \subset I$ and $r > 0$, K can be partitioned into intervals on which f is constant up to r; (c) there exists a sequence of step functions converging uniformly to f on every compact set $K \subset I$ (hence on I if I is compact). Chap. V, n° 7, Theorem 6. The Sum, product and quotient of two regulated functions are also of the same type. If f and g are of class $C^{1/2}$, the product fg is a continuous function which admits a derivative outside some countable set, the regulated function $f'(t)g(t) + f(t)g'(t)$; fg is, therefore, a primitive for $f'g + fg'$. This allows us to apply the integration by parts formula to functions of class $C^{1/2}$ defined later.

(ii) *Differential calculus in* $\mathbb{R}^2$. Let U be an open subset of $\mathbb{R}^2$ and f a map from U to $\mathbb{R}$ or $\mathbb{R}^2$. It is said to be differentiable at a point $c \in U$ if, $h \in \mathbb{R}^2$ being a varying vector, $f(c + h) - f(c)$ is "approximately linear" in h for sufficiently small h; to be precise, there needs to be a linear map from $\mathbb{R}^2$ to $\mathbb{R}$ or $\mathbb{R}^2$, the tangent map to f (or derivative, or differential of f) of f at c, written $f'(c)$, such that

$$f(c + h) = f(c) + f'(c)h + o(h)$$

as the length $|h|$ of the vector h tends to 0; hence

(1.1) $$f'(c)h = \frac{d}{dt} f(c + th) \qquad \text{for } t = 0 .$$

If $c = (a, b)$, then

$$f(a + u, b + v) = f(a, b) + pu + qv + o(|u| + |v|) ,$$

where the coefficients p, q, elements of $\mathbb{R}$ or $\mathbb{R}^2$ as the case may be, do not depend on u, v. These are the partial derivatives[3]

$$p = D_1 f(c) = \lim_{u=0} \frac{f(a + u, b) - f(a, b)}{u} ,$$

$$q = D_2 f(c) = \lim_{v=0} \frac{f(a, b + v) - f(a, b)}{v}$$

of f at c. Hence, if $h = (u, v) \in \mathbb{R}^2$, then

(1.2) $$f'(c)h = D_1 f(c)u + D_2 f(c)v .$$

In the case of a function with values in $\mathbb{R}^2$, if we set $f(x, y) = (f_1(x, y), f_2(x, y))$, then for $c = (a, b)$, the function $f'(c)$, therefore, maps $h = (u, v)$ to the vector

(1.3) $$f'(c)h = D_1 f(c)u + D_2 f(c)v =$$
$$= (D_1 f_1(c), D_1 f_2(c)) u + (D_2 f_1(c), D_2 f_2(c)) v =$$
$$= (D_1 f_1(c)u + D_2 f_1(c)v, D_1 f_2(c)u + D_2 f_2(c)v) .$$

Conversely, if the partial derivatives exist for all $c \in U$ and are continuous on U, in which case f is said to be of class C^1 in U, then f is differentiable at all points of U. If $D_1 f$ and $D_2 f$ are also of class C^1, f is said to be of class C^2, and so on. We then have

(1.4) $$D_1 D_2 f = D_2 D_1 f .$$

---

[3] A notation such as $D_1 f(c)$ will always denote the value of the function $D_1 f$ at c.

We often write

$$df(c; h) = f'(c)h$$

instead of $f'(c)h$; for $h = (u, v)$, $df(c; h) = u$ clearly holds if f is the coordinate function $(x, y) \mapsto x$, $df(c; h) = v$ if f is $(x, y) \mapsto y$ and $df(c; h) = h$ if f is[4] the identity map $z = (x, y) \mapsto (x, y)$. We can, therefore, write

$$df[c; dz(x; h)] = D_1 f(c) dx(c; h) + D_2 f(c) dy(c; h)$$

or

(1.5) $$df(c; dz) = D_1 f(c) dx + D_2 f(c) dy$$

for short.

We will also need the chain rule in two cases.

(a) Suppose that f is of class C^1 on U and let $\mu : I \longrightarrow U$ be a function defined on an interval I of $\mathbb{R}$, whence a composite function $p = f \circ \mu : t \mapsto f[\mu(t)]$. The same holds for the function p at each point t at which μ is differentiable and

(1.6) $$p'(t) = f'[\mu(t)] \mu'(t)$$

is the image of the vector[5] $\mu'(t) \in \mathbb{R}^2$ under the linear map tangent to f at $\mu(t)$. If f and μ are of class C^1, so is p; if f is of class C^1 and μ of class $C^{1/2}$, the obviously continuous function p is of class $C^{1/2}$ since $p'(t)$ exists outside some countable set and as the product of the continuous function $f'[\mu(t)]$ and the regulated function $\mu'(t)$ is regulated. The function p is, therefore, a primitive for $f'[\mu(t)] \mu'(t)$.

(b) If g is a map from an open set $V \subset \mathbb{R}^2$ to U, whence again a composite map $p = f \circ g : V \longrightarrow \mathbb{R}^2$, then, at each point $c \in V$ where g is differentiable, so is p, and

(1.7) $$p'(c) = f'[g(c)] \circ g'(c)$$

is the composite or product of linear maps tangent to g at c and to f at $g(c)$.

---

[4] Here the letter z represents the point with x, y coordinates in $\mathbb{R}^2$ rather than the complex number $x + iy$. It is in the theory of holomorphic functions that it is essential to regard points in the plane as complex numbers. Having said that, using the letter z to represent a point in $\mathbb{R}^2$ or in any other set is not forbidden.

[5] If $\mu(t) = (\mu_1(t), \mu_2(t))$, $\mu'(t)$ is the vector $(\mu_1'(t), \mu_2'(t))$. If we do not distinguish between a point or a vector $(u, v) \in \mathbb{R}^2$ and the complex number $u + iv \in \mathbb{C}$, the complex number $\mu'(t)$ becomes the usual derivative of the complex valued function $\mu(t)$. However, this interpretation is not generally compatible with formula (6), since in the latter $f'[\mu(t)]$ is a linear map from $\mathbb{R}^2$ to $\mathbb{R}^2$ and not a mere complex number. It is only if f is holomorphic that that three derivatives occurring in (6) can be interpreted as complex numbers. The possibility of interpreting elements of $\mathbb{R}^2$ in these two different ways often leads to confusion that, for good reason, does not occur in $\mathbb{R}^n$, when $n \geq 3$.

This result can easily be recovered by writing that

$$p(c+h) = f[g(c+h)] \sim f[g(c) + g'(c)h]$$
$$\sim f[g(c)] + f'[g(c)]g'(c)h = p(c) + f'[g(c)]g'(c)h,$$

but this is not a *proof*. The previous formula can also be written as

(1.8) $$dp(z; dz) = df[g(z); dg(z; dz)] :$$

in the differential $df(z; dz)$ of f, z and dz are replaced by $g(z)$ and by the differential of g at z, as was already known by Leibniz.

When the points of $\mathbb{R}^2$ are identified with complex numbers, any function $f(x,y) = (f_1(x,y), f_2(x,y))$ with values in $\mathbb{R}^2$ is identified with the complex valued function

$$z \longmapsto f_1(z) + if_2(z),$$

the partial derivatives being then identified with the functions

$$D_1 f = D_1 f_1 + iD_1 f_2, \quad D_2 f = D_2 f_1 + iD_2 f_2.$$

In case (a), the composite function $p(t) = f_1[\mu(t)] + if_2[\mu(t)]$ is complex valued; setting $\mu(t) = \mu_1(t) + i\mu_2(t)$ and $D = d/dt$,

$$p'(t) = D_1 f_1[\mu(t)] D\mu_1(t) + D_2 f_1[\mu(t)] D\mu_2(t) +$$
$$+ i\{D_1 f_2[\mu(t)] D\mu_1(t) + D_2 f_2[\mu(t)] D\mu_2(t)\} =$$
$$= \{D_1 f_1[\mu(t)] + iD_1 f_2[\mu(t)]\} D\mu_1(t) +$$
$$+ \{D_2 f_1[\mu(t)] + iD_2 f_2[\mu(t)]\} D\mu_2(t);$$

the same formula

$$p'(t) = D_1 f[\mu(t)] D\mu_1(t) + D_2 f[\mu(t)] D\mu_2(t)$$

is, therefore, recovered, but this time, the derivatives are the usual complex valued derivatives of complex valued functions. In case (b), it is necessary to assume that two holomorphic functions f are g are being composed to obtain a simple formula; see below.

(iii) *Holomorphic functions.* Let f be a complex valued function defined in the open subset U of $\mathbb{C}$ and suppose that as a map from U to $\mathbb{R}^2$ it is differentiable at $c \in U$. It, therefore, has a derivative $f'(c) : \mathbb{R}^2 \longrightarrow \mathbb{R}^2$ which is linear over the field $\mathbb{R}$. It may be $\mathbb{C}$-linear, i.e. of the form $h \mapsto ah$, where $a \in \mathbb{C}$ is a constant (namely the value of the map for $h = 1$); this means that then

$$f(c+h) = f(c) + ah + o(h)$$

as $|h|$ tends to 0, a relation where, this time, c, h, a, $f(c)$, etc. are complex numbers. The number a, which characterizes $f'(c)$, is then given by the relation

$$(1.9) \qquad a = \lim_{h=0} \frac{f(c+h) - f(c)}{h}$$

where h is made to approach 0 through non-zero *complex* numbers. f is then said to be differentiable in the complex sense at the point c of U and we write $a = f'(c)$. Hence the notation $f'(c)$ represents both a complex number and a linear map from $\mathbb{R}^2$ to $\mathbb{R}^2$; this apparent ambiguity is due to the fact that, in $\mathbb{R}^2$, maps of the form $h \mapsto ah$, where $a \in \mathbb{C}$ is a constant, are just $\mathbb{C}$-linear maps; it is, therefore, natural to make no distinction between such a map and the coefficient a which it is determined by. This generalizes to maps from a field K into itself that are linear over K : these are the schoolboys' functions $x \mapsto ax$.

The function f is said to be *holomorphic* on U if the limit $f'(z)$ exists for all $z \in U$ and is a continuous function of z (Chap. II, § 3, n° 19); or equivalently if f is C^1 as a function of (x, y) and satisfies Cauchy's relation

$$(1.10) \qquad D_1 f = -i D_2 f \quad (= f')$$

(Chap. III, § 5, n° 20), which conveys exactly the $\mathbb{C}$-linearity of the differential (2).

There is a chain formula for holomorphic functions. It is formally identical to the one in the theory of functions of a real variable and is used in two cases.

(a) Consider first an interval $I \subset \mathbb{R}$, an open set $U \subset \mathbb{C}$, a map $\mu : I \longrightarrow U$ and a function f defined and *holomorphic* on U, whence a composite map $p : t \mapsto f[\mu(t)]$ from I to U. If μ is differentiable at a point t in I, so is p and

$$(1.11) \qquad p'(t) = f'[\mu(t)]\, \mu'(t) ,$$

where f' denotes the function defined by the limit (9). This result generalizes immediately to the case of a composite function of the form $p = f \circ \mu$ where μ is a function of several real variables $s_1, \ldots, s_p$: denoting by D_i the partial differential operator with respect to s_i,

$$(1.11') \qquad D_i p(s_1, \ldots, s_p) = f'[\mu(s_1, \ldots, s_p)]\, D_i \mu(s_1, \ldots, s_p) ,$$

because in order to differentiate with respect to s_i, the other variables are kept fixed, which reduces to (11).

(b) If I is now replaced by an open subset V of $\mathbb{C}$ and μ by a function $g : V \longrightarrow U$ *holomorphic* on V, the composite map p from V to $\mathbb{C}$ is also holomorphic and

$$(1.12) \qquad p'(z) = f'[g(z)]\, g'(z)$$

for all $z \in V$.

Indeed, by (7), formula (12) is exact if the derivatives f', g' and p' are interpreted as linear maps from $\mathbb{R}^2$ to $\mathbb{R}^2$ and the right hand side as the composition of functions $f'\,[g(z)]$ and $g'(z)$. By assumption, these maps are $\mathbb{C}$-linear. But the composition of two $\mathbb{C}$-linear maps $h \mapsto ah$ and $h \mapsto bh$ on $\mathbb{C}$ gives a map $h \mapsto abh$; formula (12) can, therefore, be obtained by substituting the corresponding complex numbers to the maps $p'(z)$, etc..

As shown in Chap. VII, saying that a function f is holomorphic on an open set G amounts to saying that it is *analytic* on G, i.e. that, for each $a \in G$, it has a power series expansion $f(z) = \sum_{n \geq 0} c_n\,(z - a)^n$ which converges and represents it on a disc centered at a and in fact on the largest disc centered at a contained in G; the power series which represents f in a neighbourhood of a is just its Taylor series

$$f(z) = \sum_{n \geq 0} f^{(n)}(a)(z - a)^{[n]},$$

where, we remind the reader that $z^{[n]} = z^n/n!$. The terms "holomorphic" and "analytic" are, therefore, synonymous.

Nonetheless, all the results that we will prove in this chapter are based only on the initial definition of holomorphic functions, in other words, do not use their analyticity, a result that we will recover by Cauchy's traditional method. Hence, we will maintain the strict distinction between "holomorphic" functions and "analytic" functions until we again prove the equivalence of these two notions.

2 – The Problem of Primitives

(i) *Local primitives of a holomorphic function.* One of the basic problems in the theory of holomorphic functions is to find a function f holomorphic on an open subset U of $\mathbb{C}$ a *primitive* of f on U, i.e. a holomorphic function F such that $F' = f$. If U is a disc centered at a, the problem always has a solution since a power series can be differentiated term by term (Chap. II, n° 19):

$$(2.1)\; f(z) = \sum_{n \geq 0} c_n(z - a)^n \iff F(z) = c + \sum_{n \geq 0} c_n(z - a)^{n+1}/(n + 1),$$

where c is an arbitrary constant. A proof which does not use analyticity and which generalizes to differential forms consists in observing that if f is holomorphic on the disc $D : |z| < R$ and if $F' = f$, then, by (11), the derivative of the function $t \mapsto F(tz)$, defined at least on $[0, 1]$ for a given $z \in D$, is $F'(tz)z = f(tz)z$; The FT then shows that, when $F(0) = 0$,

$$(2.2) \qquad\qquad F(z) = \int_0^1 f(tz)z\,dt$$

for all $z \in D$.

Conversely, if F is defined on D using this formula, then F is holomorphic and satisfies $F' = f$. To see this, observe that the function of (t, x, y) under the sign $\int$ is C^1. This allows us to differentiate under the sign $\int$ with respect to x or y; setting $D_1 = d/dx$ and $D = d/dt$, omitting the limits of integration and noting that $D_1 z = 1$, we get

$$D_1 F(z) = \int D_1 \left[f(tx, ty)z \right] dt = \int \left[D_1 f(tz).tz + f(tz) \right] dt \, ;$$

since f is holomorphic, $D_1 f = f'$ and $f'(tz)z$ is the derivative of $f(tz)$ with respect to t; hence, integrating by parts, it follows that

$$D_1 F(z) = \int f(tz) dt + \int t.D \left[f(tz) \right] dt = \int f(tz) dt + tf(tz) \Big|_0^1 - \int f(tz) dt \, ,$$

and so $D_1 F = f$. If D_1 is now replaced by $D_2 = d/dy$, the calculation remains the same except that $D_2 f = if'$. Then $D_2 F = if$; as a result, F satisfies Cauchy's condition and $F'(z) = D_1 F(z) = f(z)$, q.e.d.

This method applies more generally to any star G, i.e. which does not have any point a such that, for all $z \in G$, the line segment $[a, z]$ is contained in G: replace tz by $a + t(z - a)$ in (2). It is for example the case of an open convex subset of $\mathbb{C} - \mathbb{R}_+$ by choosing a on the negative real axis, etc. We will, however, find further down a less restrictive result regarding G.

Let us return to the general case. The results obtained above mean that every holomorphic function f on an open set G has a primitive *in the neighbourhood* of each point of G; but, as already seen (Chapter IV, § 4) for $1/z$ and its pseudo-primitive $\mathscr{L}og\, z$, this local result in no way implies the existence of a global primitive, i.e. valid on all of G; we will return to this point later.

(ii) *Integration along a path. Admissible paths.* Let f be a function defined and holomorphic on an open connected subset G of $\mathbb{C}$, i.e. a *domain*, and suppose that f has a primitive F in G. If we were on $\mathbb{R}$, the FT

$$(2.3) \qquad F(z) - F(a) = \int_a^z f(\zeta) d\zeta \iff F'(z) = f(z)$$

where, despite the notation, z and the integration variable ζ are reals, would allow us to calculate F up to an additive constant. But, at first sight, integrating from a point $a \in G$ to another point $z \in G$ is not well-defined on $\mathbb{C}$.

Nevertheless, we know – this is (1.11) – that if $\mu : I \longrightarrow G$ is a map from an interval $I \subset \mathbb{R}$ to G, i.e. a *path*[6] in G, then

$$(2.4) \qquad \frac{d}{dt} F\left[\mu(t)\right] = F'\left[\mu(t)\right] \mu'(t) = f\left[\mu(t)\right] \mu'(t)$$

at all points where the derivative $\mu'(t)$ exists. If μ is of class C^1, if $I = [u,v]$ and if $\mu(u) = a$, $\mu(v) = z$, the simplest version of FT, therefore, shows that

$$(2.5) \qquad F(z) - F(a) = \int_I f\left[\mu(t)\right] \mu'(t)dt$$

since the right hand side of (4) is a continuous function of t; formula (2) is obtained for $\mu(t) = tz$. This result still holds if μ is of class $C^{1/2}$: formula (4) holds at all points where μ has a derivative, hence outside some countable subset of I, and the function $f\left[\mu(t)\right] \mu'(t)$ is regulated; the continuous function $F\left[\mu(t)\right]$ is, therefore, a primitive for the latter, and so (5) follows. A path of class $C^{1/2}$ will also be said to be *admissible*.

If we set $\zeta = \mu(t)$, *à la* Leibniz, then $d\zeta = \mu'(t)dt$, so that on the right hand side of (5) we integrate the expression $f(\zeta)d\zeta$; this leads to define the *integral of f along a path μ* by

$$(2.6) \qquad \int_\mu f(\zeta)d\zeta = \int_I f\left[\mu(t)\right] \mu'(t)dt$$

in the same way as in Chap. V, eq. (5.16) for Cauchy's formula for a circle. The notation introduced on the left hand side of (6) could be justified by observing that choosing a subdivision of $I = [u,v]$ by the points $u = t_0 < t_1 < \ldots < t_n = v$ and setting $\zeta_i = \mu(t_i)$, integral (5) is approximately equal to

$$\sum f\left(\zeta_i\right) \mu'\left(t_i\right) \left(t_{i+1} - t_i\right)$$

and hence, by the mean value formula, no less approximately, to $\sum f\left(\zeta_i\right) \left(\zeta_{i+1} - \zeta_i\right)$; whence notation (6). The reader will easily be able to add the ε necessary to correct this simplified argument by subdividing I so that the functions considered are piecewise constant up to ε; see point (iv) further down.

Note that integral (6) does not depend exclusively on the "curve" $\mu(I)$ defined by $\mu(t)$ as t varies in I; indeed, the latter does not change if the map

---

[6] Everyone uses the letter γ to denote a path. I will use the letter μ because (i) computer keyboard "designers" have had the good idea to include one and only one Greek letter, namely μ, (ii), more seriously, as we will see a bit further down, the function $\mu(t)$ occurs by way of the Radon (or Stieltjes) measure $d\mu(t) = \mu'(t)dt$ which it defines.

μ is replaced by $\nu(t) = \mu\left(\varphi(t)\right)$, where φ is a *surjective* map from an interval J to I; supposing, for simplicity's sake, that φ is of class C^1,

$$\int_\nu f(z)dz = \int_J f\left[\mu\left(\varphi(t)\right)\right]\mu'\left[\varphi(t)\right]\varphi'(t)dt$$

by the chain rule; hence, setting $F(t) = f\left[\mu(t)\right]\mu'(t)$,

$$\int_\mu f(z)dz = \int_I F(t)dt, \quad \int_\nu f(z)dz = \int_J F\left[\varphi'(t)\right]\varphi'(t)dt,$$

As $I = \varphi(J)$, the equality of these two integrals seems to follow from the change of variables formula for integrals (Chapter V, § 6, n° 19). But the latter concerns *oriented* integrals. Hence, the equality

$$\int_\mu f(z)dz = \int_\nu f(z)dz$$

supposes that φ maps the initial (resp. terminal) point of J onto the initial (resp. terminal) point of I. Otherwise, the previous relation only holds up to sign. In practice, only strictly increasing "changes of parameter" φ are considered. This avoids difficulties and reduces the question to paths for which $I = [0,1]$. From now on, we will suppose this to be the case, unless stated otherwise.

Admissible paths or those of class $C^{1/2}$ cover all cases that might arise. In practice, I can almost always be divided into intervals on which the function μ is C^1, if not linear. But, once the notion of a primitive for a regulated function has been understood, using these "piecewise" paths of class C^1 (or linear), as they are called, is not any easier than using paths of class $C^{1/2}$. Applying Theorem 12 bis of Chap. V, n° 13, we see that if a path is considered to be the trajectory of a moving object, then admissible paths can be characterized by imposing the following conditions upon them:

(a) the map μ is continuous,
(b) it has a right and left derivative at every point t,
(c) these are equal outside some countable subset D of I (for example, it may be that $D = I \cap \mathbb{Q}$, but it is better to avoid this kind of paths in practical calculations...),
(d) The right (or left) derivative is a *regulated* function of t, i.e. has right-hand and left-hand limits for all t or, equivalently, is the uniform limit on I of step functions.

The trajectory defined by $\mu(t)$, possibly passing over the same point several times, therefore, admits a "velocity vector" $\mu'(t)$ outside D as it can change direction ("angular points") at points of D. It has a tangent at each point where $\mu'(t)$ exists and is non-zero. As any driver attempting to park between two cars knows, the case $\mu'(t) = 0$ can entail a cusp point.[7]

---

[7] Example: $t \mapsto (t^2, t^3)$ at $t = 0$, with $I = [-1, 1]$.

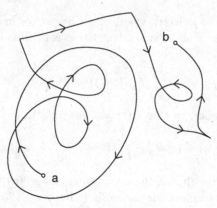

Fig. 2.1.

(iii) *Integral along a path as a Stieltjes integral.* It is sometimes convenient to interpret integral (6) as a Stieltjes integral (Chap. V, § 9, n° 32) with respect to a Radon or a Stieltjes complex measure defined on I by the function $\mu(t)$. In Chap. V, we only defined Stieltjes integrals over an interval I of $\mathbb{R}$ with respect to real increasing functions in order to obtain positive measures, but this method can easily be generalized to linear combinations with complex coefficients of increasing functions.[8] This is the case of every function $C^{1/2}$ μ since the standard formula $\mu' = \mathrm{Re}(\mu')^+ - \mathrm{Re}(\mu')^- + \ldots$, transforms μ into a linear combination of increasing functions as they are primitives of positive functions. *Complex* Radon measures are thus obtained on I in the sense of Chap. V, § 9, i.e. continuous linear functionals on the space $C^0(I)$ equipped with the norm of uniform convergence, at least if I is compact, which is the only case that interests us here.

As all $C^{1/2}$ functions are continuous, formula (32.1) of Chap. V, § 9 defining the measure of an interval $J = (u, v) \subset I$ with respect to μ becomes $\mu(J) = \mu(v) - \mu(u)$ regardless of the nature of J. Then the integral $\int f(t)d\mu(t)$ of a continuous or more generally of a regulated function f can be defined as the usual Riemann integral: the sum $\sum f(t_p)\,\mu(J_p)$ is associated to every finite partition $I = J_1 \cup \ldots \cup J_n$ of I into intervals, where $t_p \in J_p$; the integral $\int f(t)d\mu(t)$ is the limit of these sums when the partition considered becomes finer. If the J_p are chosen so that f is constant up to r on each J_p (characterization of regulated functions), then

---

[8] The classical terminology is functions *of bounded variation.* They are directly characterized as follows: there exists a positive finite constant M such that

$$\sum |\mu(t_{i+1}) - \mu(t_i)| \leq M$$

for all points $t_1 < t_2 < \ldots < t_n$ of the interval considered. See for example Rudin, Chap. 6.

$$(2.7) \qquad \left| \int f(t) d\mu(t) - \sum f(t_p) \mu(J_p) \right| \leq r \|\mu\|$$

since for all t, $f(t)$ is, up to r, equal to the value at t of the step function equal to $f(t_p)$ on each J_p. Recall that the notation $\|\mu\|$ for the *norm* or *total mass* of the measure μ is the smallest positive number such that

$$(2.8) \qquad \left| \int f(t) d\mu(t) \right| \leq \|\mu\| \cdot \|f\|_I$$

for all continuous and in fact regulated functions. If the measure μ is positive, i.e. if the function $\mu(t)$ is real and increasing, then $\|\mu\| = \mu(I)$. In fact, the main point of (8) is not the exact value of $\|\mu\|$; any constant independent of f will do. But see n° 4, (ii).

To show that the curvilinear integral (6) is also a Stieltjes integral, first recall that in Chap. V we obtained a formula (32.15) which says that, for any real, increasing function $\mu(t)$ of class C^1 on I and for any continuous function f on I,

$$(2.9) \qquad \int f(t) d\mu(t) = \int f(t) \mu'(t) dt \, ;$$

it trivially generalizes to the case when μ is complex valued. In fact, formula (9) still holds when $\mu(t)$ is $C^{1/2}$. To see this, suppose that μ' is positive, i.e. that μ is increasing. As μ is a primitive for μ', first of all

$$\mu(J) = \mu(v) - \mu(u) = \int_J \mu'(t) dt$$

for any interval $J = (u, v) \subset I$. Using as above a sufficiently fine partition of I, the regulated function μ' may be assumed to be constant up to r on each J_p. Hence, for all $t_p \in J_p$,

$$(2.11) \qquad |\mu(J_p) - \mu'(t_p) m(J_p)| \leq m(J_p) r \, ,$$

where m is the usual Lebesgue measure. Replacing each term $\mu(J_p)$ by $\mu'(t_p) m(J_p)$ in the Riemann sum $\sum f(t_p) \mu(J_p)$, the error made is less than $\|f\|_I \sum m(J_p) r = \|f\|_I m(I) r$. So

$$\left| \int f(t) d\mu(t) - \sum f(t_p) \mu'(t_p) m(J_p) \right| \leq \mu(I) r + \|f\|_I m(I) r \, ,$$

qed.

In all cases, (6) can, therefore, be written as

$$(2.10) \qquad \int_\mu f(\zeta) d\zeta = \int_I f[\mu(t)] \, d\mu(t)$$

in line with Leibniz's ideas.

(iv) *A necessary and sufficient condition for a primitive.* Let us return to a holomorphic function f on a domain G of $\mathbb{C}$. If it admits a primitive F on G if a point a of G is chosen, then, as seen at the start of this n°,

$$(2.12) \qquad F(z) - F(a) = \int_\mu f(\zeta)d\zeta$$

for any admissible path μ connecting a to z in G. Hence, the integral of f along such a path depends only on its endpoints.

In the general case, it may be tempting to construct a primitive by applying the previous formula and choosing its value at a arbitrarily, for example $F(a) = 0$. But this definition of F is totally ambiguous: the value of the integral can very well depend on the choice of the path μ connecting a to z in G as the case[9] of the function $1/z$ already shows. Hence, the notation $F(z)$ used is, *a priori*, not well-defined; the only reasonable notation is to set

$$(2.13) \qquad F(\mu) = \int_\mu f(\zeta)d\zeta$$

for every integration path[10] μ. As in the case of the logarithm of a complex number $z \neq 0$ (Chap. IV, §4 or Chap. VII, n° 16), we get a *set* $\mathcal{F}(z)$ of possible values of the function sought, namely all the numbers obtained by integrating f along a path μ connecting a to z in G or, in the case of the logarithm, all the numbers obtained by using a uniform branch of $\mathcal{L}og$ along a path connecting a fixed point a to the point z (as we will see, this amounts to integrating $1/\zeta$ along this path). But as in the case of the logarithm, the problem is to construct a truly holomorphic function F, and in particular continuous, such that $F(z) \in \mathcal{F}(z)$ for all $z \in G$. In the case of the logarithm, we have seen this to to be possible if and only if the following condition is satisfied: if, for some given $a \in G$ and a varying path $\mu : I \longrightarrow G$ with initial

---

[9] If it was independent of the path in this case, the integral of $1/\zeta$ along the path $t \mapsto \exp(2\pi i t)$ connecting the point $a = 1$ to the point $z = 1$ would be equal to that obtained by integrating along the "constant" path $t \mapsto 1$, i.e. to 0. However, the integral over the circle is obtained by integrating the function $2\pi i$ over $[0,1]$ and hence is not zero. Integrals in $1/z$ will be discussed in detail later.

[10] Mathematicians who invented the "calculus of variations" almost three centuries ago already had the idea of considering functions of a curve varying in the plane, on a surface or in space, a curve along which a given function is integrated; a century ago, the mathematician Vito Volterra used to call these *line functions*. For a "smooth" surface S in $\mathbb{R}^3$, we can, for example, try to find the curves of minimal length drawn on S connecting two given points: the geodesics; the length of a curve μ is given by (4.8) and by comparing it to a curve "infinitely near" to it we get a differential equation characterizing the geodesics. Quite an old problem in mechanics consists in find a curve connecting A to B for two given points A and B such that the time taken to go from A to B by an object moving along it under the action of gravity is minimum. Fermat already knew that the trajectory of a light beam going from point A to a point B through a medium whose retractive index varies is the one that minimizes travel time. Etc.

point a in G, we consider the uniform branch $t \mapsto L(t)$ from $\mathcal{L}og\,z$ along μ which takes a given value in the set $\mathcal{L}og\,a$ at $t = 0$, then the value of this branch should only depend on the endpoint $z = \mu(1)$ of μ.

As we will see, the problem of primitives has a similar solution. First, if f has a primitive F on G, integral (13) only depends on z. Conversely, suppose that for fixed a, the value of integral (13) is, for all z, independent of the path μ; we can then talk unequivocally of the function $F(z)$. The function F is then a global primitive for f.

Indeed, consider an arbitrary point $b \in G$ and let $D \subset G$ be an open disc centered at b. Then, formula (2) adapted to the point b gives a primitive F_D of f on D, and $F_D(z) - F_D(b) = \int f(\zeta)d\zeta$ where integration is along the line segment $[b, z]$. Since a constant can be added to F_D, we may assume that $F_D(b) = F(b)$. To calculate $F(z)$ at a point $z \in D$, f must be integrated along an arbitrary path connecting a to z in G; for example, we can choose a path connecting a to b in G, then a path from b to z in D; by definition, integration along the arc connecting a to b gives $F(b)$, and, as seen above, the arc connecting b to z, for example the radius, gives $F_D(z) - F_D(b) = F_D(z) - F(b)$; adding, we find $F(z) = F_D(z)$ on D. It follows that F is holomorphic and satisfies $F' = f$ on D, and hence globally on G since b is arbitrary. As a result:

Theorem 1. *A holomorphic function f on a domain G has primitive on G if and only if its integral along any admissible path in G depends only on the latter's endpoints.*

In particular, the integral of f along a *closed* path, i.e. such that $\mu(0) = \mu(1)$, is zero. In fact this condition is sufficient for ensuring the existence of a primitive. Indeed, if

$$\mu_1, \mu_2 : [0, 1] \longrightarrow G$$

are two paths connecting a given point a to the same point z, we get a closed path $[0, 1] \longrightarrow G$ by following first the path $[0, 1/2] \longrightarrow G$ given by $t \mapsto \mu_1(2t)$, then the path: $[1/2, 1] \longrightarrow G$ given by $t \mapsto \mu_2(2 - 2t)$; Clearly, the integral of f along the first path is equal to the integral along μ_1 and the integral along the second path is the opposite of the integral along the first one. The integral along the total path:[11] $[0, 1] \longrightarrow G$ is, therefore, the difference between the integrals along μ_1 and μ_2. As a result, these are equal for all μ_1 and μ_2. Hence the result follows from Theorem 1: *A holomorphic function f on a domain G has primitive on G if and only if its integral along any admissible closed path in G is zero.*

For another proof of theorem 1, take an open disc $D \subset G$ centered at z. To go from a to a point $z + h \in D$, we can follow a path connecting a à z and then the radius $[z, z + h]$, i.e. the path $t \mapsto z + th$; then $F(z + h) - F(z)$ is

---

[11] which may not be C^1 even if that is the case of μ_1 and μ_2. Hence it is necessary to include paths that are... piecewise admissible or at least C^1.

clearly the integral along this radius. Hence $F(z+h) - F(z) = \int f(z+th)hdt$, where, as usual, integration is over $[0,1]$. It follows that

$$F(z+h) - F(z) - f(z)h = \int [f(z+th) - f(z)]\, hdt;$$

f being continuous, $|f(z+th) - f(z)| \leq r$ for all $t \in I$ provided $|h| \leq r'$ (uniform continuity on a compact set); so

$$F(z+h) = F(z) + f(z)h + o(h)$$

as h approaches 0, which proves the existence of $F'(z) = f(z)$, qed.

(v) *The case of a contractible domain.* A naive attempt at constructing a primitive without using theorem 1 would be to arbitrarily choose for every $z \in G$ a path μ_z connecting a to z and to set $F(z) = F(\mu_z)$. Albeit strange, this leads to the result *provided* μ_z depends on z in not too... arbitrary a manner. This, we will see, implies a drastic restriction on G. Formula (2) used in the case of a star domain clearly falls within this framework, but is based on an all too providential choice of μ_z.

So let us assign to every $z \in G$ a path

$$\mu_z : t \in [0,1] \longrightarrow \mu_z(t)$$

connecting a to $z = x + iy = (x,y)$ in G and set

$$H(z,t) = \mu_z(t).$$

So $H(z,0) = a$, $H(z,1) = z$ for all $z \in G$. To show that the function

$$(2.14) \quad F(z) = F(\mu_z) = \int_{\mu_z} f(\zeta)d\zeta = \int_0^1 f\left[H(z,t)\right] DH(z,t).dt,$$

where $D = d/dt$, is a primitive for f, it would suffice to show that, with respect to x and y, it has partial derivatives D_1F and D_2F equal to f and if respectively. For this, let us *assume* that differentiation is possible under the $\int$ sign without any difficulty– this would be miraculous if μ_z was arbitrarily chosen – and calculate as Euler or Cauchy would have done; the calculation is similar to the one done for function (2) – only slightly harder. Using the product and chain rules for differentiation and the relations $DD_1 = D_1D$, $D_1f = f'$, the FT gives[12]

---

[12] In a notation such as $DH(z,t).D_1H(z,t)$, the point means that the operator D is applied to $H(z,t)$ and not to the product $H(z,t)D_1H(z,t)$.

$$(2.15) \quad D_1 F(z) = \int D_1 \{f [H(z,t)] \, DH(z,t)\} \, dt =$$

$$= \int \{f' [H(z,t)] \, D_1 H(z,t).DH(z,t) +$$

$$+ f [H(z,t)] \, D_1 DH(z,t)\} \, dt =$$

$$= \int \{f' [H(z,t)] \, DH(z,t).D_1 H(z,t) +$$

$$+ f [H(z,t)] \, DD_1 H(z,t)\} \, dt =$$

$$= \int D \{f [H(z,t)] \, D_1 H(z,t)\} \, dt = f [H(z,1)] \, D_1 H(z,1) -$$

$$- f [H(z,0)] \, D_1 H(z,0)$$

But since $H(z,0) = \mu_z(0) = a$ is independent of z and in particular of x, $D_1 H(z,0) = 0$; and since $H(z,1) = \mu_z(1) = z = x + iy$, $D_1 H(z,1) = 1$. So (15) becomes $D_1 F(z) = f(z)$. If $D_1 = d/dx$ is replaced by $D_2 = d/dy$, the calculation is similar except that $D_2 f = if'$ and $D_2 H(z,1) = i$; hence $D_2 F(z) = if(z)$. The function F is, therefore, holomorphic and is a primitive for f.

This is all formal calculation. To justify it, the theorem on differentiation under the $\int$ sign (Chap. V, n° 9, Theorem 9) needs to be applied. Leaving aside subtleties unnecessary for the time being, this supposes that the function $f [H(z,t)] \, DH(z,t)$ integrated in (14) has continuous functions of the couple $(z,t) \in G \times I$ as partial derivatives with respect to x and y. Since f does not present any problems, the derivatives of $H(z,t)$ and $DH(z,t)$ with respect to x and y must, therefore, exist and be continuous on $G \times I$. The formula $DD_i = D_i D$ has also been used; this is justified if H is of class C^2 on[13] $G \times I$, in which case the previous conditions are obviously satisfied.

Calculation (15) and the relation $F' = f$ are, therefore, justified, provided there is a map

$$H : G \times I \longrightarrow G$$

---

[13] This is problematic since functions of class C^2 have only been defined on an open Cartesian space; however, I is compact and G is open, so that the product $G \times I \subset \mathbb{C} \times \mathbb{R} = \mathbb{R}^3$, a vertical cylinder having G as base and height 1, is neither open nor closed in $\mathbb{R}^3$. The solution is to constrain H to be C^2 on the open set $G \times]0,1[$ and H and its derivatives of at most second order to be the restrictions of functions defined and *continuous* on $G \times I$ to this set. Then derivatives at points of the form $(z,0)$ or $(z,1)$ are well-defined and the relation $D_1 D = DD_1$ which holds at (z,t) for $0 < t < 1$, by passing to the limit, also holds for $t = 0$ or 1. It would be simpler to assume that H is defined and of class C^2 on $G \times J$, where J is an open interval containing I. In practice, this does not change the results in any way.

satisfying the following conditions:

(i) $H(z,0) = a$, $H(z,1) = z$ for all $z \in G$,
(ii) H is of class C^2

in the sense specified in the previous footnote. It is the case of the map $(z,t) \mapsto tz$ for a star domain about the origin.

The existence of a *continuous*, but not necessarily C^2, map H from $G \times I$ to G satisfying (i) for some point $a \in G$ is expressed by saying that the domain G is *contractible* onto a. Setting $H_t(z) = H(1-t,z)$, we then get a one-parameter family of continuous maps H_t indexed by t from G into itself starting with the identity map $z \mapsto z$ and, at the end of the process, taking G onto the point a; under the "contraction", each $z \in G$ describes a trajectory $t \mapsto H(1-t,z)$ which takes it from its initial position to the point a. It can be shown that if there is a contraction of class C^0 from G onto a point, then there also is one that is C^2 and even C^∞; while not being easy, it is not very difficult to prove. Hence, if we admit this point,[14] we get a more general result than that of n° 1 regarding star domains, but it will in its turn be generalized (?) further down:

Theorem 2. *Every holomorphic function defined on a contractible domain $G \subset \mathbb{C}$ has a primitive on G.*

Corollary. An annulus $r < |z| < R$ is not contractible (which is physically obvious), since the function $1/z$ does not have primitives: its integral along the circle centered at 0 is obtained by integrating $2\pi i$ over $[0,1]$, and so is equal to $2\pi i$ despite the closure of the integration path. However, $\mathbb{C} - \mathbb{R}_-$ is contractible and even a star domain (consider the homotheties with centre 1), which explain why the function $1/z$ has a primitive on the domain, namely any uniform branch of the pseudo-function $\mathcal{L}og\, z$.

3 – Homotopy Invariance of Integrals

(i) *Homotopic paths.* Computation (2.15) to differentiate under $\int$ sign would continue to hold if the function $H(x,y,t)$ was replaced by a function of multiple real variables with values in G. The simplest case is that of a C^2 map,[15]

---

[14] It is in fact unnecessary as theorem 2 is a consequence of theorem 3 which will be proved later. The relevance of theorem 2 as stated here only lie in its proof and, as such, is only a calculus exercise.

[15] For reasons stated in chapter 9 – similarities between curvilinear integrals (dimension 1) and surface integrals (dimension 2) –, σ may be called a 2 *dimensional path* in $\mathbb{C}$; the reader will easily generalize to all dimensions. There is no orthodox terminology; some, like Serge Lang, talk wrongly of a 2 dimensional *simplex* as in algebraic topology. Ours suggests that such a "path" takes us in a continuous manner from the usual path $\mu_0 : t \mapsto \sigma(0,t)$ to another one, $\mu_1 : t \mapsto \sigma(1,t)$, just like an usual one-dimensional path takes us in a continuous manner from a point, a 0-dimensional path, to another one.

$$\sigma : I \times I \longrightarrow G$$

i.e. satisfying the following conditions:

(a) σ is of class C^2 on the open interior of $I \times I \subset \mathbb{R}^2$;
(b) the partial derivatives of order ≤ 2 of σ can be extended by continuity[16] to $I \times I$.

Such a map defines two families of C^2 paths on G, namely

(3.1)
$$\mu_s : t \longmapsto \sigma(s,t)$$

and

(3.2)
$$\nu_t : s \longmapsto \sigma(s,t) .$$

As σ is continuous, the family of paths μ_s may be regarded as a "deformation" of μ_0 to μ_1. The fact that a path can be deformed into another one in this way, or even by a merely *continuous* function σ from $I \times I$ to G, is expressed by saying that the two paths considered are *homotopic*. A first useful case is that of a *fixed-endpoint homotopy* of μ_0 to μ_1: then suppose that

$$\mu_s(0) = \sigma(s,0) \quad \text{and} \quad \mu_s(1) = \sigma(s,1)$$

are independent of s. Another case occurs when, μ_0 and μ_1 being closed, the intermediate paths μ_s stay closed during the deformation:

$$\sigma(0,t) = \sigma(1,t) \text{ for all } t;$$

μ_0 is said to be *homotopic through closed paths* to μ_1.

Apart from these cases, the homotopy condition is always fulfilled (hence uninteresting) because, on the one hand, every path μ is homotopic to a "constant" path by $\sigma(s,t) = \mu[(1-s)t]$, and because, on the other, two "constant" paths are always homotopic as can be seen by connecting the former to the latter by a path in G and by shifting the former along it to take it onto the latter.

Despite being somewhat abstract, fixed-endpoint homotopy can be interpreted in an interesting way. Remark first that the set $C^0(I)$ of all continuous paths $I \longrightarrow \mathbb{C}$,[17] equipped with the norm $\|\mu\|_I = \sup |\mu(t)|$ and the obvious algebraic operations (addition, multiplication by a complex number), is a complete normed vector space (Cauchy's criterion for uniform convergence), i.e. a Banach space (Chap. III, Appendix, no 5). Paths can, therefore, be defined in $C^0(I)$ as in any topological space: they are continuous maps $h : [0,1] = I \longrightarrow C^0(I)$. Hence, for any $s \in I$, $h(s) = \mu_s$ is a path in $\mathbb{C}$, and setting $\sigma(s,t) = \mu_s(t)$, we get a map σ from $I \times I$ to $\mathbb{C}$; $t \mapsto \sigma(s,t) = \mu_s(t)$ is clearly continuous for all s.

---

[16] As $I \times I$ is compact, this means precisely that they are *uniformly* continuous on the open set $]0,1[\times]0,1[$: Chap. V, § 2, n° 2, Corollary 2 of Theorem 8.

[17] Up to vocabulary, a continuous "path" is just a complex valued function defined and continuous on I.

Having said this, let us show that the map $h : I \longrightarrow C^0(I)$ is continuous if and only if so is the map $\sigma : I \times I \longrightarrow \mathbb{C}$. If indeed the latter is continuous, it is so uniformly since $I \times I$ is compact (Chap. V, § 2, no 8); in particular, this means that for all $r > 0$, there exists $r' > 0$ such that

$$|s - s'| \leq r' \Longrightarrow |\sigma(s,t) - \sigma(s',t)| \leq r \quad \text{for all } t \in I \,;$$

but since $h(s) \in C^0(I)$ is just the path $t \mapsto \sigma(s,t)$, this relation can be written

(3.3) $$|s - s'| \leq r' \Longrightarrow \|h(s) - h(s')\|_I \leq r \,.$$

Hence h is continuous. Proving the converse amounts to showing that (3) implies continuity of $\sigma(s,t)$ at all points $(s,t) \in I \times I$. To see this, let us start with the inequality

$$\left| \sigma\left(s',t'\right) - \sigma(s,t) \right| \leq \left| \sigma\left(s',t'\right) - \sigma\left(s,t'\right) \right| + \left| \sigma\left(s,t'\right) - \sigma(s,t) \right|$$

and choose some $r > 0$. If $|s - s'| \leq r'$, by (3), the first term on the right hand side is $\leq r$ for all t'; but as the function $t \mapsto \sigma(s,t)$ is continuous for given s, for given (s,t), $\leq r$ if $|t - t'|$, the second term on the right hand side is sufficiently small, qed.

Let us now consider an open subset G of $\mathbb{C}$ and let $C^0(I, G)$ be the subset of in $C^0(I)$ consisting of paths $I \longrightarrow G$; it is *open* in $C^0(I)$ since if $\mu \in C^0(I, G)$, the image $\mu(I)$ is a compact subset of G whose distance R to the border of G is strictly positive;[18] It is then obvious that any path $\nu : I \longrightarrow \mathbb{C}$ such that $\|\mu - \nu\|_I < R$ remains a path in G that is in fact homotopic to μ as the line segment

$$\sigma(s,t) = (1 - s)\mu(t) + s\nu(t)$$

connects μ and ν in $C^0(I)$. On the other hand, it is obvious that for given $a, b \in G$, the set $C^0_{a,b}(G)$ of continuous points $I \longrightarrow G$ connecting a to b in G is a closed subset of the open set $C^0(I, G)$. The same holds for the set of closed paths in G.

In conclusion, two paths with given endpoints a and b in G are fixed-endpoint homotopic if and only if they can be connected by a continuous path in the space $C^0_{a,b}(G)$ of all such paths. A similar result holds for a homotopy of closed paths.

(ii) *Differentiation with respect to a path.* A norm can be defined in the vector space $C^{1/2}(I)$ of admissible paths $I \longrightarrow \mathbb{C}$ by setting

$$\|\mu\| = \|\mu\|_I + \|\mu'\|_I \,;$$

equipped with it, $C^{1/2}(I)$ is *complete*. Indeed, if (μ_n) is a Cauchy sequence, the functions $\mu_n(t)$ and $\mu'_n(t)$ converge uniformly to some limits μ and ν that are respectively continuous and regulated; the relation

---

[18] Let F be this border; the function $d(z, F)$ is continuous on the compact set $\mu(I)$, and so reaches its minimum at some point $a \in \mu(I)$; were this minimum zero, there would be a sequence of points of F converging to a. This would imply that $a \in F$ since F is closed, a contradiction.

$$\mu(s) - \mu(t) = \int_s^t \nu(x)dx$$

proving that μ is a primitive for ν is then obtained by passing to the limit. Besides, it is obvious that in the space $C^0(I)$, the set $C^{1/2}(I;G)$ of $\mu \in C^{1/2}(I)$ such that $\mu(I) \subset G$ is open in $C^{1/2}(I)$.

Let us now return to formula (2.13)

$$(3.4) \qquad F(\mu) = \int_\mu f(\zeta)d\zeta = \int_0^1 f[\mu(t)]\,\mu'(t)dt$$

defining a function on $C^{1/2}(I;G)$. It may seem strange to differentiate it with respect to μ, but as it it defined on the open subset $C^{1/2}(I;G)$ of the Banach space $C^{1/2}(I)$, definition (1.1) which holds in $\mathbb{R}^2$ can be imitated. Here, c and h will be replaced by some $\mu \in C^{1/2}(I;G)$ and some $\nu \in C^{1/2}(I)$. So the expression

$$F(\mu + s\nu) = \int f[\mu(t) + s\nu(t)][\mu'(t) + s\nu'(t)]\,dt =$$
$$= \int f[\mu(t) + s\nu(t)]\,d\mu(t) + s\int f[\mu(t) + s\nu(t)]\,d\nu(t)$$

must be differentiated with respect to s. For any ν, it is well-defined for sufficiently small $|s|$. To differentiate under the $\int$ sign [Chapter V, § 2, Theorem 9 or § 9, formula (30.15)], it is sufficient to check that $f[\mu(t) + s\nu(t)]$ is a continuous function of (s,t), which is obvious, and that its derivative with respect s is a continuous function of (s,t); its existence is obvious – it is $f'[\mu(t) + s\nu(t)]\nu(t)$ – and so is its continuity since the functions f', μ and ν are continuous. Hence in telegraphic style,

$$\int f'(\mu + s\nu)\nu d\mu + \int f(\mu + s\nu)d\nu + s\int f'(\mu + s\nu)\nu d\nu =$$
$$= \int f'(\mu + s\nu)\nu(d\mu + sd\nu) + \int f(\mu + s\nu)\nu'dt\,.$$

Integration by parts can justifiably be used to compute the last integral since the functions $f(\mu + s\nu)$ and ν are of class $C^{1/2}$. So it can also be written

$$f(\mu + s\nu)\nu\Big|_{t=0}^{t=1} - \int f'(\mu + s\nu)(\mu' + s\nu')\nu dt\,.$$

Since $[\mu'(t) + s\nu'(t)]\,dt = d\mu(t) + sd\nu(t)$, the last integral can be written $\int f'(\mu + s\nu)\nu(d\mu + sd\nu)$, canceling out the first term of the penultimate formula. Hence finally,

$$(3.5) \qquad \frac{d}{ds}F(\mu + s\nu) = f[\mu(1) + s\nu(1)]\nu(1) - f[\mu(0) + s\nu(0)]\nu(0)\,.$$

All this obviously supposes that only values of s such that $\mu+s\nu \in C^{1/2}(I;G)$ are used. The reader will probably be under the impression of having come across similar calculations above. A more or less vague form of this idea is due to Cauchy: in his work on the series expansion of $f[\mu(t) + s\nu(t)]$ with respect to s, he calculated the coefficient of s. His calculations have long since disappeared from textbooks. This is a good reason for reintroducing them by rectifying his simplistic, but ultimately correct idea, since it reappears in a very generalized form in the version of the calculus of variations for example found in H. Cartan, *Differential Calculus*, where a function of the following form is differentiated with respect to μ:

$$F(\mu) = \int_a^b f\left[t, \mu(t), \mu'(t)\right] dt .$$

On the other hand, note that in the presence of a formula such as (5), the idea of deducing an expression for $F(\mu+s\nu)$ by applying the FT immediately arises. This will be done a bit later.

Exercise 1. Let μ_0 and μ_1 be two paths on G and $\sigma : I \times I \longrightarrow G$ a homotopy from μ_0 to μ_1; write $F(s)$ for the integral of the function $f(z)$ along the path μ_s. assuming σ to be of class C^2, find a formula similar to (5) for $F'(s)$.

(iii) *Effects of a linear homotopy on an integral.* We can now return to the behaviour of an integral when an integration path μ_0 is deformed into a path μ_1 without leaving the domain G where the function f to be integrated is defined and holomorphic. The difficulty is that, for $0 < s < 1$, the intermediary paths μ_s are continuous, but not necessarily admissible. We can get around it by altering the homotopy so that the μ_s become admissible or, equivalently, by showing that it is possible to go from μ_0 to μ_1 by a succession of *linear* homotopies between admissible paths, i.e. of the form

$$(3.6) \qquad \sigma(s,t) = (1 - s)\mu_0(t) + s\mu_1(t) = \mu_s(t),$$

where $s, t \in I = [0, 1]$. There is always such a homotopy when μ_1 is sufficiently near μ_0 in the sense of uniform convergence: if $R > 0$ is the distance from $\mu_0(I)$ to the border of G, then $\sigma(s,t) \in G$ for all $s, t \in I$ provided $\|\mu_1-\mu_0\|_I < R$.

However, path (6) is of the form $\mu_0 + s\nu$ with

$$\nu(t) = \mu_1(t) - \mu_0(t).$$

It is, therefore, possible to apply (5) to the function $F(\mu_s)$, whence

$$(3.7) \qquad \frac{d}{ds} \int_{\mu_s} f(\zeta)d\zeta = f\left[\mu_s(t)\right]\left[\mu_1(t) - \mu_0(t)\right]\Big|_{t=0}^{t=1} .$$

Suppose first that it is a fixed-endpoint homotopy. Then, $\mu_1(t) - \mu_0(t) = 0$ for $t = 0$ or 1, the derivative is zero for all s and the integral is, therefore, independent of $s \in [0, 1]$. In other words, in this case

(3.8)
$$\int_{\mu_0} f(\zeta)d\zeta = \int_{\mu_1} f(\zeta)d\zeta .$$

Suppose now that all paths μ_s are closed, i.e. that $\mu_s(1) = \mu_s(0)$ for all s. A short calculation shows that the right hand side of (7) is still zero for all s. Hence (8) follows again.

Without using these assumptions, the FT applied to relation (7) shows that, by integrating over $[0,1]$,

(3.9)
$$\int_{\mu_1} f(\zeta)d\zeta - \int_{\mu_0} f(\zeta)d\zeta = \int_0^1 f[\mu_s(1)][\mu_1(1) - \mu_0(1)]\,ds -$$
$$- \int_0^1 f[\mu_s(0)][\mu_1(0) - \mu_0(0)]\,ds .$$

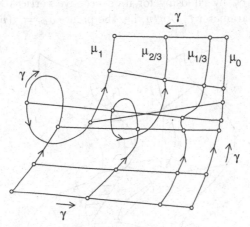

Fig. 3.2.

As $\mu_1(1) - \mu_0(1)$ is the derivative of $\mu_s(1)$ with respect to s, the first term is just the integral of f along the path $s \mapsto \mu_s(1)$, the second being the integral of f along the path $s \mapsto \mu_s(0)$. Therefore, the relation obtained means that, the integral of f along the coherently oriented closed path γ drawn in figure 2 above is zero. This would be obvious if f had a primitive on G, which is, however, not assumed. We will not generalize to any closed path : as a closed path, γ is homotopic to the path consisting in traveling μ_0 twice in opposite directions, and so is homotopic to a point. As for the fact that the integral of f along γ is zero, it follows from Theorem 3 which will be proved shortly.

(iv) *The homotopy invariance theorem.* Paths μ_0 and μ_1 always being admissible, suppose only that the homotopy σ deforming μ_0 into μ_1 is C^0; it is no longer possible to differentiate integrals or even to write them. But a given homotopy can be approached by linear homotopies and the preceding point can be used, which, as will be seen, again leads to the same results.

The image $\sigma(I \times I) \subset G$ being compact, as already stated, its distance R from the border of G is > 0. Choose $r < R$. As σ is uniformly continuous on the compact set $I \times I$, there exists $r' > 0$ such that

$$(3.10) \quad |s - s'| \leq r' \quad \& \quad |t - t'| \leq r' \Longrightarrow |\sigma(s,t) - \sigma(s',t')| \leq r.$$

For $t = t'$, in particular this shows that

$$(3.11) \qquad\qquad |s - s'| \leq r' \Longrightarrow \|\mu_s - \mu_{s'}\|_I \leq r;$$

see remarks at the end of (i).

Having said that, let us choose an integer n, give s values of the form $s_p = p/n$ with $0 < p < n$ and let ν_p be the path μ_s for $s = p/n$. It may not be admissible, but it can be approached by a piecewise linear, and hence admissible, path γ_p by choosing for its successive vertices the points of ν_p indexed by the parameter $t_q = q/n$, i.e. the points $a_{pq} = \sigma(p/n, q/n)$.

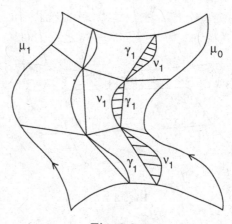

Fig. 3.3.

If n is sufficiently large, the diameter of the square K_{pq} with vertices $(s_p, t_q), (s_{p+1}, t_q), (s_{p+1}, t_{q+1}), (s_{p+1}, t_{q+1})$ in $I \times I$ is $< r'$. Therefore, by (10), its image under σ is contained in the disc D_{pq} centered at a_{pq} and of radius r. However, the latter is convex and contained in G since $r < R$. As a result:

(a)   the line segment connecting a_{pq} to $a_{p,q+1}$ is contained in G for all q, so that the same holds for the path γ_p obtained by juxtaposing these segments for different values of q;

(b)   for any $t \in [t_q, t_{q+1}]$, the line segment connecting $\sigma(s_p, t)$ to $\sigma(s_{p+1}, t)$ is contained in G since its endpoints are in D_{pq}. Hence, there is a linear deformation that takes us from γ_p to γ_{p+1} without leaving G.

The same arguments show that there are linear deformations taking us from μ_0 to γ_1 and from γ_{n-1} to μ_1.

Hence, if the definition of the γ_p, given for $0 < p < n$ is completed by setting $\gamma_0 = \mu_0$ and $\gamma_n = \mu_1$, we get a sequence of admissible (and even piecewise linear except for the first and the last one) paths in G

$$\gamma_0 = \mu_0, \quad \gamma_1, \ldots, \quad \gamma_n = \mu_1$$

such that there are linear deformations taking us from each of them to the next one without leaving G. If the deformation σ we started with is a fixed-endpoint homotopy, the intermediate paths γ_p clearly also have the same endpoints as the two given paths. As seen in (iii), a fixed-endpoint linear homotopy leaves the integral invariant. So the integrals along γ_p and γ_{p+1} are equal for all p. Similarly, if μ_0 and μ_1 are closed and remain so during the given deformation σ, γ_p is closed for all p and remains so during the linear deformation taking it onto γ_{p+1}. Hence, once again the integrals are equal.

In conclusion:

Theorem 3. *Let G be a domain in $\mathbb{C}$, f a holomorphic function on G and μ_0, μ_1 two admissible paths in G. If one of the two following conditions is satisfied, then the integrals of f along μ_0 and μ_1 are equal:*

(a) *there is a fixed-endpoint homotopy on G joining μ_0 and μ_1*
(b) *μ_0 and μ_1 are closed and homotopic in G as closed paths.*

Exercise 2 (direct proof of Theorem 3). We keep the above construction and notation by, for example, supposing that σ is a fixed-endpoint homotopy; proving this result amounts to showing that integrals along γ_p and γ_{p+1} are equal for all p. For this, take a closed path γ_{pq} made of line segments connecting a_{pq}, $a_{p,q+1}$, $a_{p+1,q+1}$, $a_{p+1,q}$ and a_{pq} in the given order; using the existence of a primitive for f on D_{pq}, show that the integral of f along this path is zero. Show that the difference between integrals of f along γ_p and γ_{p+1} is equal to the sum, extended to q, of integrals along γ_{pq} and conclude.[19]

Theorem 3 provides a new existence theorem for primitives. For this it suffices to suppose that condition (b) is satisfied for all μ_0 and μ_1; in this case, any closed path μ is indeed homotopic as a closed path to a "constant" path $t \mapsto a$, where $a \in G$ is arbitrarily chosen, so that the integral of f along μ is zero. Then theorem 1, or its equivalent in terms of closed paths, shows that f has a primitive on G.

In the next chapter it will be shown that conditions (a) and (b) are equivalent in a more general framework. Domains in which they are satisfied for all continuous closed paths μ_0 and μ_1 are called *simply connected*.

Corollary 1. *Any holomorphic function on a simply connected domain G of $\mathbb{C}$ has a global primitive on G.*

Corollary 2. *Let f be a holomorphic function on a simply connected domain G; suppose that f does not vanish on G. Then, there is a holomorphic*

---

[19] For similar proofs, see Dieudonné, *Éléments d'analyse*, vol. 1, (9.6.3) or Remmert, *Funktionentheorie 2*, Chap. 8, § 1, n° 5 and 6.

function g on G such that $e^{g(z)} = f(z)$ for all $z \in G$; it is unique up to addition of a multiple of $2\pi i$.

Since f does not vanish on G, the function f'/f is defined and holomorphic on G, and hence admits a primitive g; then $(e^g)' = g'e^g = e^g f'/f$, i.e. $(e^g)' f - e^g f' = 0$, hence $(e^g/f)' = 0$, so that the function e^g is proportional to f. Adding a constant to g, $e^g = f$ may be assumed. Any other holomorphic solution g_1 must satisfy the condition $g_1(z) - g(z) \in 2\pi i \mathbb{Z}$ for all z, which obviously requires the left hand side to be a constant, qed.

The relation $e^g = f$ means that $g(z) \in \mathcal{L}og\, f(z)$ holds for all $z \in G$, where $\mathcal{L}og\, w$ denotes the set $z \in \mathbb{C}$ such that $\exp(z) = w$ (Chapter IV, § 4). Such a function g is called a *uniform branch* of the pseudo-function $\mathcal{L}og\, f(z)$. Then

$$g(z) = \log |f(z)| + i.\operatorname{Arg} f(z),$$

where, at all points, the argument of $f(z)$ must be chosen so that it is a continuous function of z.

The definition of uniform branches of the no less pseudo-functions $f(z)^s$, where $s \in \mathbb{C}$ is given and is not an integer, follows from such branches g: these are the functions $e^{s \cdot g(z)}$. For $s = 1/p$ with p integer, we thus get holomorphic solutions of the equation $h(z)^p = f(z)$; they can be deduced from any one of them by taking its product with a pth root of unity.

All this supposes that G is simply connected. The case of the function $f(z) = 1/z$ on $G = \mathbb{C} - \{0\}$ shows that this assumption is essential. In fact, Corollary 1 can be shown to *characterize* simply connected domains, but this result is rarely used.

Corollary 3. *Let f be a holomorphic functions on a domain G. The integral of f along any null-homotopic closed path, i.e. homotopic to a point in G, is zero.*

This result explains Theorem 2 whose proof used a homotopy of class C^2, an assumption now unnecessary.

Any *contractible* domain G is simply connected. Indeed, if σ is a contraction to a point $a \in G$ and μ a closed path in G, then μ is homotopic through closed paths to the constant path $t \mapsto a$ under the map $(s, t) \mapsto \sigma[1 - s, \mu(t)]$.

This trivial result has a far less trivial converse: any simply connected domain G in $\mathbb{C}$ is not only contractible, but also homeomorphic to the unit disc $|z| < 1$; except when $G = \mathbb{C}$, a case excluded by Liouville's theorem on integral functions, even in G, there is a *holomorphic* function mapping G bijectively onto the open disc $D : |z| < 1$ and whose inverse is holomorphic (Riemann). Any holomorphic bijection $f : U \longrightarrow V$ from an open set onto another whose inverse $g : V \longrightarrow U$ is holomorphic is called a *conformal representation* of U on V. The existence of such a representation means that U and V are "isomorphic" from the point of view of analytic function theory: everything that holds for holomorphic or harmonic functions on U can be immediately transferred to holomorphic or harmonic functions on V. As relation $g\left[f(z)\right] = z$ shows that the derivatives of f and g are mutually inverse at corresponding points, it follows that $f'(z) \neq 0$ for all $z \in G$. Conversely, if this condition is satisfied, then f, though not necessarily a global homeomorphism (for this f would need to be injective) transforms every open subset of U, and in particular U itself, into an open subset of $\mathbb{C}$. This was proved in Chapter III, § 5, n° 24 using the local inversion theorem: set $f = p + iq$ to be the map taking the point $(x, y) \in \mathbb{R}^2$ to the point (ξ, η) such that $\xi + i\eta = f(x + iy)$ can also be written

$$\xi = p(x, y), \quad \eta = q(x, y),$$

so that, by Cauchy's equations, its Jacobian

$$J_f(x, y) = D_1 p(x, y) D_2 q(x, y) - D_2 p(x, y) D_1 q(x, y)$$

is equal to

$$D_1 p(x, y)^2 + D_1 q(x, y)^2 = \left|f'(z)\right|^2 ,$$

and hence is non-zero. This proves the result. Moreover, if $f : U \longrightarrow V = f(U)$ is assumed to be injective, then f is a homeomorphism [since the inverse image of an open set $U' \subset U$ under f^{-1} is then $f(U')$, and so is open] and the local inversion theorem shows that, like f, the inverse map $g : V \longrightarrow U$ is of class C^1 as a function of two real variables. It is *holomorphic* since the relation $g\left[f(z)\right] = z$ implies that the Jacobian matrix of g at $\zeta = f(z)$ is the inverse of that of f at z. Now, holomorphic functions are characterized by the fact that, at all points, their Jacobian matrix is of the form

$$\begin{pmatrix} a & b \\ -b & a \end{pmatrix} .$$

It is, therefore, sufficient to check that the inverse of such a matrix is also of the same type. More simply: the inverse of any $\mathbb{C}$-linear map is $\mathbb{C}$-linear. All this has been proved in Chapter III, § 5, but it is worth recalling here. We will see in n° 5 (Theorem 7) that the relation $f'(z) \neq 0$ is in fact a consequence of the injectivity of f, in other words that conformal representations are just bijective holomorphic maps.

The fact that a simply connected domain G other than $\mathbb{C}$ is isomorphic to the unit disc is one of the most famous results of Riemann; while being based on a method going far beyond the framework of holomorphic functions (PDE specialists' "Dirichlet's principle"), his proof was not really satisfactory. Simpler proofs have since then been found[20] and the behaviour of a conformal representation f of G on the unit disc in the neighbourhood of the boundary of G has been widely studied; if, for example, G is bounded and

---

[20] See for instance Chap. 14 in Rudin and for examples Chap. X in Dieudonné, *Analyse infinitésimale*.

if its boundary consists of a finite number of simple arcs of curve, or if G is bounded and convex, then f can be extended to a homeomorphism from the closure $\bar{G}$ of G onto the closed disc $|z| \leq 1$.

If f is a conformal representation of G on the unit disc D, then any conformal representation of G on D is obviously of the form $h \circ f$, where $h = g \circ f^{-1}$ is a conformal representation of D on itself, and conversely. This leads us to determine the *conformal representations of D on D*, which is much easier than proving Riemann's theorem: these are precisely the maps given by

$$(3.11) \qquad h(z) = (az + b)/(\bar{b}z + \bar{a}) = \zeta \quad \text{où} \quad a\bar{a} - b\bar{b} = 1 \; .$$

Exercise 3. (i) Show that (11) is defined for $|z| < 1$. (ii) Show that $\zeta\bar{\zeta} - 1 = (z\bar{z} - 1)/|\bar{b}z + \bar{a}|^2$ and deduce that $h(D) \subset D$. (iii) Observing that h^{-1} is also of the form given in (11), show that $h(D) = D$. (iv) Let f be a conformal representation of D on D such that $f(0) = 0$; show that $f'(0) \neq 0$ and $f(z) = zg(z)$ where g is holomorphic and verify that $|g(z)| \leq |z|^{-1}$ in D. (vi) Using the maximum principle, show that $|g(z)| \leq 1/r$ for $|z| \leq r < 1$ and deduce that $|g(z)| < 1$ in D (particular case of Schwarz's lemma: Chap. VII, §4, n° 15, cor. 3 of theorem 11). (vii) Show that if $f(0) = 0$, then $|f(z)| \leq |z|$ and $|f^{-1}(z)| \leq |z|$. Deduce that $f(z) = az$, where $|a| = 1$. (viii) Show that, for any conformal representation f of D on D, there is a function (11) such that $h \circ f$ has 0 as fixed point. Deduce that f is of the form given in (11).

Exercise 4. Let P be the half-plane $\text{Im}(z) > 0$. (i) Show that the map $z \mapsto (z - i)/(z + i)$ is a conformal representation of P on D. (ii) Deduce that the conformal representations of P on P are the maps

$$(3.12) \quad z \longmapsto (az + b)/(cz + d) \quad \text{with} \quad a, b, c, d \in \mathbb{R}, \quad ad - bc = 1 \; .$$

For Riemann, a simply connected domain was a domain partitioned into two disjoint ones by all "cuttings" – simple curve segments connecting two boundary points. For example this is clearly not the case of an annulus. The equivalence of these two definitions is intuitively obvious, but proving it is another matter...

Another "obvious" idea can be justified, namely that a domain is simply connected if its complement does not have any *compact*, connected component, in other word if there are no "holes" in G. It is even possible to go much further[21] and consider domains whose complements have a finite number of compact connected components $K_i (1 \leq i \leq n)$. A first result that is also made "obvious" by figure 4 is that there are then closed paths μ_i in G such that K_i is in the "interior" of μ_i and K_j in "exterior" of μ_i for all $j \neq i$; further explanations for the meaning of these terms will be given a bit later (n° 4, (i)).

---

[21] See chapters 8 and 14 in Remmert, *Funktionentheorie 2*, in particular the historical statements about Riemann's theorem in chapter 8, and especially vol. 2 of Conway, where everything is proved.

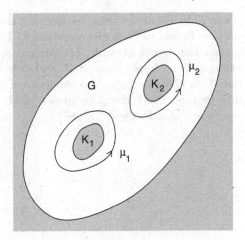

Fig. 3.4.

A second far less "obvious" result, but which trivially implies the first, is that a bounded domain with n holes has a conformal representation on the domain obtained by removing from an open disc, could be all of $\mathbb{C}$, n adequately chosen pairwise disjoint compact discs and possibly reduced to a point. This generalization of Riemann's theorem proved at the start of the century by Paul Koebe, presents enough difficulties that even Remmert only mentions it at the end of about 700 pages of general theorems on analytic functions. The number of holes can also be shown to be the same for two homeomorphic domains, but this is a very particular case of much more general theorems in algebraic topology. In fact, this entire subject is characterized by an amalgamation of methods from analytic function theory and topology that are sometimes difficult to separate out. Their generalization to functions of several complex variables gave rise to remarkable Franco-German discoveries after the war. In some sense, they are easier to understand than those from theory in one variable: as they are more general, they do not use "elementary" ad hoc arguments that hide the real reasons for these phenomena.

There are also close links between these theories and the problem of approximating holomorphic functions on a given domain G by simple polynomial functions or, when this is not possible, by rational functions without poles in G. If for example G is simply connected, and only in this case, any holomorphic function f on G is the limit of a sequence of polynomials in z converging uniformly to f on every compact subset of G. This result and more general ones are due to Carl Runge[22] (1885).

---

[22] In fact, Runge did not see that the approximation by rational functions led to the result in question in the case of a simply connected domain. On another note, Runge was interested in atomic spectra in the hope of finding simple formulas that would allow their frequencies to be computed, as had already been done by Balmer for the hydrogen atom. We now know that this amounts to calculating the corresponding eigenvalues of the Schrödinger operator. This problem is still too hard, the helium atom, and all the more the following ones, continuing to resist all *exact* solutions. After 1900, when Felix Klein created the first applied mathematics team in Göttingen, one of his recruits would be Runge, the first leading expert of numerical analysis. He also recruited Ludwig Prandtl, who remained until 1945 the greatest German expert of aerodynamics, perhaps even the greatest world expert, while Prandtl's first brilliant student, the Hungarian Theodor von Kármán who emigrated to CalTech at the end of the 1920s, would play the same role in the USA until the end of the 1950s. See Paul A. Hanle, *Bringing Aerodynamics to America* (MIT Press, 1982).

§ 2. Cauchy's Integral Formulas

4 – Integral Formula for a Circle

(i) *Integrals in* $1/z$. Integrals of functions $1/(z-a)$ occur everywhere in the theory of holomorphic functions and knowing how to compute them is important. The integration path $\mu : I \longrightarrow G$ is obviously assumed not to pass through a. Supposing $a = 0$ for simplicity's sake, the definition of $\int dz/z$ reduces to the integral of $\mu'(t)/\mu(t)$ over the interval I. Same as μ', this function is regulated. So admits a primitive $L(t)$ and, by the FT, the expected result will be the variation of $L(t)$ between the endpoints of I. But if we set $h(t) = \exp[L(t)]$, then $h'(t) = L'(t)h(t)$. As $L'(t) = \mu'(t)/\mu(t)$, the derivative of the continuous function $h(t)/\mu(t)$ is seen to be identically zero outside a countable set of values of t. Hence the function is constant. Adding an adequate constant to L, it may, therefore, be assumed that

$$(4.1) \qquad\qquad \exp[L(t)] = \mu(t)$$

for all $t \in I$. As $L(t)$ is continuous, this means that, by definition, $L(t)$ is a uniform branch of the pseudo function $\mathcal{L}og\, z$ along μ in the sense of Chap. IV, § 4, (vii) and (viii). Recall again that for us the notation $\mathcal{L}og\, z$ does not describe a clearly determined complex number, but, on the contrary, the *set* of $\zeta \in \mathbb{C}$ such that $\exp(\zeta) = z$.

As an aside, recall also (Chap. VII, end of n° 16) that a uniform branch of $\mathcal{L}og\, z$ on a domain $G \subset \mathbb{C}^*$ is similarly a (genuine) holomorphic function L – continuity would suffice – defined on G and satisfying $L(z) \in \mathcal{L}og\, z$, i.e.

$$(4.2) \qquad\qquad \exp[L(z)] = z,$$

for all $z \in G$. Unlike what happens in the case of a path, such a branch does not always exist, in particular if $G = \mathbb{C}^*$. It does so if and only if for any path μ in G, the variation of a uniform branch of $\mathcal{L}og\, z$ along μ, or, equivalently, of the argument of z, only depends on the endpoints of the path considered. Verified by the means available in Chap. IV, § 4, (ix), this result is just theorem 1 of § 1 applied to $1/z$.

If, instead of integrating $1/z$, we integrate $1/(z-a)$ for a point a not located on μ, the result would obviously be the same. In conclusion:

Theorem 4. *The integral of* $1/(z-a)$ *along an admissible path* μ *in* $\mathbb{C} - \{a\}$ *is equal to the variation of a uniform branch of* $\mathcal{L}og(z-a)$ *along* μ.

Such a branch is of the form

$$(4.3) \qquad\qquad L(t) = \log|\mu(t) - a| + i.A(t)$$

where log is the elementary function defined on $\mathbb{R}_+^*$ and where, in its turn, $t \mapsto A(t)$ is a uniform branch along μ of the no less pseudo-function $\mathcal{A}rg(z-a)$, i.e. a *continuous* function such that

(4.4) $$\mu(t) - a = |\mu(t) - a| \cdot \exp\left[i.A(t)\right]$$

for all t. Assuming μ to be *closed*, the term $\log|\mu(t) - a|$ of (3) has the same values at $t = 0$ and $t = 1$ since it only depends on $\mu(t)$. Its variation along μ is, therefore, zero, so that, up to a factor i, that of $L(1) - L(0)$ is equal to the variation $A(1) - A(0)$ of the argument of $\mu(t) - a$. But as the various possible values of the argument of a complex number differ by multiples of 2π,

(4.5) $$A(1) - A(0) = 2\pi . \operatorname{Ind}_\mu(a),$$

where $\operatorname{Ind}_\mu(a)$ is an integer called the *index of a with respect to μ*, unless it is the index of μ with respect to a. As explained at the end of Chap. IV, § 4, physically, it is the number of rotations carried out by the ray with initial point a and passing through $\mu(t)$ as t varies from 0 to 1. It is a positive or negative number calculated by taking into account the direction of the rotations. This will be justified in Chap. X, no 3, (iii). Clearly, $\operatorname{Ind}_\mu(a)$ *only depends on the homotopy class of μ in $\mathbb{C} - \{a\}$*. As a result, setting

$$\operatorname{Supp}(\mu) = \mu(I)$$

for the *support* of the path μ, the next result follows:

Corollary. *For any closed path μ in $\mathbb{C}$ and all $a \in \mathbb{C} - \operatorname{Supp}(\mu)$,*

(4.6) $$\int_\mu \frac{dz}{z - a} = \int_0^1 \frac{\mu'(t)}{\mu(t) - a} dt = 2\pi i . \operatorname{Ind}_\mu(a)z, .$$

Note that the left hand side of (6) – and hence the right hand side – is a continuous function of a outside the compact set $\mu(I) = \operatorname{Supp}(\mu)$, i.e. outside the image of I under μ, since the function of the (t, a) integrated over $[0, 1]$ is continuous (Chap. V, no 9, Theorem 9). From the fact that $a \mapsto \operatorname{Ind}_\mu(a)$ has values in $\mathbb{Z}$, it can be deduced that *the index of a point a with respect to a closed path μ only depends on the connected component of a in the open set $\mathbb{C} - \operatorname{Supp}(\mu)$.* By (6), it clearly approaches 0 as $|a|$ increases indefinitely. Hence it is zero in the unbounded connected component of $\mathbb{C} - \operatorname{Supp}(\mu)$. The latter is unique because it at least contains the exterior of any disc D having $\operatorname{Supp}(\mu)$ as a subset, so that all other components are contained in D, and hence are bounded.

We sometimes write $\operatorname{Ext}(\mu)$, *exterior* of μ, for the set of $z \in \mathbb{C} - \operatorname{Supp}(\mu)$ where $\operatorname{Ind}_\mu(z) = 0$ and $\operatorname{Int}(\mu)$, *interior* of μ, for the set of z where $\operatorname{Ind}_\mu(z) \neq 0$. The exterior of μ contains the non-compact connected component of $\mathbb{C} - \operatorname{Supp}(\mu)$, but may be strictly larger. These notions are not at all related to those defined in Chap. III, no 1 with respect to an arbitrary subset of $\mathbb{C}$, but refer instead to what has been said at the end of Chap. III, § 4 (Jordan curve theorem) in the case of a "simple" curve, i.e. homeomorphic to the unit circle $\mathbb{T}$. This supposedly simple case being already quite

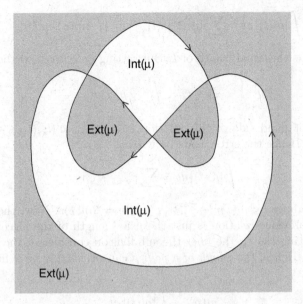

Fig. 4.5. Freitag-Busam, p. 240

subtle, it would be wrong to thing that the general case is less so even if, in practice, everything is always more or less obvious.

(ii) *Length of a path.* It is often necessary to find an upper bound for an integral

$$\int f(\zeta)d\zeta = \int_I f\left[\mu(t)\right]\mu'(t)dt\,.$$

Setting

(4.7)
$$\|f\|_\mu = \sup_{t\in I} |f\left[\mu(t)\right]|\,,$$

for the *uniform norm of f along μ*, i.e. the uniform norm of f on the set of points $\text{Supp}(\mu) = \mu(I)$ of the "curve" described by $\mu(t)$ in the sense of Chap. III, no 7,

$$\left|\int f(\zeta)d\zeta\right| \le \|f\|_\mu \cdot \int |\mu'(t)|\,dt$$

obviously follows. The integral occurring on the right hand side has a geometric interpretation. Indeed, if a subdivision $u = t_0 < t_1 < \ldots < t_n = v$ of $I = [u, v]$ is chosen to be sufficiently fine so that μ' is constant up to $r > 0$[23] on each partial interval, then

---

[23] Recall our language conventions (Chapter III, no 2). A numerical function f is *constant up to r* on a set E if $|f(x) - f(y)| \le r$ for all $x, y \in E$. An equality $a = b$ holds up to r if $|a - b| \le r$.

$$\int |\mu'(t)|\, dt = \sum |\mu'(t_i)| (t_{i+1} - t_i) \quad \text{upto } m(I)r\,,$$

where $m(I)$ is the usual length of I. But setting $\zeta_i = \mu(t_i)$, by the FT,

$$\zeta_{i+1} - \zeta_i = \int_{t_i}^{t_i+1} \mu'(t) dt$$

and the right hand side of this relation is equal to $\mu'(t_i)(t_{i+1} - t_i)$ up to $r(t_{i+1} - t_i)$. Hence the error made by writing

$$\int |\mu'(t)|\, dt = \sum |\zeta_{i+1} - \zeta_i|$$

is bounded above by $m(I)r + \sum(t_{i+1} - t_i)r = 2m(I)r$. Now, the left hand side of the previous relation is just the usual length of the piecewise linear path connecting the ζ_i; the finer the subdivision considered, the closer this path is to μ. Hence, the *length of a path* μ can reasonably be defined by the formula

$$(4.8) \qquad\qquad m(\mu) = \int |\mu'(t)|\, dt\,,$$

where integration is over I: the scalar (not the vectorial) velocity of the moving object along a path is integrated with respect to time. The letter m suggests an analogy with the usual length or measure of an interval in $\mathbb{R}$.

The conclusion of these arguments is the inequality

$$(4.9) \qquad\qquad \left| \int_\mu f(\zeta) d\zeta \right| \le m(\mu).\|f\|_\mu\,.$$

This result supersedes the almost trivial inequality in real variables we came across and is constantly used.

Note that if $\mu(t)$ is replaced by $\nu(t) = \mu\,[\varphi(t)]$, where φ is a C^1 map from an interval $J \subset \mathbb{R}$ to the interval I where μ is defined, which does not change $\mathrm{Supp}(\mu)$, then

$$m(\nu) = \int |\mu'\,[\varphi(t)]\,\varphi'(t)|\, dt = \int |\mu'\,[\varphi(t)]| \,.\, |\varphi'(t)|\, dt\,.$$

Hence, the change of variable formula for integrals (Chap. V, §6, no 19) involving the function $\varphi'(t)$, but not its absolute value, gives the equality $m(\mu) = m(\nu)$ only if the sign of φ' is constant, i.e. if φ is monotone: it is generally accepted that the direct route from Paris to Marseille is shorter than the Paris-Lyon-Dijon-Lyon-Marseille one. Despite its terminology, the length of a path is, therefore, a kinematic rather that a geometric notion applicable to the set $\mathrm{Supp}(\mu) = \mu(I)$. In fact, it would be better to call "route" what everyone calls "path", but it is too late.[24]

---

[24] *Path*: any road that can be taken to go from one place to another. *Route*: action of crossing space from one place to another. (Littré).

(iii)   *Cauchy's integral formula for a circle.* The calculations done in no 3, (iii) explain Cauchy's integral formula for a circle (Chap. V, no 5), namely

(4.10) $$2\pi i f(a) = \int_\mu \frac{f(\zeta)}{\zeta - a} d\zeta \,,$$

where a is in the interior of the circle $\mu : t \mapsto R.\exp(2\pi it)$ centered at 0 along which we integrate and where f is holomorphic on an open disc D of radius $> R$. The map

$$(s, \zeta) \longmapsto a + (1 - s)(\zeta - a)$$

is indeed a contraction from D onto the point a. For given s, it is the homothety with centre a and ratio $1 - s$, which transforms μ into a circle μ_s surrounding a and whose radius approaches 0 as s tends to 1: an example of a linear deformation. Hence, the right hand side of (10), where a holomorphic function is integrated over the open set $D - \{a\}$, does not change if μ is replaced by μ_s with $s < 1$, where this is a strict inequality. But for every $r > 0$, there exists $r' > 0$ such that

$$|\zeta - a| < r' \Longrightarrow |f(\zeta) - f(a) - f'(a)(\zeta - a)| < r|\zeta - a|$$

since f is differentiable at the point a. Therefore, if s is sufficiently near 1 for μ_s to be contained in the disc $|\zeta - a| < r'$, and if $f(\zeta)$ is replaced by $f(a) + f'(a)(\zeta - a)$ in the integral along μ_s, for all ζ, the error made on the function $f(\zeta)/(\zeta - a)$ to be integrated is bounded above by r. So, in view of the standard upper bound (9), the error on the right hand side (10) is bounded above by $m(\mu_s)r$, where $m(\mu_s)$ is the length of μ_s. If it is assumed that for one complete circuit of a circumference, our scholarly definition of the length coincides with that of Archimedes – it is anyhow his since he approximated the circle with inscribed polygons, without, however, following it through...–, then it is clear that $s \mapsto m(\mu_s)$ is bounded on I; in fact, $m(\mu_s)$ is the product of the length of the initial circle and of the homothety ratio $1 - s$. Hence the error made on the right hand side of (10) has upper bound r up to a constant factor. In other words,

(4.11) $$\int_\mu \frac{f(\zeta)}{\zeta - a} d\zeta = \lim_{s=1-0} \int_{\mu_s} \left[\frac{f(a)}{\zeta - a} + f'(a) \right] d\zeta =$$

$$= \lim f(a) \int_{\mu_s} \frac{d\zeta}{\zeta - a} + \int_{\mu_s} f'(a) d\zeta \,.$$

The contribution of $f'(a)$ to the second expression is zero since a constant function is being integrated along a closed path. It remains to evaluate the integral of $1/(\zeta - a)$ along μ_s; for this, μ_s may be replaced by a closed contour homotopic to it as a closed path in the open set $\mathbb{C} - \{a\}$ where the function $1/(\zeta - a)$ is holomorphic; for example, by a circle with centre a. By the corollary to Theorem 4, this is the product of $2\pi i$ and of the index of a with respect to such a circle, which is obviously equal to 1. In view of the factor $f(a)$, the final result is, therefore, indeed $2\pi i f(a)$. This is one of Cauchy's own proofs.

(iv) *Modes of convergence of holomorphic functions.* Recall that formula (10), an immediate consequence of the elementary theory of Fourier series, is the essential tool in the proof of Weierstrass' theorem on limits of holomorphic functions (Chap. VII, § 4, no 19). Indeed, if f is holomorphic on the open set G, if K is a compact subset of G and if $r > 0$ is *strictly* less than the distance from K to the boundary of G, then, taking for μ the circle $|\zeta - w| = r$, (9) may be applied for all $w \in K$. Set $\zeta = w + r\mathbf{e}(t)$, whence $d\zeta = 2\pi i r \mathbf{e}(t) dt$ and

$$f(\zeta) = \sum_{n \geq 0} f^{(n)}(w) r^n \mathbf{e}(t)^n / n! \,,$$

Then,

(4.12) $$f^{(n)}(w)/n! = \int f(w + r\mathbf{e}(t)) \, r^{-n} \mathbf{e}(t)^{-n} dt \,,$$

where integration is over $[0, 1]$. Hence setting $K(r)$ to be the compact set contained in G, consisting of points at a distance $\leq r$ from K, and maximizing of the left hand side over K, we get

(4.13) $$\left\| f^{(n)} \right\|_K \leq n! r^{-n} \|f\|_{K(r)} \,,$$

which proves that, *in the vector space of holomorphic functions on G, the map $f \mapsto f^{(n)}$ is continuous with respect to the topology of compact convergence.*[25] Indeed, if a sequence (f_p) converges uniformly to a limit f on every compact subset of G, the same holds for the successive derivatives of f in the real sense, which is identical up to constant factors to the complex sense. The limit function is, therefore, C^∞ and satisfies Cauchy's condition by passing to the limit. Its limit is, therefore, holomorphic, and as the derivatives in the real sense converge uniformly to those of f on every compact subset, the expected result follows (Chap. VII, no 19, Weierstrass' theorem).

In fact, Weierstrass' conclusion can be reached by making seemingly much weaker assumptions than compact convergence on the convergence of the sequence (f_n). Let us in particular consider $L^p (1 \leq p < +\infty)$ convergence in the theory of integration. It is defined by the norm

(4.14) $$\|f\|_p = \left(\iint_G |f(z)|^p \, dm(z) \right)^{1/p} \,,$$

---

[25] Recall (Chapter III, Appendix, no 8) that it is defined by the seminorms $f \mapsto \|f\|_K$, where $K \subset G$ is an arbitrary compact set. The inequality obtained shows that if f converges uniformly on $K(r)$, then $f^{(n)}$ converges uniformly on K.

where integration is over the given open set G^{26} with respect to the usual measure $dm(z) = dxdy$. We need to show that *for holomorphic functions on G, the relation*

$$(4.15) \qquad \lim \|f - f_n\|_p = 0 \quad implies \quad \lim f_n(z) = f(z)$$

uniformly on any compact subset K of G, i.e. that there is an upper bound

$$(4.16) \qquad \|f\|_K \leq M_K \|f\|_p$$

valid for any compact subset $K \subset G$ and any function f holomorphic on G. In fact, since under these conditions, for all r, $\lim f_n^{(r)}(z) = f^{(r)}(z)$ uniformly on every compact subset, we also get upper bounds

$$(4.16) \qquad \left\| f^{(r)} \right\|_K \leq M_{K,r} \|f\|_p$$

for all $r \in \mathbb{N}$.

To prove (16), let us again consider the compact set $K(r) \subset G$. For all $a \in K$, the disc $D(a,r)$ is contained in $K(r)$ and

$$f(a) = \int_0^1 f\left[a + r\mathbf{e}(t)\right] dt.$$

If we assume that in polar coordinates $x = \rho \cos 2\pi t$, $y = \rho \sin 2\pi t$, the measure $dm(z) = dxdy$ is given by $dxdy = 2\pi \rho d\rho dt$, then

$$\iint_{D(a,r)} f(z) dxdy = 2\pi \int_0^r \rho d\rho \int_0^1 f\left[a + \rho\mathbf{e}(t)\right] dt = \pi r^2 f(a).$$

Since πr^2 is the area of the disc $D(a,r)$, this means that *the value of a holomorphic function at the centre of a disc is equal to its mean value over the disc*. As $D(a,r) \subset K(r)$ for all $a \in K$,

$$(4.17) \qquad \pi r^2 \|f\|_K \leq \iint_{K(r)} |f(z)| dm(z) \leq \iint_G = \|f\|_1,$$

which proves (16) for $p = 1$. For $1 < p < +\infty$, applying Hölder's inequality (Cauchy-Schwarz for $p = 2$) to the functions f and 1 on $K(r)$, we get

$$\pi r^2 \|f\|_K \leq \left(\iint_{K(r)} |f(z)|^p dm(z) \right)^{1/p} \left(\iint_{K(r)} dm(z) \right)^{1/q},$$

---

[26] To define integral (14), which pertains to a positive continuous function, the method holding for lower semicontinuous functions must be applied (Chap. V, § 9, no 33, theorem 31): consider functions $\leq |f(z)|^p$ in $\mathbb{R}^2$, everywhere continuous positive on G and zero outside compact subsets of G. The integral of $|f(z)|^p$ is then the supremum of the integrals of these functions. Considering the supremum of extended integrals of $|f(z)|^p$ over compact sets $K \subset G$ would amount to the same. This presupposes that we know how to integrate over an arbitrary compact set (same reference).

where $1/p + 1/q = 1$ (Chap. V, § 3, no 14); the first integral is less that $\|f\|_p$ and the second does not depend on f, hence (16). (16') can be similarly obtained from (12).

The reader will easily verify that relation (16) continues to hold if, instead of the usual measure $dxdy$, we use a measure $d\mu(z) = \rho(z)dm(z)$ to define $\|f\|_p$, i.e. the formula

$$(4.18) \qquad \|f\|_p = \left(\iint_G |f(z)|^p \rho(z)dxdy \right)^{1/p},$$

where the given "density" ρ is continuous and has *strictly* positive values. It suffices to note that the minimum m over $K(r)$ of ρ is > 0. Hence

$$\iint_{K(r)} |f(z)|^p dm(z) \leq \frac{1}{m} \iint_{K(r)} |f(z)|^p \rho(z)dm(z)$$

for any function f.

From this result, it follows that, for any measure $d\mu(z)$ of the form (18), the normed vector space $\mathcal{H}^p(G, \mu)$ of holomorphic functions such that

$$\iint |f(z)|^p d\mu(z) < +\infty$$

is *complete*:[27] Indeed, inequality (16) transforms every Cauchy sequence with respect to the norm L^p into a Cauchy sequence with respect to compact convergence. Hence we get a holomorphic limit which can easily be inferred to also be the L^p- limit of the functions f_n. This is obvious *if* we have the benefit of the simplest results from Lebesgue theory.[28]

In particular, $\mathcal{H}^2(G, \mu)$, equipped with the scalar product

$$(f|g) = \iint f(z)\overline{g(z)}d\mu(z),$$

is a *Hilbert space* which plays an important role in some questions, especially complex Fourier transformations, the theory of modular functions, conformal representations, etc.

In fact, as observed by Laurent Schwartz half a century ago, much more can be proved. Let G be an open subset of $\mathbb{C}$ and $\mathcal{D}(G)$ the vector space of

---

[27] As we will see in n° 12, it can be reduced to zero if the open set G is not bounded, especially in the trivial case when $G = \mathbb{C}$ since (17) then applies for all $r > 0$.

[28] Given a Cauchy sequence (f_n) in an L^p-space , if $\lim f_n(x) = f(x)$ exists almost everywhere, then $f \in L^p$ and $\lim \|f - f_n\|_p = 0$ (Chap. XI). In the case at hand, the functions f_n and f are continuous and the sequence converges uniformly on every compact subset. Hence the theorem is in fact about Riemann integrals. But a proof based on elementary arguments requires far more ingenuity than the use of Lebesgue's sledgehammer.

C^∞ functions with compact support in G. Like on $\mathbb{R}$ (Chapter V, § 10), it can be used to define *distributions* on G. For all $r \in \mathbb{N}$, define a semi-norm on $\mathcal{D}(G)$ (Appendix of Chap. III, end of no 8)

$$N_r(\varphi) = \sum_{p,q \leq r} \sup_{z \in G} |D_1^p D_2^q \varphi(z)| = \sum \|D_1^p D_2^q \varphi\|_G \,,$$

where D_1 and D_2 are the differential operators with respect to the real coordinates x, y of z. Writing $\mathcal{D}(G, K)$ for the subspace of the functions $\varphi \in \mathcal{D}(G)$ which vanish outside a given compact set $K \subset G$, a distribution of G is then a linear functional $\varphi \mapsto T(\varphi)$ in $\mathcal{D}(G)$, continuous in the following sense : for any compact set $K \subset G$, there exists $r \in \mathbb{N}$ and a constant $M_K(T) \geq 0$ such that

$$|T(\varphi)| \leq M_K(T) N_r(\varphi) \quad \text{for all} \quad \varphi \in \mathcal{D}(G, K).$$

Like on $\mathbb{R}$, the successive derivatives of the distributions can be defined by iterating the formulas defining the first derivatives

$$D_1 T : \varphi \longmapsto -T(D_1 \varphi) \,, \quad D_2 T : \varphi \longmapsto -T(D_2 \varphi)$$

of T. If T is defined by a function f of class C^∞ on G, i.e. if

$$T(\varphi) = \iint \varphi(z) f(z) dm(z) \,,$$

it can be immediately checked that $D_i T$ is defined by a function $D_i f$: like over $\mathbb{R}$, integrate by parts[29] with respect to x or y. The notion of a holomorphic function can then be generalized by saying that a distribution T on G is holomorphic if it satisfies Cauchy's condition

$$D_2 T = i D_1 T$$

or, equivalently, if T vanishes for all functions of the form $\partial \varphi / \partial \bar{z}$.

Having set this, (i) *any holomorphic distribution is defined by a holomorphic function*, in other words this is not a generalization; (ii) if a sequence T_n of holomorphic distributions converges to a distribution T, i.e. if

$$\lim T_n(\varphi) = T(\varphi) \quad \text{for all} \quad \varphi \in \mathcal{D}(G) \,,$$

then T is holomorphic (obvious from the definition); (iii) if f_n and f are holomorphic functions defining T_n and T, then $f_n(z)$ converges uniformly to $f(z)$ on any compact subset of G.

Regarding *holomorphic* functions, any definition of convergence more restrictive than of convergence in the sense of distributions implies, therefore,

---

[29] As φ is zero outside a compact subset $K \subset G$, the function under the $\int$ sign is the restriction to G of a C^∞ function on $\mathbb{R}^2$ and vanishes outside K, hence outside a square $I \times I$, where I is a compact interval of $\mathbb{R}$.

compact convergence. This result covers every type of convergence encountered in practice, in particular L^p convergence considered above.

(v) *Analyticity of holomorphic functions.* Cauchy's formula for a circle assumes only that f is holomorphic in the sense initially defined in Chap. II, n° 19. Using the theory of Fourier series, it was directly shown in Chap. VII, no 14 that in fact all holomorphic functions are analytic, i.e. have power series expansions, and (10) was deduced. Cauchy's formula provides under proof of analyticity, that of Cauchy which everyone reproduced and which proceeds in reverse order. It suffices to replace a by a variable z in (10), and to write that

$$1/(\zeta - z) = \zeta^{-1}/(1 - z/\zeta) = \sum_{n \geq 0} z^n \zeta^{-n-1}.$$

This geometric series expansion is justified as long as $|z| < |\zeta| = R$. For given z, this series of functions of ζ converges normally (Chap. III, n° 8) on the circle $|\zeta| = R$ since it is dominated by the series $\sum q^n/R$, with $q = |z|/R < 1$. As the function f is continuous on the circle, it can be integrated term by term. Hence

(4.19) $2\pi i f(z) = \sum c_n z^n$ with $c_n = \int f(\zeta) \zeta^{-n-1} d\zeta,$

where integration is along the circle of radius R, qed.

Like the one of Chap. VII, this proof shows that expansion (19) holds in the largest disc D centered at 0 contained in the domain G where f is holomorphic. As indeed all circles centered at 0 are pairwise homotopic as closed paths in the domain $G - \{0\}$ where the function $f(\zeta)\zeta^{-n-1}$ being integrated is holomorphic, the integral representing the c_n is independent from R as long as the closed disc $|z| \leq R$ is contained in G. Since the power series then converges for $|z| \leq R$, the result follows. This argument also shows that Cauchy's integral formula for a disc characterizes holomorphic functions on the disc since it implies a power series expansion or else because the function being integrated on a compact set depends holomorphically on the parameter z, which allows theorems on differentiation under the $\int$ sign to be applied.

(vi) *Laurent Series.* Consider a function f holomorphic on an annulus $G : r < |z| < R$ and let z be a point of G. Choose numbers r' and R' such that $r < r' < |z| < R' < R$ and a ray D with initial point 0 which does not pass through z; let A and B be the points where it meets circles of radius r' and R'. Consider the closed path μ consisting of AB, followed by the circumference $|\zeta| = R'$ oriented positively, then by BA followed in turn by the circle $|\zeta| = r'$ oriented clockwise. It is obviously homotopic in $G - \{z\}$ to a circumference γ centered at z contained in G. The integrals of

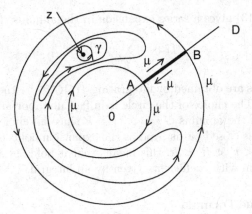

Fig. 4.6.

$f(\zeta)/(\zeta - z)$ along these two paths being equal, Cauchy's formula for a circle shows that

$$(4.20) \qquad f(z) = \frac{1}{2\pi i} \int_\mu \frac{f(\zeta)}{\zeta - z} d\zeta .$$

The contribution from the line segment AB is clearly zero since it is followed twice in reverse directions. Hence, this leaves the difference between the extended integrals along the circumferences $|\zeta| = R'$ and $|\zeta| = r'$ oriented positively.

For $|\zeta| = R'$, $|z/\zeta| < 1$ and so

$$1/(\zeta - z) = 1/\zeta(1 - z/\zeta) = \sum_N z^n \zeta^{-n-1} .$$

The contribution from the circumference $|\zeta| = R'$ to integral (13) is, therefore, like in (v), the power series

$$\sum_{n \geq 0} c_n z^n \quad \text{with} \quad 2\pi i c_n = \int_{|\zeta|=R'} f(\zeta) \zeta^{-n-1} d\zeta .$$

For $|\zeta| = r'$, $|\zeta/z| < 1$, which allows us to write

$$1/(\zeta - z) = -1/z(1 - \zeta/z) = - \sum_{n \geq 0} \zeta^n z^{-n-1} .$$

The product with $f(\zeta)$ is again integrable term by term: this is clearly a normally convergent series. Replacing n by $-n - 1$ where, this time, $n < 0$, its contribution is easily seen to be equal to

$$\sum_{n < 0} c_n z^n \quad \text{with} \quad 2\pi i c_n = \int_{|\zeta|=r'} f(\zeta) \zeta^{-n-1} d\zeta .$$

Hence formula (13) gives a series expansion in the annulus $r' < |z| < R'$

(4.21) $$f(z) = \sum_{\mathbb{Z}} c_n z^n$$

whose coefficients are obtained by integrating $f(\zeta)\zeta^{-n-1}$ either over $|\zeta| = R'$, or over $|\zeta| = r'$. The choice of the circle is in fact unimportant provided it is in the interior of the annulus $G : r < |z| < R$ since all circles are obviously homotopic in G as closed paths. And as the expansion holds for $r' < |z| < R'$ provided that $r < r' < R' < R$, this means that it holds in all of G. This is Laurent's theorem with coefficients given by an integral.

5 – The Residue Formula

(i) *The residue formula.* A formula (4.10) valid for any closed path μ in G, any function f holomorphic on G and any $z \in G$, as well as a more general and extremely classical result, can be proved in a simply connected domain: Cauchy's residue formula, an inexhaustible source of exercises and examination questions that, though sometimes subtle, have long become stale. Indeed, everything can be proved at the same time and G need not be assumed to be simply connected provided μ is taken to be null-homotopic in G.

First some remarks about the Laurent series of a function f holomorphic for $0 < |z - a| < R$, a being a singular isolated point of f (Chapter VII, §4, n° 16). We saw above that it is given by

(5.1) $f(z) = \sum c_n (z - a)^n$ with $2\pi i c_n = \int f(\zeta)(\zeta - a)^{-n-1} d\zeta$

where integration is over any circle $t \mapsto a + r.\exp(2\pi i t)$ of radius $r < R$ centered at a. The sum of the terms of degree $n < 0$, i.e. the *polar* or *singular part* of the Laurent series, is a power series in $w = 1/(z - a)$; it converges for $0 < |z - a| < R$, i.e. for $|w| > 1/R$; Now, the domain of convergence of a power series is the interior of a disc; if it converges outside a disc, then it converges everywhere. Therefore, the series given by the terms of negative degree in (1) converges for all $z \neq a$. So, in the neighbourhood of a, f *is the sum of a power series and of a holomorphic function on* $\mathbb{C} - \{a\}$.

Having said this, let us consider a function f which, instead of being holomorphic everywhere on a domain G, is so on $G - S = G'$, where S is a temporarily finite set. Let $\rho_a = \text{Res}(f, a)$ be the *residue* of f at $a \in S$, i.e. the coefficient of $1/(z - a)$ in its Laurent series at a. Write $g_a(z)$ for the sum of the terms of degree ≤ -2 – as seen above, it is in fact defined and holomorphic on $\mathbb{C} - \{a\}$ – and consider the function

(5.2) $$g(z) = f(z) - \sum [g_a(z) + \rho_a/(z - a)] \ ;$$

each g_a being holomorphic on $\mathbb{C} - \{a\}$, g is at least defined on G'. Its only singular points in G are among the points $a \in S$, but, since like in (2) the

terms indexed by $b \neq a$ are holomorphic with respect to a, the polar part of
the Laurent series of g with respect to a is obtained by removing from the
series of f that of $g_a(z) + \rho_a/(z-a)$, i.e. the sum of terms of degree < 0
of the Laurent series of f. Therefore, in reality, the Laurent series of g with
respect to a is a *power* series, and denoting the constant term of the Laurent
series of f with respect to a by $g(a)$, g is transformed into a function defined
and holomorphic on all of G.

To compute the integral of f along a closed path μ in G', it then suffices
to compute those of the functions g_a, the functions $1/(z-a)$ and of g.

By definition of the index, to begin with we have

$$(5.3) \qquad \int_\mu \frac{d\zeta}{\zeta - a} = 2\pi i.\,\mathrm{Ind}_\mu(a) \ .$$

As, on the other hand, g is holomorphic on all of G and as by assumption, μ
is null-homotopic, its integral along μ is zero (Corollary 3 of Theorem 3). As
for the function g_a, for $\zeta \neq a$, it is represented by a series

$$g_a(\zeta) = \sum_{n \leq -2} c_n(\zeta - a)^n$$

converging in $\mathbb{C} - \{a\}$ and without any terms of degree -1; it, therefore,
admits a primitive

$$\sum_{n \leq -2} c_n(\zeta - a)^{n+1}/(n+1)$$

on $\mathbb{C} - \{a\}$ (Chap. VII, n° 16: possibility of differentiating a Laurent series
term by term). So irrespective of whether μ is null-homotopic or not, the
integral of g_a along μ is zero.

The only terms in sum (2) contributing effectively to the computation of
the integral are, therefore, the fractions $\rho_a/(\zeta - a)$. Hence, in view of (3), the
relation

$$(5.4) \qquad \int_\mu f(\zeta)d\zeta = 2\pi i \sum_{a \in S} \mathrm{Ind}_\mu(a)\,\mathrm{Res}(f, a)$$

follows.

This result assumes that f has finitely many singular points in G. In fact,
it remains valid when S is a possibly infinite *closed and discrete* subset of
the topological space G or, equivalently, has a neighbourhood V for all $z \in G$
such that $V \cap S$ is finite, or else is such that $K \cap S$ is *finite* for any compact[30]

---

[30] "closed in G" means that every point of G (and not of $\mathbb{C}$) is a limit point of S
and is in S; "discrete in G" means that every $z \in S$ has a neighbourhood V such
that $V \cap S = \{z\}$. Supposing that $K \subset G$ is compact, if $K \cap S$ is infinite, then
there exists (use BL) $a \in K$ such that $V \cap S$ is infinite for every neighbourhood V
of a. Hence, we get a sequence of pairwise distinct points of S converging to a;
since S is closed in G, it follows that $a \in S$, which contradicts the discreteness
assumption on S.

set $K \subset G$. To see this, let us first prove the following result which holds for far more general spaces:

Lemma. *For any open subset G of $\mathbb{C}$, there exist a sequence of compact subsets K_n and open subsets G_n such that*

$$G = \bigcup K_n, \quad K_n \subset G_n \subset K_{n+1}.$$

Every compact set contained in G is then contained in some K_n.

For every $z \in \mathbb{C}$, set $d(z) = d(z, \mathbb{C} - G)$. As $\mathbb{C} - G$ is closed, relation $d(z) = 0$ is equivalent to $z \in \mathbb{C} - G$, and so G is the set of $z \in \mathbb{C}$ such that $d(z) > 0$. Besides, it is obvious that

$$(*) \qquad\qquad |d(z') - d(z'')| \leq d(z', z'') = |z' - z''|$$

for all z' and z''. So the function d is continuous. Then define K_n by

$$z \in K_n \iff d(z) \geq 1/n \quad \& \quad |z| \leq n.$$

The subsets K_n are closed and bounded, and hence compact; they form an increasing sequence and G is obviously their union. Besides, every $a \in K_n$ is in the interior of K_{n+1} because, by $(*)$,

$$d(a, z) \leq 1/n - 1/(n+1) \implies d(z) \geq d(a) - d(a, z) \geq 1/(n+1),$$

so that K_{n+1} contains a disc centered at a. The set G_n consisting of the interior points of K_{n+1} is, therefore, suitable. Finally, if K is a compact subset of G, it is covered by the open subsets G_n, hence by a finite number of them. So $K \subset K_n$ for large n, qed.

Having done this, let us return to formula (4) for a holomorphic function on $G - S$ and a closed path μ in $G - S$, null-homotopic in G. As the path μ contracts to a point under a homotopy, it describes a compact compact set $K \subset G$ contained in one of the open subsets G_n of the Lemma. As $G_n \cap S = S_n$ is finite and as μ is null-homotopic in G_n, (18) applies to G_n provided only the points $a \in S_n$ are included in it. But as the result holds for sufficiently large n, passing to the limit is trivial (in fact, there are finitely many non-zero residues), qed. In conclusion:

Theorem 5 (Cauchy's residue formula). *Let G be a domain in $\mathbb{C}$, S a closed discrete subset of G and f a holomorphic function on $G' = G - S$. Then*

$$(5.5) \qquad\qquad \int_\mu f(\zeta)d\zeta = 2\pi i \sum_{a \in S} \mathrm{Ind}_\mu(a)\,\mathrm{Res}(f, a)$$

for any admissible closed path μ in G' null-homotopic in G.

If G is simply complex, the formula can be applied to every closed path in $G - S$. Hence, if the residues of f are all zero, then the integral of f along any closed path in G' is zero. As a result:

Corollary. *A holomorphic function f on G − S,where G in simply connected, has a primitive on G − S if and only if* $\mathrm{Res}(f,a) = 0$ *for all* $a \in S$.

The figure below is considered in the classical theory where $S = \{a_1, \ldots, a_n\}$ is finite. It contains the path μ and a path ν consisting, on the one hand, of

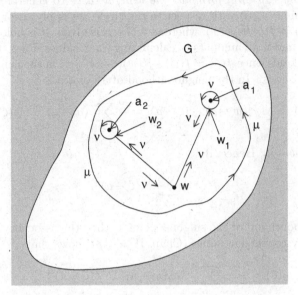

Fig. 5.7.

paths connecting in $G - S$ an arbitrarily chosen point w to points w_p in the neighbourhood of a_p, and on the other, of circular paths μ_p centered at a_p. The path ν consists of a loop surrounding the point a_1, followed by a loop surrounding the point a_2 and so on until the point a_n, the first loop consisting of the path from w to w_1, followed by a path μ_1 from w_1 to w_1, followed in turn by the first path in the reverse direction from w_1 to w, the other loops being defined similarly. When f is integrated along ν, the contributions from the paths connecting w to w_p cancel out, leaving the sum of integrals along small circles surrounding the points a_p. But if $f(\zeta) = \sum c_n(\zeta - a_p)^n$ is integrated along a sufficiently small circle μ_p centered at a_p, the formula for the calculation of the coefficients when $n = -1$ gives $2\pi i c_{-1}$. The integral of f along ν is, therefore, $2\pi i \sum \mathrm{Res}(f, a_p)$. On the other hand, ν and the initially given path μ are "obviously" homotopic in G as closed paths. Hence the integrals of f along μ and ν are equal. This proves the residue formula.

Except for one detail: this argument leaves out the factors $\mathrm{Ind}_\mu(a_p)$. To obtain them, the above figure needs to be made more complicated when μ circles the points a_p several times, in both possible directions.

This type of argument, heavily exploited by Riemann in his theory of algebraic functions – in his days, he did not have any choice – and by his many successors, has, nonetheless, supplied several useful formulas as we will have occasion to see later.

(ii) *Cauchy's integral formula : the general case.* To generalize Cauchy's integral formula (4.10) with respect to the circle, we apply theorem 5 to the function $g(\zeta) = f(\zeta)/(\zeta - w)$, where $w \in G - S$ is given. It is holomorphic on $G - \{w\} \cup S$ and it all amounts to calculating the residues. First, $\mathrm{Res}(g, w) = f(w)$ clearly holds since $f(\zeta) = f(w) + f'(w)(\zeta - w) + \ldots$ in the neighbourhood of w (Taylor series). In the neighbourhood of a point $a \in S$,

$$1/(\zeta - w) = -1/\left[(w - a) - (\zeta - a) \right] = - \sum_{m \in \mathbb{N}} (w - a)^{-m-1} (\zeta - a)^m$$

if $|\zeta - a| < |w - a|$; Hence, if

$$(5.6) \qquad f(\zeta) = \sum_{n \in \mathbb{Z}} c_n (\zeta - a)^n$$

in the neighbourhood of the singular point a, then the associativity theorem for absolutely convergent series (Chap. II, n° 18) shows that

$$g(\zeta) = - \sum_{n \in \mathbb{Z}, \, m \geq 0} c_n (w - a)^{-m-1} (\zeta - a)^{m+n} .$$

The residue of g is obtained by grouping together the terms for which $m + n = -1$ (same reference), and so

$$(5.7) \quad \mathrm{Res}(g, a) = - \sum_{m \in \mathbb{N}} c_{-m-1} (w - a)^{-m-1} = - \sum_{n < 0} c_n (w - a)^n .$$

This is the value up to sign at w of the polar part[31] of the Laurent series of f with respect to a.

This formula simplifies if f has a *simple pole* at a, i.e. if its Laurent series with respect to a reduces to its term of degree -1, leaving $\mathrm{Res}(g, a) = -c_{-1}/(w - a) = -\mathrm{Res}(f, a)/(w - a)$. Hence formula (5) applied to g gives the following result:

Theorem 6. *Let G be a domain $\mathbb{C}$, S a closed and discrete subset of G and f a holomorphic function on $G - S$ with simple poles at every point of S. If $w \in G - S$ and if μ is a closed path in $G - S \cup \{w\}$ null-homotopic in G, then*

$$(5.8) \qquad \frac{1}{2\pi i} \int_\mu \frac{f(\zeta)}{\zeta - w} d\zeta = \mathrm{Ind}_\mu(w) f(w) + \sum_{a \in S} \mathrm{Ind}_\mu(a) \frac{\mathrm{Res}(f, a)}{a - w} .$$

---

[31] It was shown above that this polar part converges in $\mathbb{C} - \{a\}$.

This type of argument has several versions.

Theorem 6 bis. *Let G be a* simply connected *domain, f a holomorphic function on G and μ a closed admissible path in G. Then*

(5.9)
$$\frac{1}{2\pi i}\int_{\mu}\frac{f(\zeta)}{(\zeta-w)^{n+1}}d\zeta = \mathrm{Ind}_{\mu}(w).f^{(n)}(w)/n!$$

for all $w \in G - \mathrm{Supp}(\mu)$ and for all $n \in \mathbb{N}$.

This result is the residue theorem applied to the function $f(\zeta)/(\zeta-w)^{n+1}$; indeed, it has at most one singular point in G: a pole at $z = w$, with residue $f^{(n)}(w)/n!$, since Taylor's formula

$$f(z)/(z-w)^{n+1} = (z-w)^{-n-1}\sum(z-w)^p f^{(p)}(w)/p!$$

shows that the coefficient of $1/(z-w)$ is equal to $f^{(n)}(w)/n!$.

(iii) *The number of zeros and poles of a function.* The residue formula has several other immediate consequences. Consider for example a *meromorphic* function f on a domain G; f being holomorphic and without any zeros on $G - S$, this means that there exists a discrete and closed subset S of G such that the points of S are zeros or poles, but not essential singularities, of f. The function $f'(z)/f(z)$ is then holomorphic on $G - S$ and its only singularities are the points $a \in S$. In the neighbourhood of any point $a \in G$, there is a series expansion

(5.10)
$$f(z) = c_p(z-a)^p + c_{p+1}(z-a)^{p+1} + \dots$$

with $c_p \neq 0$; the integer p is ≥ 0 if f is holomorphic at a; it is > 0 if $f(a) = 0$, and it is < 0 if a is a pole of f. Denote it by $v_a(f)$, so that if S is finite, then

$$f(z) = u(z)\prod(z-a)^{v_a(f)},$$

where the function u is holomorphic everywhere and $\neq 0$ everywhere on G, and so has an inverse in the ring of meromorphic functions on G. This is the analogue of the decomposition of an integer into primes.[32] Having said this, let us apply the residue formula to the function $f'(z)/f(z)$, holomorphic outside S. By (10),

$$f(z) = (z-a)^p g(z),$$

where g is holomorphic and non-zero at a, and so

---

[32] for extension to meromorphic functions in divisibility theory , see Remmert, *Funktionentheorie 2*, Chap. 3 and 4, where it is in particular shown that every meromorphic function on a domain G is the quotients of two holomorphic functions on G having no common zeros.

$$f'(z)/f(z) = p/(z-a) + g'(z)/g(z).$$

The function g'/g being holomorphic at a, $\mathrm{Res}(f'/f, a) = p = v_a(f)$ and the residue formula then shows that

(5.11)
$$\int_\mu \frac{f'(z)}{f(z)} dz = 2\pi i \sum \mathrm{Ind}_\mu(a) v_a(f) = 2\pi i v_\mu(f),$$

where the sum, extended to all $a \in G$, only has a finite number of non-zero terms. If for example f is holomorphic everywhere on G and if μ is a simple closed curve whose interior is contained in G and on which f does not vanish, then integral (11) allows us to calculate the total number of zeros of f in the interior of μ; this number takes into account the order or *multiplicity* $v_a(f)$ of each zero.

As a function of f, the left hand side of (11) satisfies a remarkable continuity property resulting from that of the map $f \mapsto f^{(n)}$ studied above. The open set G and the path μ being fixed, first note that if f does not vanish on $\mathrm{Supp}(\mu)$, then neither do functions g defined on G and sufficiently near f in the sense of compact convergence. Then choose a number $r > 0$ strictly less that the distance from $\mathrm{Supp}(\mu)$ to the boundary of G and let $K \subset G$ be the compact set of points whose distance to $\mathrm{Supp}(\mu)$ is $\leq r$. Since f does not vanish on $\mathrm{Supp}(\mu)$, for sufficiently small r, it does not vanish on K either. The lower bound d of $|f(z)|$ in K is, therefore, > 0 if r is sufficiently small, and that of $|g(z)|$ is $\geq d/2$ if $\|f - g\|_K < d/2$. If g is holomorphic, the results stated in n° 4, (iv) show that $\|f' - g'\|_{\mu(I)} \leq Md$ where M does not depend on g. If $\|f - g\|_K$ is sufficiently small, the integral

$$2\pi i v_\mu(g) - 2\pi i v_\mu(f) = \int_\mu \left[\frac{g'(z)}{g(z)} - \frac{f'(z)}{f(z)} \right] dz$$

is well defined and considering an upper bound for it,[33] it follows that for $\|f - g\|_K$ sufficiently small, $|v_\mu(g) - v_\mu(f)| < 1$, and so $= 0$. As a result, *the number $v_\mu(f)$ of roots of a holomorphic function f in the interior of a closed path μ is a continuous function of f with respect to the topology of compact convergence.* In particular, for every functions f holomorphic on G and all $a \in G$, there exist $r > 0$, $\rho > 0$ and a compact set $K \subset G$ such that the number of zeros in the disc $|z - a| < r$ of any function g holomorphic on G and satisfying $\|g - f\|_K < \rho$ is the same as that of f.

If f and g are replaced by $f - c$ and $f - c'$, where c and c' are constants, so that $\|g - f\|_K = |c' - c|$, then, for given c, the equations $f(z) = c$ and $f(z) = c'$ have the same number of solutions in the interior of μ, provided $|c' - c|$ is sufficiently small. A direct argument: if $f(z) - c$ does not vanish on

---

[33] It all amounts to finding an upper bound for an expression of the form $|f/g - p/q|$ when lower bounds > 0 of g and q and upper bounds for f, p, $|f - p|$ and $|g - q|$ are known . See rules of Chapter III, § 2, n° 7 with respect to algebraic operations on uniformly convergent sequences.

Supp(μ), neither does $f(z) - c'$ for c' sufficiently near c; when c' approaches c, $f'/(f - c')$ obviously converges uniformly to $f'/(f - c)$ on Supp(μ), implying the result. This also means that *the number $v_\mu(f - c)$ of solutions of $f(z) = c$ in the interior of μ only depends on the connected component of $\mathbb{C} -$ Supp(μ) containing c.*

If for example f has a zero of order p at a and does not vanish elsewhere on a disc $|z - a| < r$, then, for any sufficiently small $c \in \mathbb{C}$, the equation $f(z) = c$ has p solutions in the disc. This shows that *the image under f of a disc centered at a contains a disc centered at $f(a)$* and, as a is arbitrary, that f maps every open subset of G to an open subset of $\mathbb{C}$. Moreover, the roots of $f(z) = c \neq 0$ in $|z - a| < r$ are *pairwise distinct* for sufficiently small r since even if $f'(a) = 0$, $f'(z) \neq 0$ for $0 < |z - a| < r$ if r is sufficiently small, which prevents $f(z) = c$ from having multiple roots in this disc.

For $p = 1$, this argument shows that f is injective in the neighbourhood of a if and only if $f'(a) \neq 0$. If f is globally injective on G, then $f'(z) \neq 0$ everywhere and as shown above (and in Chapter III, § 5, n° 24, by using the real-variable version of the local inversion theorem), f is a conformal representation on an open subset as remarked earlier. In conclusion:

Theorem 7. *Any holomorphic and non-constant function f on G maps open subsets of G to open subsets of $\mathbb{C}$. It is a conformal representation of G on $f(G)$ if and only if f is injective on G.*

(iv) *Residues at infinity.* The simplest functions the residue formula can be applied to are rational functions $f(z) = p(z)/q(z)$, where p and q are polynomials without common roots. As will be seen later, using this method, the integral over $\mathbb{R}$ of any function of this type can be computed, at least when it converges. But an important theoretical result about residues of rational functions can now be proved.

Indeed, integrate f over a circle $|z| = R$ going around it once counter-clockwise. If R is sufficiently large, we get, up to a factor of $2\pi i$, the sum of the residues of f at all its poles, which are the roots of q. Now, for large z, there is an asymptotic estimate of the form $f(z) \sim cz^n$, with $c \neq 0$ and $n = d°(p) - d°(q)$, and in particular $|f(z)| \leq M|z|^n$, where M is a constant. The integral over the circle, equal to

$$(5.12) \qquad \int_0^1 2\pi i f\left[\mathbf{Re}(t)\right] \mathbf{Re}(t) dt\,,$$

where $\mathbf{e}(t) = \exp(2\pi i t)$, is, therefore, $O(R^{n+1})$. So it approaches 0 if $n \leq -2$, i.e. if

$$(5.13) \qquad d°(q) \geq d°(p) + 2\,.$$

Hence assumption (13) implies the relation

$$(5.14) \qquad \sum \mathrm{Res}(f, a) = 0\,.$$

Otherwise, the argument falls apart otherwise and in fact (14) no longer holds. As seen in n° 4, (vi), on the exterior of a disc of very large radius R, the function f has a Laurent series expansion $\sum c_p z^p$ with

$$2\pi i c_p = \int_{|\zeta|=R} f(\zeta)\zeta^{-p-1}d\zeta .$$

For $p = -1$, we recover the integral of p along the circle, which is therefore equal to $2\pi i c_{-1}$. In the general case, (14) needs to be replaced by the relation

(5.15) $$\sum \mathrm{Res}(f,a) = c_{-1} .$$

If in particular $f(z) \sim c/z$ at infinity (case $n = -1$), then

(5.16) $$\sum_{a \in \mathbb{C}} \mathrm{Res}(f,a) = c = \lim_{z \infty} z f(z) .$$

Setting (without forgetting the sign!)

$$\mathrm{Res}(f,\infty) = -c_{-1}$$

to be the *residue of f at infinity*, instead of (14), we get

(5.14') $$\sum_{a \in \mathbb{C} \cup \{\infty\}} \mathrm{Res}(f,a) = 0.$$

This tautology is justified by its extension to much more general situations (compact Riemann surfaces) where it is far less obvious.

This residue at infinity can be defined for any function f, both rational and otherwise, defined and holomorphic for large $|z|$. First of all, what will be meant by the behaviour of f "in the neighbourhood of infinity" needs to be specified. Such a function, whether rational or not, has a Laurent series expansion $f(z) = \sum c_n z^n$ which converges for large $|z|$, with possibly infinitely many terms of negative or positive degree. It is then natural to say that f is *holomorphic at infinity* or *on a neighbourhood of infinity* if $c_n = 0$ for all $n > 0$ – in other words, if $f(z) = O(1)$ for large $|z|$ – and to set, by definition,

(5.17) $$f(\infty) = c_0 = \lim_{z \infty} f(z) .$$

If $f(\infty) = 0$, f will be said to have a *zero of order p at infinity* if

(5.18') $$f(z) = \ldots + c_{-p-1}z^{-p-1} + c_{-p}z^{-p} \quad \text{with} \quad c_{-p} \neq 0,$$

in other words if $f(z) \asymp 1/z^p$ for large z.

If f is not holomorphic at infinity, f will be said to have a *pole of order p at infinity* if

(5.18") $$f(z) = \ldots + c_{p-1}z^{p-1} + c_p z^p \quad \text{with} \quad c_p \neq 0,$$

in other words, if $f(z) \asymp z^p$ large large z. Otherwise, the expression *essential singular point at infinity* is used, for example in the case of e^z. Then, expansion (18") leads to setting

$$\operatorname{Res}(f, \infty) = -c_{-1}$$

as above, whence

(5.19) $$\int_{|z|=R} f(z)dz = -2\pi i \operatorname{Res}(f, \infty) \quad \text{for large } R$$

follows, attention needing to be paid to the sign on the right hand side...

A rational function only has polar singularities in $\hat{\mathbb{C}} = \mathbb{C} \cup \{\infty\}$, in other words is meromorphic on $\hat{\mathbb{C}}$, and no other function satisfies this property. First of all, such a function f can only have finitely many poles because, even if it has a pole at infinity, it is holomorphic outside a compact disc D. However, it can only have finitely many poles in D. Multiplying f by a polynomial chosen so as to remove the poles of f in D, we get a function whose only possible singularity in $\hat{\mathbb{C}}$ is a pole at infinity; it is, therefore, an entire function on $\mathbb{C}$ whose order of magnitude at infinity is that of a power of z, and hence, by Liouville's theorem (Chapter VII, § 4, n° 18, theorem 15), is a polynomial. The result follows.

(v) *The conformal invariance of a residue.* At first sight, the definition of the residue of f at infinity seems strange; apart from the sign chosen, in the series expansion of $f(z)$, the residue is the coefficient of a power of z approaching 0 as z approaches infinity, whereas the exact opposite holds for residues at points $\neq \infty$. This requires some explanation, found by replacing the variable z by $1/z$.

Indeed, computing *à la* Leibniz, the change of variable $z = 1/\zeta$ transforms the expression[34] $\omega = f(z)dz$ into $\varpi = f(1/\zeta)d(1/\zeta) = g(\zeta)d\zeta$, where

$$g(\zeta) = -f(1/\zeta)\zeta^{-2} = -\left(\ldots + c_{-1}\zeta + \ldots\right)\zeta^{-2} = \ldots - c_{-1}/\zeta + \ldots.$$

Therefore,

$$\operatorname{Res}(f, \infty) = -c_{-1} = \operatorname{Res}(g, 0).$$

This suggests that the residue of a function f at a point a in fact involves the differential form $\omega = f(z)dz$ rather than the function f itself; it would therefore be better to write it $\operatorname{Res}(\omega, a)$. To justify this, let us generalize the situation by considering a holomorphic function f on $U - S$, where S is closed and discrete in an open subset U of $\mathbb{C}$, and let us investigate what happens

---

[34] It is some sort of differential in the sense of Chapter IX; the transformation it is made to undergo here consists in computing its "inverse image" under $z \mapsto 1/z$.

to $\omega = f(z)dz$ under a conformal representation $z \mapsto \varphi(z) = \zeta$ of U on an open subset V of $\mathbb{C}$. If the inverse of φ is the map $\psi : V \longrightarrow U$, a formal calculation shows that ω is transformed into

$$\varpi = f\left[\psi(\zeta)\right]\psi'(\zeta)d\zeta .$$

The function $f\left[\psi(\zeta)\right]\psi'(\zeta)$ is holomorphic on $V - \varphi(S)$, and the more general result we have in mind is the formula

$$(5.20) \qquad\qquad \mathrm{Res}(\omega, a) = \mathrm{Res}\left[\varpi, \varphi(a)\right] ,$$

which holds for all $a \in S$ and expresses the *invariance of the residue of a holomorphic differential form under a conformal representation*. On the other hand, résidues at a and $b = \varphi(a)$ of the functions $f(z)$ and $f\left[\psi(\zeta)\right]$ are obviously not equal: for $f(z) = 1/z$, $a = 0$, $\varphi(z) = 2z$, $\psi(\zeta) = \zeta/2$, $f\left[\psi(\zeta)\right] = 2/\zeta \neq 1/\zeta$, but $\omega = dz/z$ and $\varpi = d\zeta/\zeta$.

To prove (20), we may assume that $a = \varphi(a) = 0$ and that U is an open disc centered at 0 containing no singular points of f other than 0. Then – use the Laurent series– there is a holomorphic function F on $U - \{a\}$ such that $f(z) = F'(z) + c/z$, where $c = \mathrm{Res}(f, a)$. Then,

$$\omega = F'(z)dz + cdz/z = dF + cdz/z ,$$

and so

$$\varpi = F'\left[\psi(\zeta)\right]\psi'(\zeta)d\zeta + c\psi'(\zeta)d\zeta/\psi(\zeta) .$$

As $F'\left[\psi(\zeta)\right]\psi'(\zeta)$ is the derivative of $F\left[\psi(\zeta)\right]$, the contribution of the first term to the residue of ϖ at $b = 0$ is zero; $\mathrm{Res}(\varpi, b)$ is, therefore, up to a factor c, the coefficient of $1/\zeta$ in the Laurent series of

$$\psi'(\zeta)/\psi(\zeta) = \left[\psi'(0) + \ldots\right] / \left[\psi'(0)\zeta + \ldots\right] .$$

As φ and ψ are mutually inverse, $\psi'(0) \neq 0$, and so $\mathrm{Res}(\psi'/\psi, b) = 1$ and $\mathrm{Res}(\varpi, b) = c = \mathrm{Res}(\omega, a)$, which proves (20).

If a conformal representation leaves the residues of a holomorphic differential form $\omega = f(z)dz$ invariant, it may be assumed that the integrals of ω are also left invariant. To see this, consider a path μ in $U - S$ and its image $t \mapsto \nu(t) = \varphi\left[\mu(t)\right]$ under φ; to compute the integral

$$(5.21) \qquad\qquad \int_\nu \varpi = \int_\nu f\left[\psi(\zeta)\right]\psi'(\zeta)d\zeta ,$$

by definition, ζ needs to be replaced by $\nu(t)$ and $d\zeta$ by $\nu'(t)dt$; $\psi(\zeta)$ is thereby replaced by $\psi\left\{\varphi\left[\mu(t)\right]\right\} = \mu(t)$ since ψ and φ are mutually inverse, $f\left[\psi(\zeta)\right]$ by $f\left[\mu(t)\right]$, and $\psi'(\zeta)d\zeta$ by $\psi'\left[\nu(t)\right]\nu'(t)dt$; but ψ being holomorphic, definitions obvious imply what we know, namely that

$$\psi'\left[\nu(t)\right]\nu'(t) = \frac{d}{dt}\psi\left[\nu(t)\right] = \mu'(t)\,;$$

so finally

(5.22) $$\int_\nu \varpi = \int f\left[\mu(t)\right]\mu'(t)dt = \int_\mu \omega$$

as expected.

This result is independent from residue theory, but suppose that μ is a closed path in $U - S$ null-homotopic in U. In this case, ν is a closed path in $V - \varphi(S)$, clearly null-homotopic in V. Then the residue formula (Theorem 5) shows that

$$\sum \mathrm{Ind}_\nu(b)\,\mathrm{Res}(\varpi, b) = \sum \mathrm{Ind}_\mu(a)\,\mathrm{Res}(\omega, a)$$

follows from (22). This result can be applied to the case $S = \{a\}$, $\varphi(a) = b$, where a is any point in U, and where f is a holomorphic function on $U - \{a\}$, for example $1/(z - a)$. The residues being equal, we conclude that

(5.23) $$\mathrm{Ind}_\nu\left[\varphi(a)\right] = \mathrm{Ind}_\mu(a)\,.$$

This may seem obvious geometrically but is not so, especially as φ could a priori transform a counterclockwise path μ around a into a counterclockwise path ν around $b = \varphi(a)$.

Hence a conformal representation φ leaves the "rotation direction around a point" of a closed path invariant; as will be seen in the next chapter, in a far more general situation, this is due to the fact that the Jacobian of $\varphi = p + iq$, regarded as a map $\mathbb{R}^2$ to $\mathbb{R}^2$, namely

$$J_\varphi(z) = D_1 p.D_2 q - D_2 p.D_1 q = |\varphi'(z)|^2\,,$$

is positive.

(vi) *Functions on the Riemann sphere.* A new set

$$\hat{\mathbb{C}} = \mathbb{C} \cup \{\infty\}\,,$$

obtained by adding to $\mathbb{C}$ an element written ∞, whose choice and nature matter little, was introduced above – reread Hardy in Chapter II, end of n° 2. Having done this, define a topology on $\hat{\mathbb{C}}$ by setting $U \subset \hat{\mathbb{C}}$ to be open if $U \cap \mathbb{C}$ is open in $\mathbb{C}$ in the usual sense and if the exterior of a disc is contained in U when $\infty \in U$. Open subsets containing the point ∞ are, therefore, precisely the complements of the compact subsets of $\mathbb{C}$ in $\hat{\mathbb{C}}$. Axioms about unions and intersections of open sets are immediately seen to be satisfied. This topology of $\hat{\mathbb{C}}$ allows us to define the notions of limit and continuity; for example, saying that a sequence of points $z_n \in \mathbb{C}$ approaches ∞ with respect to the topology of $\hat{\mathbb{C}}$ means that $|z_n|$ increases indefinitely since we need to intimate that for any compact set $K \subset \mathbb{C}$, $z_n \in \mathbb{C} - K$ for large n. With respect to this topology, $\hat{\mathbb{C}}$ is *compact*. If indeed $\hat{\mathbb{C}}$ is the union of a family of open subsets U_i, one of them, say U_j, contains the point at infinity, and hence also the exterior of a compact set $K \subset \mathbb{C}$; as K can be covered (BL) by finitely may sets $\mathbb{C} \cap U_i$, these U_i, together with U_j, form a finite cover of $\hat{\mathbb{C}}$, qed. BW can also be checked.

To understand the topology of $\hat{\mathbb{C}}$ "geometrically", consider the classical unit sphere S^2 in $R^3 = \mathbb{C} \times \mathbb{R}$ and, denoting its north pole by $\nu = (0, 0, 1)$, consider the map p associating to any $\zeta \in S^2$ other than ν the point $z \in \mathbb{C} = \mathbb{R}^2$, where the line passing through ν and ζ meets the equatorial plane $\mathbb{C}$; this is the "stereographic projection" from the north pole used to map regions not too close to it. We thus get a homeomorphism from $S^2 - \{\nu\}$ onto $\mathbb{C}$ transforming the exterior of a disc of radius R centered at 0 in $\mathbb{C}$ into a set of points $\zeta \in S^2 - \{\nu\}$ whose third coordinate satisfies a relation $a < \zeta_3 < 1$. Therefore, if we generalize the definition of p by setting $p(\nu) = \infty$, we obtain a continuous bijection from S^2 onto $\hat{\mathbb{C}}$, hence a homeomorphism since the sphere is compact. $\hat{\mathbb{C}}$ is generally called the la *Riemann sphere*; I do not know whether he would have appreciated this tribute: it is like congratulating an Olympic cycling champion for having won the amateur criterion of his home town.

$\hat{\mathbb{C}}$ is indeed the only trivial example of a compact "Riemann surface" or, in today's terminology, of a "compact complex analytic manifold of dimension 1" (Chap. X): A reasonable definition of the notion of a holomorphic function on an open subset U of S^2 follows by stipulating that such a function should depend holomorphically, possibly also at infinity, on the point $p(\zeta) \in p(U)$, where $p : S^2 \longrightarrow \hat{\mathbb{C}}$ is the stereographic projection. Without resorting to a cartography unlikely to yield useful generalizations in this type of context,[35] a function f defined on an open subset U of $\hat{\mathbb{C}}$ with values in $\mathbb{C}$ is said to be holomorphic if it is so in the usual sense when $U \subset \mathbb{C}$, and in case $\infty \in U$, if it is so in the usual sense on $U \cap \mathbb{C}$ and approaches a finite limit $f(\infty)$ at infinity; as the value $f(\infty)$ is defined by (17), this amounts to saying that $f(z) = g(1/z)$, where g is holomorphic in the neighbourhood of 0. Besides, classical definitions together with those given in (iv) for behaviour at ∞ makes it possible to give meaning to the notion of a "pole" of a function when it is holomorphic in the neighbourhood of a point $a \in \hat{\mathbb{C}}$, except

---

[35] The construction of $\hat{\mathbb{C}}$ can be generalized to any locally compact space X: the open subsets of $\hat{X} = X \cup \{\infty\}$ containing the point ∞ are set to be the complements of the compact subsets of X. Thus X becomes the complement of a point in the *compact* space $\hat{X}$, the *Alexandrov one-point compactification* of X. For $X = \mathbb{R}$, the space obtained is homeomorphic to the unit circle $\mathbb{T}$. This can be seen by using the map $t \mapsto (t - i)/(t + i)$ from $\mathbb{R}$ to $\mathbb{T} - \{1\}$; it can be extended by continuity to $\hat{\mathbb{R}}$ if a value of 1 is set for $t = \infty$, and as it is then bijective and continuous, it is necessarily a homeomorphism. This construction transforms functions approaching a limit at infinity into functions on $\hat{X}$ continuous at ∞. This is low level, but sometimes useful, general topology.

at a itself. In particular, any rational function $f(z)$ can be interpreted on $\hat{\mathbb{C}}$ as a function whose only singularities are its poles, of which are there are necessarily finitely many since $\hat{\mathbb{C}}$ is compact; and we have seen that they are characterized by this property: *rational functions are identical to meromorphic function on $\hat{\mathbb{C}}$*.

To give the reader a less trivial example connected to the theory of elliptic functions, we choose a lattice L in $\mathbb{C}$ (Chapter II, § 3, n° 23) and we consider the set $\mathbb{C}/L$ of equivalence classes mod L, obtained by regarding identical two numbers z', $z'' \in \mathbb{C}$ such that $z' - z'' \in L$. Writing p for the map $\mathbb{C} \longrightarrow \mathbb{C}/L$ associating to each $z \in \mathbb{C}$ its class mod L, a topology can be defined on $\mathbb{C}/L$ by setting a subset $U \subset \mathbb{C}/L$ to be open if and only if $p^{-1}(U)$ is open in $\mathbb{C}$: to ensure the continuity of p, we confine ourselves to the minimum required. The space $\mathbb{C}/L$ is compact[36] since, if we choose a compact subset K in $\mathbb{C}$ meeting all the classes mod L (for example a closed parallelogram generated by two basis vectors[37] for L), then $p(K) = \mathbb{C}/L$. As p is continuous, since BW or BL holds for K it holds for $\mathbb{C}/L$ (see Theorem 11 of Chapter III, § 3, n° 9, whose proof generalizes immediately). Having said this, a function f defined on such an open set is, by definition, holomorphic if and only if the function $z \mapsto f[p(z)]$, defined on the open subset $p^{-1}(U)$ of $\mathbb{C}$, is holomorphic in the usual sense (and doubly periodic since it is constant on the classes mod L). In this case, it would be easy to explain what is meant by the pole of a function defined in the neighbourhood of a point of $\mathbb{C}/L$, but not at the point itself; transposing what we are doing to $\mathbb{C}$ would be sufficient. In particular, a meromorphic function f on $\mathbb{C}/L$ only has finitely many poles since $\mathbb{C}/L$ is compact; composing it with p, we get a doubly periodic and meromorphic function on $\mathbb{C}$: as we will see in Chap. XII, this is precisely what is called an *elliptic function* of the lattice L and we will show that, in this case, the meromorphic functions on $\mathbb{C}/L$ are just the rational functions in $\wp_L(z)$ and $\wp'_L(z)$, where $\wp_L$ is the Weierstrass function of L (Chapter II, § 3, n° 23).

Hence all this is only a màtter of translation, but this point of view has proved itself exceedingly fruitful in much more general cases in most of which the construction is not at all as simple as in the last two examples, starting with the case of *algebraic functions* of one variable studied by Riemann, i.e.

---

[36] It is also necessary to show that $\mathbb{C}/L$ satisfies the *Hausdorff axiom*: that two distinct points have disjoint neighbourhoods; this is equivalent to saying that if $a, a' \in \mathbb{C}$ do not belong to the same class mod L, then there are discs D and D' with centered respectively at a and a' such that $(D + L) \cap (D' + L) = \varnothing$, which is immediate. I have not mention this axiom in the Appendix to Chap. III in order not to steer the reader towards situations rarely encountered in everyday mathematics, but it is nonetheless fundamental. It obviously holds in all metric spaces.

[37] Topologically, $\mathbb{C}/L$ can, therefore, be obtained by taking a period-parallelogram P and by "gluing" its parallel sides pairwise; by gluing two parallel sides, we get a cylinder and by gluing the end circles, a ring. In other words, $\mathbb{C}/L$ is homeomorphic to the surface of a torus in $\mathbb{R}^3$.

given, like the "function" $\zeta = z^{1/3}$, by an equation $P(z, \zeta) = 0$, where P is a polynomial (see chap. X).

6 – Dixon's Theorem

One may wonder in what cases Cauchy's integral formulas continues to hold when μ is not null-homotopic in G. Though the proof[38] forms a double-sided page of ingenious but perfectly elementary arguments, it is was not until 1971 that this was known without recourse to heuristic arguments. To state it, we need to again use notions defined at the end of n° 4, (i), namely the interior and exterior of a closed path μ. Hence

$$\mathbb{C} = \text{Ext}(\mu) \cup \text{Int}(\mu) \cup \text{Supp}(\mu),$$

these sets being pairwise disjoint and, except the first one, bounded.

Theorem 8 (Dixon). *Let G be a domain and μ a closed path in G. The following statements are equivalent:*

(i)   *The integral along μ of any holomorphic function on G is zero.*
(ii)   *The interior of μ is contained in G.*

(iii)   *For any holomorphic function on G and any $a \in G - \text{Supp}(\mu)$,*

(6.1)
$$\int_\mu \frac{f(z)}{z - a} dz = 2\pi i \, \text{Ind}_\mu(a) f(a)$$

The proof consists in showing easy logical implications except for the second one. We will provide more details for it than its inventor.

(i) $\Longrightarrow$ (ii). For some $a \notin G$, consider the function $f(z) = 1/(z - a)$; by (i), its integral along μ is zero, but, up to a factor $2\pi i$, it is also the index of a with respect to μ; the latter is, therefore, zero, proving (ii).

(ii) $\Longrightarrow$ (iii). This is Dixon's contribution. Define a function g on $G \times G$ by setting

(6.2)
$$g(\zeta, z) = [f(\zeta) - f(z)] / (\zeta - z) \quad \text{if} \quad \zeta \neq z,$$
$$= f'(z) \quad \text{if} \quad \zeta = z.$$

By definition of the index, (1) is equivalent to

(6.1')
$$\int_\mu g(\zeta, a) d\zeta = 0.$$

To prove (1') for a given path μ, we proceed step-by-step by showing that

---

[38] J.D. Dixon, *A brief proof of Cauchy's integral theorem* (Proc. Amer. Math. Soc., **29**, 1971, pp. 625–626), reproduced in Remmert, *Funktionentheorie 1*, Chap. 9, §5, which I follow except for a few details.

(a) $g(\zeta, z)$ is a continuous function of $(\zeta, z) \in G \times G$,
(b) $h(z) = \int_\mu g(\zeta, z) d\zeta$ is a holomorphic function on G,
(c)   it can be analytically extended to all of $\mathbb{C}$,
(d)   the entire function thus obtained approaches 0 at infinity.

Liouville's theorem (Chap. VII, §4, n° 18) will then show that $h(z) = 0$ everywhere, proving (1') and the implication (ii) $\Longrightarrow$ (iii).

(a) g is clearly a continuous function of the couple (ζ, z) on the subset of $G \times G$ defined by the relation $\zeta \neq z$. Continuity at each point (a, a) of the "diagonal" of the Cartesian product $G \times G$ is less obvious.

To simplify the notation, suppose that $a = 0$. The Taylor series $f(z) = \sum c_n z^n$ de f at $a = 0$ converges and represents f on a disc of radius $R > 0$. Let D be a disc $|z| < r$ with $r < R$; hence

$$f(\zeta) - f(z) = \sum c_n \left(\zeta^n - z^n \right) = (\zeta - z) \sum_{n \geq 1} c_n \left(\zeta^{n-1} + \zeta^{n-2} z + \ldots + z^{n-1} \right)$$

in $D \times D$. As $g(0,0) = f'(0) = c_1$, it follows that

$$|g(\zeta, z) - g(0,0)| = \left| \sum_{n \geq 2} c_n \left(\zeta^{n-1} + \zeta^{n-2} z + \ldots + z^{n-1} \right) \right| \leq$$

$$\leq \sum_{n \geq 2} n |c_n| r^{n-1}.$$

This is a convergent series in r, without a constant term. As r approaches 0, so does its sum, proving the continuity of g.

(b) It is then possible to define

$$(6.3) \qquad h(z) = \int g(\zeta, z) d\zeta = \int g[\mu(t), z] \mu'(t) dt,$$

where integration is along the given path $\mu : I \longrightarrow G$. As $g[\mu(t), z]$ is a continuous function of (t, z) on $I \times I$ by (a), and for given t, is holomorphic in z, the result is holomorphic in z by Theorem 9 stated below.

(c) $\text{Ext}(\mu)$ contains the exterior of a disc, its complement

$$K = \text{Int}(\mu) \cup \text{Supp}(\mu)$$

is compact, and by assumption (ii), contained in G. Since G is open, and hence distinct from K, the open set $U = G \cap \text{Ext}(\mu)$ is not empty. So for $z \in U$, $z \notin \text{Supp}(\mu)$, and writing

$$(6.4) \qquad h(z) = \int_\mu \frac{f(\zeta)}{\zeta - z} d\zeta - f(z) \int_\mu \frac{d(\zeta)}{\zeta - z} =$$

$$= \int_\mu \frac{f(\zeta)}{\zeta - z} d\zeta - 2\pi i f(z) \, \text{Ind}_\mu(z) = \int_\mu \frac{f(\zeta)}{\zeta - z} d\zeta$$

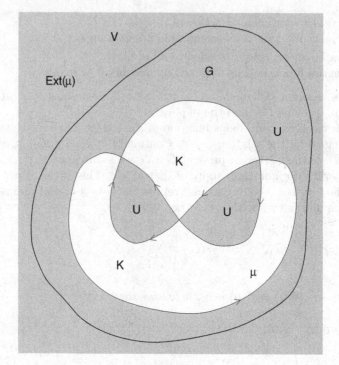

Fig. 6.8.

is justified since $\mathrm{Ind}_\mu(z) = 0$. However, the last integral obtained is well defined for all $z \notin \mathrm{Supp}(\mu)$, in particular on $\mathrm{Ext}(\mu)$, and hence is obviously a holomorphic function of z. As it coincides with $h(z)$ on the non-empty open set $G \cap \mathrm{Ext}(\mu)$, we get a holomorphic function on the open set $G \cup \mathrm{Ext}(\mu)$ by requiring it to be equal to h on G and to the integral in question on $\mathrm{Ext}(\mu)$. But as $\mathbb{C} - \mathrm{Ext}(\mu)$ is contained in G, $G \cup \mathrm{Ext}(\mu) = \mathbb{C}$. The new function h is, therefore, defined and holomorphic on all of $\mathbb{C}$.

(d) The reason why this entire function approaches 0 at infinity is obvious: if the compact subset $\mathrm{Supp}(\mu)$ is contained in the disc $|\zeta| \leq R$, with R finite, and if z is exterior to it, then $|\zeta - z| > |z| - R$ for all $\zeta \in \mathrm{Supp}(\mu)$, and so

$$|h(z)| \leq m(\mu)/\left(|z| - R\right),$$

which gives the result. Liouville's Theorem then shows that $h(z) = 0$ for all $z \in \mathbb{C}$, proving (ii) $\Longrightarrow$ (iii).

(iii) $\Longrightarrow$ (i). If (1) is satisfied for any f, it also is for $g(z) = f(z)(z - a)$; as $g(z)/(z - a) = f(z)$ and as $g(a) = 0$, we get $\int f(z)dz = 0$. This ends the proof.

We still need to justify fully point (b) of the previous proof. This is the aim of the next theorem, also very useful in many other circumstances. Basically, it is almost always applied to a measure of the form $d\mu(t) = \mu'(t)dt$ where the function $\mu'(t)$ is regulated, but the general case is not hard since, in all questions of this type, the only properties of measures really used are their definitions – linearity of $f \mapsto \mu(f)$ and upper bound in terms of the uniform norm of f – and elementary theorems about passing to the limit under the $\int$ sign that follow directly from the definition.

7 – Integrals Depending Holomorphically on a Parameter

Theorem 9. *Let I be an interval, μ a measure on I, U an open subset of $\mathbb{C}$ and $f : I \times U \longrightarrow \mathbb{C}$ a function satisfying the following conditions:*

(a) *f is continuous on $I \times U$,*
(b) *$f(t, z)$ is a holomorphic function of z for all $t \in I$,*
(c) *for any compact set $H \subset U$, there exists a positive μ-integrable[39] function $p_H(t)$ on I such that*

$$(7.1) \qquad |f(t, z)| \leq p_H(t) \text{ for all } t \in I \text{ and } z \in H .$$

Let $f^{(r)}(t, z)$ be the derivative of order r of $z \mapsto f(t, z)$. Then $f^{(r)}(t, z)$ satisfies conditions (a), (b) and (c) for all r, the function

$$(7.2) \qquad g(z) = \int f(t, z) d\mu(t)$$

is holomorphic on U, the functions $f^{(r)}(t, z)$ are μ-integrable and

$$(7.3) \qquad g^{(r)}(z) = \int f^{(r)}(t, z) d\mu(t) \text{ for all } r \in \mathbb{N} .$$

It is sufficient to prove the statements with respect to $r = 1$: the general case will follow by a repeated application of the result.

First proof. Theorem 24 bis of Chap. V, § 7 is analogous to the result we need to prove but based on different assumptions: f' (instead of f) was assumed to satisfy conditions (a) and (c). At that point, the only tool at our disposal was indeed the formula for differentiation of an integral with respect to a *real* parameter; it assumes that the derivative being integrated is continuous and that its integral converges normally on compact sets. We then obtained the holomorphy of (1) and formula (2) by differentiating integral (1) with respect to coordinates x and y of z and by checking Cauchy's condition

---

[39] If $d\mu(t) = \mu'(t)dt$ with $\mu'(t)$ regulated, this means that $\int p_H(t)|\mu'(t)|dt < +\infty$. In the case of an arbitrary positive measure, $p_H(t)$ can be assumed to be lsc (and in practice, even continuous) because a (Lebesgue) integrable positive function is always dominated by an integrable lsc function. Besides, note that (c) is always satisfied when I is compact because f is bounded on the compact set $I \times H$, so that it suffices to choose the constant function $p_H(t) = \sup|f(s, z)|$, sup being extended to $(s, z) \in I \times H$.

$D_2g = iD_1g$. To prove theorem 9, it will, therefore, suffice to show that f' satisfies (a) and (c).

To prove that f' is continuous, take a point a of G. Let R be the distance from a to the boundary of G. Assuming $a = 0$ to simplify formulas, the open disc $D : |z| < r$ is contained in G for $r < R$ and Cauchy's integral formula shows that

$$(7.4) \qquad f'(t,z) = \int f\,[t, re(u)]\,[re(u) - z]^{-2}\,re(u)du$$

for $|z| < r$, where integration is over $J = [0,1]$. The function under the $\int$ sign depends on variables $u \in J$, $z \in D$ and $t \in I$ and, in view of the simplest results on integrals depending on parameters [Chap. V, n° 9, Theorem 9, (i)], it all amounts to showing that this function of (u, z, t) is continuous on $J \times D \times I$, which is clear.

To show that f' satisfies (c) for any compact set $H \subset U$, it suffices (Borel-Lebesgue) to show this in the neighbourhood of all $a \in U$, for example $a = 0$. As the points $re(u)$ remain in a compact set U, by (c), there is a positive, μ-integrable function p such that

$$|f\,(t, re(u))| \le p(t) \text{ for all } t \in I \text{ and all } u.$$

If z remains in the disc $D' : |z| \le r/2$, then $|re(u) - z| \ge r/2$, whence $|re(u) - z|^{-2} \le 4r^{-2}$ and so $|f'(t, z)| \le Mp(t)$ for all $t \in I$ and $z \in D'$, with a constant M independent of t and z; thus (c) follows for f'.

It remains to apply Theorem 24 bis of Chap. V. For the reader's convenience, we recall its proof. For this suppose μ positive, a case it can be reduced to. For any compact interval $K \subset I$, set

$$g_K(z) = \int_K f(t, z)d\mu(t).$$

Regarded as a function of t, $x = \mathrm{Re}(z)$ and $y = \mathrm{Im}(z)$, the function f has derivatives $D_1f(t, z) = f'(t, z)$ and $D_2f(t, z) = if'(t, z)$ with respect to x and y, and, like f', these are continuous functions on $K \times U$. Since integration is over a compact set, it is possible to differentiate under the $\int$ sign with respect to x or to y (Chap. V, § 2, n° 9, Theorem 24); the derivatives are obviously

$$D_1g_K(z) = \int_K f'(t, z)d\mu(t), \quad D_2g_K(z) = i\int_K f'(t, z)d\mu(t).$$

As f' is continuous and K compact, they are continuous and satisfy Cauchy's condition. The functions g_K are, therefore, holomorphic, with

$$g_K'(z) = \int_K f'(t, z)d\mu(t).$$

To finish the proof, consider a compact subset H of U and an upper bound $|f'(t,z)| \leq p_H(t)$ valid for $t \in I$ and $z \in H$, with a positive μ-integrable function p_H. Then, for any compact set $K \subset I$,

$$\left| \int_I f'(t,z)d\mu(t) - g'_K(z) \right| \leq \int_{I-K} p_H(t)d\mu(t).$$

for all $z \in H$, and since the right hand side approaches 0 as K "approaches" I, it can be concluded that $g'_K(z)$, and hence also the partial derivatives of g_K with respect to x and y, converge uniformly, up to a factor i, to $\int f'(t,z)d\mu(t)$ on H. The function $g = \lim g_K$, therefore, has partial derivative with respect to x and y obtained by passing to the limit over those of g_K (Chap. III, § 4, Theorem 19). They are continuous and like those of g_K satisfy Cauchy's condition, so that g is holomorphic,[40] with $g'(z) = \lim g'_K(z) = \int f'(t,z)d\mu(t)$, qed.

Second proof. Theorem 9 can also be proved by using Weierstrass' theorem on uniform limits of holomorphic functions (Chap. VII, § 4, n° 19, Theorem 17).

Let us first consider the case where I is compact. In what follows, suppose that z remains in a compact subset H of U and set $f_t(z) = f(t,z)$. As $I \times H$ is compact, f is uniformly continuous on it. In particular, for every $r > 0$ there exists $r' > 0$ such that

$$(7.5) \qquad |s - t| < r' \Longrightarrow \|f_s - f_t\|_H < r .$$

Having said this, let us partition I into finitely many non-empty intervals I_k of length $< r'$, choose points $t_k \in I_k$ and compare integral (1) to the Riemann sum $\sum f(t_k, z)\mu(I_k)$. Since, by (5), $|f(t,z) - f(t_k,z)| < r$ for all $t \in I_k$ and all $z \in H$, it follows that

$$\left| g(z) - \sum f(t_k, z)\mu(I_k) \right| \leq \|\mu\| r \text{ for all } z \in H ,$$

where $\|\mu\|$ is the norm of μ. This means that the function g is a *uniform* limit of holomorphic functions on every compact set $H \subset U$; it is, therefore, holomorphic. Moreover, since a limit of holomorphic functions can be differentiated term by term according to the same theorem, $g^{(p)}(z)$ is the limit of the expressions $\sum f^{(p)}(t_k, z)\mu(I_k)$, which are just the Riemann sums with respect to integral (2), and so the theorem follows when I is compact.

In the general case, replace I by a compact interval $K \subset I$ and pass to the limit as in the previous proof.

---

[40] This is the "useless" result mentioned in Chap. III, § 5, at the end of n° 22.

Third proof. Let us directly show that $g(z)$ has a power series expansion on the interior of any compact disc $D \subset U$. This reduces to the case where D is the disc $|z| \leq R$. As $z \mapsto f(t, z)$ is holomorphic and hence analytic, there is a Taylor series expansion on D

$$(7.6) \qquad f(t, z) = \sum f^{(n)}(t, 0) z^{[n]}$$

with $z^{[n]} = z^n / n!$ and

$$(7.7) \qquad f^{(n)}(t, 0) R^{[n]} = \int f(t, \mathrm{Re}(u)) \, \mathbf{e}(-nu) du$$

for $t \in I$ (Fourier series...). Since, according to assumption (c), $|f(t, z)| \leq p_D(t)$ for $t \in I$ and $z \in D$, it follows that

$$(7.8) \qquad \left| f^{(n)}(t, 0) \right| \leq p_D(t) / R^{[n]}$$

for all n and all $t \in I$. For $|z| = qR$ with $q < 1$ and all $r \in \mathbb{N}$, the Taylor series

$$(7.9) \qquad \sum f^{(n+r)}(t, 0) z^{[n]} = f^{(r)}(t, z)$$

is, therefore, dominated for all $t \in I$, $z \in D$ and $p \in \mathbb{N}$ by the series

$$(7.10) \qquad \sum p_D(t) q^{[n]} R^n / R^{[n+r]} = p_D(t) R^{-r} \sum q^n (n + r)! / n! \,,$$

which converges since $(n + r)! / n! \asymp n^r$ for large n. If $K \subset I$ is compact, the continuous function $f(t, z)$ is bounded on $K \times D$ and $p_D(t)$ can be replaced in (10) by a constant independent of $(t, z) \in K \times D$, so that the Taylor series (9) converges normally on $K \times D$. As a result, the function $f^{(r)}(t, z)$ is continuous on $K \times D$ for any K, hence on $I \times D$, and so on $I \times U$ since the argument can be applied to any compact disc $D \subset U$.

As $p_D(t)$ is also a factor in (10),

$$(7.12) \qquad \sum_{n \geq 0} \left| f^{(n+r)}(t, 0) z^{[n]} \right| \leq M_r p_D(t) \,,$$

where M_r is a constant. Series (9) can, therefore (Chap. V, § 7, Theorem 20 generalized to an arbitrary measure), be integrated term by term on I, and so

$$\int f^{(r)}(t, z) d\mu(t) = \sum a_{n+r} z^{[n]} \quad \text{où} \quad a_n = \int f^{(n)}(t, 0) d\mu(t) \,.$$

For $r = 0$, this shows that $g(z) = \sum a_n z^{[n]}$, and, therefore, that g has a power series expansion and that, moreover,

$$\int f^{(r)}(t, z) d\mu(t) = \sum a_{n+r} z^{[n]} = g^{(r)}(z) \,.$$

This is relation (2). That $f^{(r)}(t, z)$ satisfies condition (c), a fact of local nature, follows immediately from (12).

§ 3. Some Applications of Cauchy's Method

In this § aimed at showing that even a limited knowledge of Cauchy theory allows us to do mathematics that does not merely amount to irrelevant exercises, the following notation will be systematically used:

$L^1(\mathbb{R})$ will denote the set of functions defined and absolutely integrable over $\mathbb{R}$ with respect to the usual measure dx; the reader is free to interpret this notation in the sense of Lebesgue theory. In fact, as is done in Lebesgue theory, we will often say "integrable" instead of "absolutely integrable", even if this means warning the reader when we will encounter semi-convergent integrals (Chapter V, § 7);

$F^1(\mathbb{R})$ will denote the set of continuous functions f[41] on $\mathbb{R}$ such that both $f \in L^1(\mathbb{R})$ and $\hat{f} \in L^1(\mathbb{R})$ hold; Fourier's inversion formula applies to these functions (Chapter VII, § 6, n° 30, Theorem 26).

Recall that, for us, the Fourier transform is defined by the formula

$$\hat{f}(y) = \int f(x)\mathbf{e}(-xy)dx \,,$$

where integration is over $\mathbb{R}$ and where, for every $z \in \mathbb{C}$, as in Chapter VII,

$$\mathbf{e}(z) = \exp(2\pi i z) \,.$$

So $\mathbf{e}(-x) = \overline{\mathbf{e}(x)}$ for $x \in \mathbb{R}$ and

$$|\mathbf{e}(z)| = \exp(-2\pi y) \,, \quad y = \mathrm{Im}(z)$$

for all $z \in \mathbb{C}$.

Expressions such as the following will be frequently used:

$$f(z) = O\left(g(z)\right) \quad \text{at infinity on } U \,,$$

where U is a subset of $\mathbb{C}$; this means (Chapter II, n° 3) that there exists $M > 0$ and $R > 0$ such that

$$z \in U \quad \& \quad |z| \geq R \Longrightarrow |f(z)| \leq M \, |g(z)| \,.$$

Similar conventions apply to relations o, $\asymp$ and $\sim$. We will also write

$$f(z) \asymp g(z) \quad \text{on } U$$

if there are constants $m, M \geq 0$ such that

$$m \, |g(z)| \leq |f(z)| \leq M \, |g(z)| \quad \text{for } all \; z \in U$$

---

[41] A somewhat superfluous condition: in Lebesgue theory, any $f \in L^1(\mathbb{R})$ whose Fourier transform is integrable is shown to be equal "almost everywhere" to a continuous function given by the inversion formula.

and not only large z or near to a given point. The next result often serves as an example if $U = \mathbb{C} - \mathbb{R}_-$:

Lemma. *Let $U \subset \mathbb{C}^*$ be a domain on which there is a uniform* bounded *branch[42] of* $\operatorname{Arg}(z)$. *Then*

$$z^s \asymp |z|^{\operatorname{Re}(s)} \quad on \quad U$$

for any uniform branch of z^s on U.

Indeed, $z^s = \exp(s \operatorname{Log} z)$, where $\operatorname{Log} z = \log |z| + i \operatorname{Arg} z$ for any uniform branch on U of the argument. Since

$$\operatorname{Re} [s \operatorname{Log} z] = \operatorname{Re}(s) \log |z| - \operatorname{Im}(s) \operatorname{Arg} z \,,$$

$$|z^s| = |z|^{\operatorname{Re}(s)} e^{- \operatorname{Im}(s) \operatorname{Arg} z} \,.$$

Since by assumption $\operatorname{Arg}(z)$ remains in a compact subset of $\mathbb{R}$, for any $z \in U$, the exponential lies between m and $M > 0$, qed.

We will most often write

$$\int f(x)dx \quad \text{or} \quad \int_{\mathbb{R}} f(x)dx \quad \text{instead of} \quad \int_{-\infty}^{+\infty} f(x)dx \,;$$

there will be no confusion as we will never use the absurd $\int f(x)dx$ to denote a primitive for a function f. On the other hand, d^*x will denote the positive measure on $\mathbb{R}_+^*$ or sometimes on $\mathbb{R}^*$, but not[43] on $\mathbb{R}$, defined by the formula

$$\int_{\mathbb{R}^*} f(x)d^*x = \int_{-\infty}^{+\infty} f(x)|x|^{-1}dx$$

for f continuous and zero in a neighbourhood of 0 and of infinity, and more generally for any function making the integral absolutely convergent;

---

[42] This is not always the case, even when U is simply connected. For a counterexample, take U to the complement in $\mathbb{C}$ of a spiral with initial point the origin and tending towards infinity, for example the curve $t \mapsto t\mathbf{e}(t)$, $t \geq 0$.

[43] Recall (Chap. V, §9) that if $X \subset \mathbb{C}$ is locally compact (i.e. the intersection of an open and of a closed set), then a positive measure μ on X is a linear functional $f \mapsto \mu(f)$ on the vector space $L(X)$ of continuous functions on X that are zero outside a compact subset of X, satisfying $\mu(f) \geq 0$ for $f \geq 0$. A general measure is a linear functional on $L(X)$ such that, for any compact set $K \subset X$, there is an upper bound

$$|\mu(f)| \leq M_K \|f\|_K$$

for any $f \in L(X)$ zero outside K. The measure d^*x is not a measure on $\mathbb{R}$ because the integral $\int f(x)|x|^{-1}dx$ is not defined for $f \in L(\mathbb{R})$, if we do not at least require f to be zero at 0.

the relevance of introducing such a measure lies in its invariance under "multiplicative translations" $x \mapsto ax(a \neq 0)$ and under $x \mapsto 1/x$:

$$\int f(ax)d^*x = \int f(x)d^*x, \int f(1/x)d^*x = \int f(x)d^*x.$$

In other words, d^*x plays the same role for the multiplicative group $\mathbb{R}^*$ as the Lebesgue measure dx does for the additive group $\mathbb{R}$.

8 – Fourier Transform of a Rational Fraction

(i) *Absolutely convergent integrals of rational functions.* A seemingly trivial example, but which illustrates one of the most used techniques in practice, is the computation of the integral of $f(x) = 1/(1 + x^2)$ over $\mathbb{R}$; the given function having arctg x as primitive, the result obviously equals π.

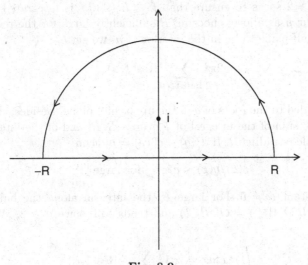

Fig. 8.9.

Consider the above path μ in $\mathbb{C}$. The function f is holomorphic on $\mathbb{C}$ except at $z = i$ or $-i$; the formula

$$2i/\left(1 + z^2\right) = 1/(z - i) - 1/(z + i)$$

shows that the residue at i is equal to $1/2i$. The value of the integral along μ is, therefore, $2\pi i/2i = \pi$.

Having done this, consider the contribution from the half circle to the computation. Its length is πR. As $1/(1 + z^2) = O(1/|z|^2)$ for large $|z|$, the general upper bound (4.9) shows that this contribution is $O(1/R)$, and so

$$\pi = \int_\mu f(\zeta)d\zeta = \int_{-R}^{R} \frac{dx}{1+x^2} + O(1/R).$$

As the integral over $[-R, R]$ approaches the integral sought, it is equal to π as expected.

It may be wondered why we choose to integrate over a half-circle rather than over other curves. The most probable reason is that, for the last two thousand five hundred years, not to go further back to homo erectus fascinated by the Moon and the Sun, the circle is justifiably an object of adoration for mathematicians. But we might as well integrate over the upper or lower part of the square bounded by the lines $\mathrm{Re}(z) = R$ or $-R$ and $\mathrm{Im}(z) = 0$ or R. The main point is that its length should be $O(R)$ as R increases indefinitely, where R denotes the minimum distance from the origin to the chosen contour points; it is even possible to go until a length of the order of $o(R^2)$.

The method generalizes to integrals of rational fractions $f(x) = p(x)/q(x)$, where q is assumed not to have any real roots and, to start with, that $d^\circ(q) - d^\circ(p) = n \geq 2$ so as to ensure that $f \in L^1(\mathbb{R})$. f is integrated along the same contour μ as above; choosing R sufficiently large for the roots of q in the upper half plane to be in the interior of μ, we get

$$(8.1) \qquad\qquad 2\pi i \sum_{\mathrm{Im}(a) > 0} \mathrm{Res}(f, a),$$

a sum extended to the roots of q in the upper half plane. Besides, the integral over μ is the sum of the integral of f over $[-R, R]$ and of the integral along the half-circle of radius R. If $d^\circ(q) - d^\circ(p) = n$, then

$$(8.2) \qquad\qquad p(z)/q(z) \sim c/z^n \quad \text{for large } |z|$$

with a constant $c \neq 0$. For large R, the integral along the half-circle is, therefore, $\pi R.O(R^{-n}) = O(R^{1-n})$ and tends to 0 since $n \geq 2$. Whence the final result:

$$(8.3') \qquad\qquad \int f(x)dx = 2\pi i \sum_{\mathrm{Im}(a) > 0} \mathrm{Res}(f, a).$$

Instead of integrating along the above contour, its symmetric with respect to the real axis would do as well; as it is followed clockwise, we get

$$(8.3'') \qquad\qquad \int f(x)dx = -2\pi i \sum_{\mathrm{Im}(a) < 0} \mathrm{Res}(f, a).$$

Comparing these results shows that

$$(8.4) \qquad\qquad \sum_{a \in \mathbb{C}} \mathrm{Res}(p/q, a) = 0 \quad \text{if } d^\circ(q) - d^\circ(p) \geq 2,$$

which has already been shown in (5.14).

If all the roots of q are simple,

$$p(z)/q(z) = [p(a) + p'(a)(z-a) + \ldots] / [q'(a)(z-a) + \ldots]$$
$$= \frac{p(a)}{q'(a)(z-a)} [1 + ?(z-a) + \ldots]$$

in a neighbourhood of such a root a since the quotient of two power series starting with 1 is a power series of the same type. Hence the formula

$$\operatorname{Res}(p/q, a) = p(a)/q'(a)$$

that allows us to compute the integral.

Let us for example choose the function $f(z) = p(z)/(1 + z^{2n})$, where p is a polynomial of degree $\leq 2n - 2$; its poles are (at most) the roots of the equation $z^{2n} = -1 = \exp(\pi i)$, i.e. the $2n$ points

$$\omega_k = \omega^{2k+1} \quad \text{where } \omega = \exp(\pi i/2n),\, 0 \leq k \leq 2n - 1.$$

The roots with positive imaginary parts are obtained for $0 < k < n - 1$ and the residue at ω_k is equal to

$$p(\omega_k)/2n\omega_k^{2n-1} = -p(\omega_k)\omega_k/2n.$$

So

$$\int \frac{p(x)}{1 + x^{2n}} dx = -\frac{\pi i}{n} \sum_0^{n-1} \omega_k p(\omega_k).$$

For $p(x) = x^{m-1}$, the sum

$$\sum \omega_k^m = \sum_{0 \leq k \leq n-1} \omega^{(2k+1)m} = \omega^m \sum \omega^{2mk} = \omega^m \frac{1 - \omega^{2mn}}{1 - \omega^{2m}} = \frac{(-1)^m - 1}{\omega^m - \omega^{-m}},$$

needs to be calculated so that the integral sought equals 0 if m is even (obvious!) and $\pi/n \sin(m\pi/2n)$ if m is odd.

(ii) *Semi-convergent integrals of rational functions.* The previous method can also be applied if $d^\circ(q) = d^\circ(p) + 1$. But some precautions are called for since the extended integral over $\mathbb{R}$ is no longer absolutely convergent. Nonetheless, we can set

$$(8.5) \qquad \int_{\mathbb{R}} f(x)dx = \lim \int_{-R}^{+R} f(x)dx = \lim \int_0^R [f(x) + f(-x)]\, dx.$$

This limit exists since

$$f(z) = c/z + O(1/z^2) \quad \text{at infinity}$$

and hence $f(z) + f(-z) = O(1/z^2)$. In the decomposition of f into simple elements, the integrals of terms of the form $A/(z-a)^r$ with $r \geq 2$ can be computed as before; computations may even be omitted because they are zero, either because that is the case of the corresponding residues or simply because the function $1/(x-a)^r$ admitting a primitive approaching 0 at infinity for $r \geq 2$, the FT solves the problem without having to invoke Cauchy.

It, therefore, remains to compute the above integral for $f(z) = 1/(z-a)$, $a \notin \mathbb{R}$. A first method consists in observing (Chapter V, §6, n° 20) that, on the simply connected open subset $G = \mathbb{C} - \mathbb{R}_-$, the function $1/z$ admits as primitive any uniform branch $L(z)$ of the pseudo-function $\mathcal{L}og\, z$, for example the one obtained by setting

$$L(z) = \log|z| + i.\operatorname{Arg} z \quad \text{with} \quad |\operatorname{Arg} z| < \pi.$$

As a is not real, the points $x - a$ are in G for $x \in \mathbb{R}$, so that $L(x-a)$ can be chosen as a primitive for $1/(x-a)$ on $\mathbb{R}$. Hence

$$\int_{-R}^{R} \frac{dx}{x-a} = L(x-a)\Big|_{-R}^{R}.$$

The variation of the real part of

$$L(x-a) = \log|x-a| + i.\operatorname{Arg}(x-a)$$

over $[-R, R]$ is equal to $\log(|R-a|/|R+a|)$ and tends to 0 since $|R-a|/|R+a|$ approaches 1 as R increases; the argument of $R - a$ tends to 0 since the half-line with initial point 0 and terminal point $R-a$ approaches the half-line $\mathbb{R}_+$; finally, the half-line with initial point 0 and terminal point $-R-a$ approaches $\mathbb{R}_-$, but is in the half-plane $\operatorname{Im}(z) < 0$ if $\operatorname{Im}(a) > 0$ and in the half-plane $\operatorname{Im}(z) > 0$ if $\operatorname{Im}(a) < 0$; the argument of $-R - a$, therefore, approaches $-\pi$ if $\operatorname{Im}(a) > 0$ and $+\pi$ if $\operatorname{Im}(a) < 0$. So

$$(8.6) \qquad \int_{\mathbb{R}} dx/(x-a) = \begin{array}{ll} \pi i & \text{if} \ \operatorname{Im}(a) > 0 \\ -\pi i & \text{if} \ \operatorname{Im}(a) < 0 \end{array}.$$

Finally, in the general case, it follows that

$$(8.7) \qquad \int_{\mathbb{R}} f(x)dx = \pi i \sum_{\operatorname{Im}(a) > 0} \operatorname{Res}(f, a) - \pi i \sum_{\operatorname{Im}(a) < 0} \operatorname{Res}(f, a).$$

When $d^\circ(q) \geq d^\circ(p) + 2$, the sum of all the residues of f in $\mathbb{C}$ are zero and (7) reduces to (3') or (3"), as desired. If on the contrary $d^\circ(q) = d^\circ(p) + 1$, it is the sum of residues in $\mathbb{C}$ and at infinity, which is zero by (5.14'). Then we for example find

$$(8.8) \qquad \int_{\mathbb{R}} f(x)dx = 2\pi i \sum_{\operatorname{Im}(a) > 0} \operatorname{Res}(f, a) + \pi i \operatorname{Res}(f, \infty).$$

À quicker second method consists in integrating over the same closed contours as in the absolute convergence case. Start with the relation $f(z) = c/z + O(1/z^2)$ where $c = - \operatorname{Res}(f, \infty)$ – careful with the sign! – and observe that the integral of $O(1/z^2)$ over the half-circle approaches 0. Therefore, the integral of f along the latter approaches the same limit as that of c/z; it is calculated by setting $z = \operatorname{Re}^{it}$, and so $dz/z = idt$, and as integration is over $(0, \pi)$, the result is equal to $\pi ic = -\pi i \operatorname{Res}(f, \infty)$. Taking into account the poles in the interior of the integration contour,

$$\int_{\mathbb{R}} f(x)dx - \pi i \operatorname{Res}(f, \infty) = 2\pi i \sum_{\operatorname{Im}(a) > 0} . \operatorname{Res}(f, a)$$

This again leads to (8).

(iii) *Absolutely convergent Fourier transforms.* For t real, let us now consider the Fourier integral

(8.9) $$\hat{f}(t) = \int f(x)\mathbf{e}(-tx)dx = \int f(x) \exp(-2\pi itx)dx,$$

where $f = p/q$ is again a rational fraction without any real roots; here too the integral is absolutely convergent if $n = d^\circ(q) - d^\circ(p) \geq 2$.

Suppose first that $t > 0$. The function

$$g(z) = f(z)\mathbf{e}(-tz)$$

is holomorphic on $\mathbb{C}$ deprived of the roots of q; $f(z) \sim c/z^n$ for large $|z|$, though $|\mathbf{e}(-tz)| = \exp(2\pi ty)$ is ≤ 1 on the half-plane $\operatorname{Im}(z) \leq 0$. Hence, if g is integrated over the contour formed by the interval $[-R, R]$ followed by the lower half-circle of radius R, its contribution is $O(1/R^{n-1})$ for large R, and so approaches 0. As the integration contour is followed clockwise, the index of a point in its interior is -1 and the residue theorem shows that

(8.10') $$\hat{f}(t) = -2\pi i \sum_{\operatorname{Im}(a) < 0} \operatorname*{Res}_{z=a} [f(z)\mathbf{e}(-tz)] \quad (t. \geq 0),$$

the notation for the residue being self-explanatory. For $t \leq 0$, we use the upper half-circle on which $\mathbf{e}(-tz)$ is bounded, and so

(8.10") $$\hat{f}(t) = +2\pi i \sum_{\operatorname{Im}(a) > 0} \operatorname*{Res}_{z=a} [f(z)\mathbf{e}(-tz)] \quad (t \leq 0).$$

For $t = 0$, we recover the results of section (i).

Suppose for example that

(8.11) $$f(x) = \left(x^2 + w^2\right)^{-1} = \frac{1}{2iw}\left(\frac{1}{x - iw} - \frac{1}{x + iw}\right)$$

with $\mathrm{Re}(w) > 0$. The function $g(z) = \exp(-2\pi itz)f(z)$ has simple poles at iw and $-iw$, and clearly[44]

$$\mathrm{Res}(g, iw) = e^{2\pi tw}/2iw, \quad \mathrm{Res}(g, -iw) = -e^{-2\pi tw}/2iw\,.$$

As $\mathrm{Im}(iw) > 0$ and $\mathrm{Im}(-iw) < 0$, multiplying by $2\pi i$, we get

$$\int \frac{\mathbf{e}(-tx)}{x^2 + w^2}dx = \begin{matrix} \pi e^{-2\pi tw}/w & \text{if } t \ge 0 \\ \pi e^{2\pi tw}/w & \text{if } t \le 0, \end{matrix} \quad \text{if } \mathrm{Re}(w) > 0;$$

in other words

$$(8.12) \qquad \int \frac{\mathbf{e}(-tx)}{x^2 + w^2}dx = \pi e^{-2\pi w|t|}/w \quad \text{if } \mathrm{Re}(w) > 0\,.$$

Note that $|e^{-2\pi w|t|}| = e^{-2\pi|t|\,\mathrm{Re}(w)}$ approaches 0 exponentially as $|t|$ increases indefinitely since $\mathrm{Re}(w) > 0$, so that $\hat{f} \in L^1(\mathbb{R})$; in other words, $f \in F^1(\mathbb{R})$. Fourier's inversion formula can, therefore, be applied, and so

$$\int e^{-2\pi w|t|+2\pi itx}dt = w/\pi\left(x^2 + w^2\right)\,. \quad \mathrm{Re}(w) > 0\,,$$

This formula easy to check directly: integrate over $t > 0$ and $t < 0$ taking into account that e^{ct} has e^{ct}/c as primitive for any $c \in \mathbb{C}^*$.

   To obtain an explicit form of $\hat{f}(t)$ in the general case, observe that any rational fraction is the sum of a polynomial and of a linear combination of functions of the form

$$(8.13) \qquad\qquad f(x) = (x - a)^{-n}\,,$$

where n is an integer ≥ 1 and a is a constant; its Fourier transform is well-defined only if its decomposition does not have any polynomial terms and if its poles are not real. Hence if an explicit formula is found for the Fourier transform of (13) for $a \notin \mathbb{R}$, the decomposition into simple elements will give the result in the general case. In this section, we will suppose that $n \ge 2$ and that $\mathrm{Im}(a) > 0$, the case $\mathrm{Im}(a) < 0$ being similar.
   The only pole of the function

$$g(z) = (z - a)^{-n}\mathbf{e}(-tz)$$

being a, it is already possible to conclude that $\hat{f}(t) = 0$ for $t \ge 0$. For $t < 0$, the residue at a of

---

[44] If φ has a simple pole at a and if ψ is holomorphic and non-zero at a, then

$$\varphi(z)\psi(z) = \left[c(z - a)^{-1} + \ldots\right][\psi(a) + \ldots]\,,$$

and so $\mathrm{Res}(\varphi\psi, a) = \psi(a)\,\mathrm{Res}(\varphi, a)$.

$$g(z) = \mathbf{e}(-tz)(z-a)^{-n} = (z-a)^{-n}\mathbf{e}(-ta)\mathbf{e}\left[-t(z-a)\right] =$$
$$= \mathbf{e}(-at)(z-a)^{-n}\sum_{\mathbb{N}}\left[-2\pi it(z-a)\right]^{[p]}$$

needs to be calculated. Here we use the notation of divided powers $x^{[n]} = x^n/n!$. As we want to find the coefficient of $(z-a)^{-1}$, it follows that

$$\mathrm{Res}(g,a) = \mathbf{e}(-at)(-2\pi it)^{[n-1]}.$$

Formulas (10') and (10") then show that

(8.14) $\displaystyle\int (x-a)^{-n}\mathbf{e}(-tx)dx = \begin{array}{ll} 2\pi i(-2\pi it)^{[n-1]}\mathbf{e}(-at) & \text{if } t \le 0 \\ 0 & \text{if } t \ge 0 \end{array}$ if $\mathrm{Im}(a) > 0$.

To avoid major mistakes, it is worth checking that $\hat{f}(t)$ approaches 0 at infinity.[45]

Allowing for notation, formula (14) was announced in Chapter VII, §6, n° 27, example 1 without then being in a position to prove it; We lacked Cauchy theory to be able to justify it. Replacing t by $-t$ and denoting by t_+ the function equal to t for $t > 0$ and to 0 otherwise, (14) can also be written as

(8.15) $\displaystyle\int (x-a)^{-n}\mathbf{e}(tx)dx = (2\pi i)^n t_+^{[n-1]}\mathbf{e}(at),\quad \mathrm{Im}(a) > 0,\quad n \ge 2.$

Formula (15) can also be written

$$\int (x-a)^{-n}\mathbf{e}\left[t(x-a)\right]dx = (2\pi i)^n t_+^{[n-1]}.$$

Setting $x-a = z$, the integral over $\mathbb{R}$ is transformed into an integral *à la* Cauchy along the (unbounded) path $\mathrm{Im}(z) = -\mathrm{Im}(a) = c < 0$:

(8.15') $\displaystyle\int_{\mathrm{Im}(z)=c\,<\,0} z^{-n}\mathbf{e}(tz)dz = (2\pi i)^n t_+^{[n-1]}.$

The change of variable $tz = \zeta$ transforms the horizontal $\mathrm{Im}(z) = c$ into a horizontal $\mathrm{Im}(\zeta) = tc = -t\,\mathrm{Im}(a) = b$. As $\zeta^{-n}d\zeta = t^{1-n}z^{-n}dz$ and as b and t are of opposite sign, relation (14) is equivalent to

$$\int_{\mathrm{Im}(\zeta)=b}\zeta^{-n}\mathbf{e}(\zeta)d\zeta = \begin{array}{ll} (2\pi i)^n/(n-1)! & \text{if } b < 0, \\ 0 & \text{if } b > 0. \end{array}$$

The function being integrated is holomorphic on $\mathbb{C}$ except at $\zeta = 0$. It,

[45] The Fourier transform of an absolutely integrable function f is continuous and approaches 0 at infinity: Chap. VII, n° 27, theorem 23.

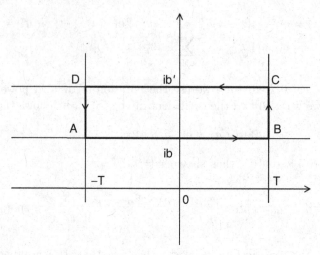

Fig. 8.10.

therefore, does not come as a surprise that the integral only depends on the sign of b. To justify this directly, we integrate along the contour of the rectangle $ABCD$ of the above figure; by Cauchy's theorems, the result is zero, and it all boils down to showing that, as T tends to $+\infty$, the contributions from the vertical sides approach 0. However, on these sides,

$$\left|\zeta^{-n}\mathbf{e}(\zeta)\right| = \left(T^2 + \eta^2\right)^{-n/2}\exp(-2\pi\eta),$$

where $\eta = \mathrm{Im}(\zeta)$ varies between b and b', and so $0 < m \le |\mathbf{e}(\zeta)| \le M < +\infty$ with constants m and M independent from T. On the other hand, $(T^2 + \eta^2)^{-n/2}$ remains between its values at $\eta = b$ and $\eta = b'$; $|\zeta^{-n}\mathbf{e}(\zeta)|$ is $\asymp T^{-n}$ on the vertical sides. Their contribution is, therefore, $O(T^{-n})$, qed.

This argument also explains why the value of the integral over the positive horizontals is different from that on the negative ones: in such a case, there is pole at $\zeta = 0$ in the interior of the rectangle $ABCD$, so that, up to a factor $2\pi i$, the integral is equal to the residue of $\zeta^{-n}\mathbf{e}(\zeta)$ at 0.

Exercise 1. Translate (15') using the change of variable $2\pi i t z = \zeta$.

Exercise 2. Applying Poisson's summation formula to the function $x \mapsto (x - z)^{-k}$, show that

$$(8.16) \qquad \sum_{\mathbb{Z}} \frac{1}{(z+n)^k} = \frac{(-2\pi i)^k}{(k-1)!} \sum_{n\ge 1} n^{k-1} e^{2\pi i n z} \quad \text{for } k \ge 2, \ \ \mathrm{Im}(z) > 0.$$

Recover (16) by differentiating the partial fractions expansion of $\mathrm{cotg}\,\pi z$.

Exercise 3. Check that Plancherel's formula holds for the Fourier transform (14).

(iv) *Semi-convergent Fourier transforms.* The calculation of the Fourier transform of a rational function $f = p/q$ assumes that $d^\circ(q) - d^\circ(p) = n \ge 2$.

As in (ii), the more subtle case when $n = 1$ can be handled; as we will see, the integral defining $\hat{f}(t)$ is then semi-convergent for all t, which we already know to be the case for $t = 0$.

Note first that

$$f(z) = c/z + O\left(1/z^2\right) \quad \text{at infinity}$$

as already remarked above. So only the term en $1/z$ poses a problem in the calculations and estimations which were successfully done for $n \geq 2$.

On the other hand, if the function f is real on $\mathbb{R}$, which may be assumed, its derivative only has finitely many real roots and hence its sign remains constant for large $|x|$; $f(x)$, therefore, monotonously tends to 0 as $x \in \mathbb{R}$ tends to $+\infty$ or $-\infty$, so if the Fourier integral is defined by the formula

$$(8.17) \qquad \hat{f}(t) = \lim \int_{-R}^{R} f(x)\mathbf{e}(-tx)dx\,,$$

it remains well-defined for $t \neq 0$ (Chapter V, n° 24, Theorem 23, integrals of "oscillating" functions), hence also for complex f by separating real and imaginary parts. In fact (17) is also well-defined for $t = 0$, but for different reasons as shown above in (ii).

Having said this, suppose first that $t > 0$ and integrate along the path already used for absolutely converging integrals. Setting $g(z) = f(z)\mathbf{e}(-tz)$, the contribution from the lower half-circle can be obtained by integrating the function $g(\mathrm{Re}^{iu})i\,\mathrm{Re}^{iu}$ from $u = 0$ to $u = -\pi$. As $|f(z)| \leq M/|z|$, where M is a constant,

$$\left|g\left(\mathrm{Re}^{iu}\right)i\,\mathrm{Re}^{iu}\right| = R\left|f\left(\mathrm{Re}^{iu}\right)\right|.\left|\mathbf{e}\left(-Rte^{iu}\right)\right| \leq M.\exp\left(2\pi Rt\sin u\right).$$

Therefore, the contribution from the lower half-circle is, in modulus, bounded above by

$$M\int_{-\pi}^{0}\exp(2\pi Rt\sin u)du = M\int_{0}^{\pi}\exp\left(-\rho\sin u\right)du\,,$$

where $\rho = 2\pi Rt$ tends to $+\infty$ since t is supposed to be strictly positive. To show that this integral tends to 0, first note that the integrated function is everywhere ≤ 1 and that it tends to 0 except for $u = 0$ or π since $-\rho\sin u$ tends to $-\infty$; the (real) Lebesgue theorem of dominated convergence takes care of the question since the measure on the set reduces to the points 0 and π. If the event of a refusal to use it, note first that, for given $t > 0$ and $\delta > 0$, the continuous function being integrated decreasingly converges to a limit function continuous on $[\delta, \pi - \delta]$, namely 0; convergence is, therefore, uniform in such an interval (Chapter V, § 2, n° 10, Dini's theorem), so that its contribution to the integral tends to 0, and so is $\leq \delta$ for large ρ; as those of the two forgotten intervals are $\leq \delta$ for all ρ, the total is $\leq 3\delta$ for large ρ, implying the result once again. If, however, you refuse to use Dini's theorem, you can check it by an explicit calculation. For this observe that

$$\sin u \geq \sin \delta = \alpha,$$

is a strictly positive constant in $[\delta, \pi - \delta]$, so that

$$\exp(-\rho \sin u) \leq \exp(-\alpha \rho)$$

and uniform convergence on the interval under consideration holds again since $\exp(-\alpha\rho)$, which does not depend on u, approaches 0 as ρ increases indefinitely.

In fact, a somewhat more precise result can be obtained. It is sometimes called Jordan's lemma by the author of a famous *Analysis Course* and professor at the École Polytechnique from 1876 to 1911 ; given there were about two hundred students per year, his students must have provided many artillerymen. It suffices to study the integral

$$I(R) = \int_0^\pi \exp(-R \sin u) du = 2 \int_0^{\pi/2}$$

and to note that between 0 and $\pi/2, u/2 \leq \sin u \leq u$.

$$I(R) \asymp \int_0^{\pi/2} \exp(-Ru)\, du \asymp 1/R.$$

immediately follows.

The previous arguments suppose that $t \neq 0$ and fall apart if $t = 0$. But in this case, we are brought back to the computations of section (i) and for example to formula (8), which shows that, one of the following holds:

$$(8.18) \qquad \hat{f}(0) = \begin{array}{l} 2\pi i \sum\limits_{\mathrm{Im}(a)>0} \mathrm{Res}(f, a) + \pi i\, \mathrm{Res}(f, \infty), \\[1em] -2\pi i \sum\limits_{\mathrm{Im}(a)<0} \mathrm{Res}(f, a) - \pi i\, \mathrm{Res}(f, \infty) \end{array}$$

where the residue at infinity is given by the relation

$$\mathrm{Res}(f, \infty) = -\lim z f(z) = -c.$$

In conclusion, (10') and (10") continue to hold for $t \neq 0$, the case $t = 0$ being a consequence of (18). For example,

$$(8.19) \qquad \int_{\mathbb{R}} \frac{\mathrm{e}(-tx)}{x - a} dx = \begin{array}{ll} 0 & \text{if } t > 0, \\ \pi i & \text{if } t = 0, \quad (\mathrm{Im}(a) > 0) \\ 2\pi i\, \mathrm{e}(-ta) & \text{if } t < 0. \end{array}$$

Replacing t by $-t$, we get

(8.20)
$$\int (x-a)^{-1}\mathbf{e}(tx)dx = 2\pi i t_+^0\mathbf{e}(at)$$

by stipulating, like Fejér, that

$$t_+^0 = 1 \text{ if } t > 0, \quad = 1/2 \text{ if } t = 0, \quad = 0 \text{ if } t < 0.$$

We could have spared ourselves all these calculations – but we did not want to – by alluding to Theorem 27 of Chapter VII, § 6, n° 30: if f is a regulated and absolutely integrable function on $\mathbb{R}$, then

$$\lim \int_{-R}^{R} \hat{f}(y)\mathbf{e}(xy)dy = \frac{1}{2}\left[f(x+) + f(x-)\right]$$

at every point where f has left and right derivatives. In the present case, it can be applied to the function f defined by the right hand sides of (19): it is obviously regulated, right and left differentiable everywhere, integrable since $\text{Im}(a) > 0$, and finally its Fourier transform is $1/(x-a)$ as shown by a most simple direct calculation. It then remains to check that $\frac{1}{2}\left[f(0+) + f(0-)\right] = \pi i$.

9 – Summation Formulas

The residue method allows us to prove several summation formulas. Let us for example show that *if $f(z)$ is an entire function satisfying an inequality of the form*

(9.1)
$$|f(z)| \le M.e^{\pi a|y|} \quad \text{with} \quad 0 < a < 1,$$

then

(9.2)
$$\pi \frac{f(z)}{\sin \pi z} = \sum_{\mathbb{Z}} (-1)^n \frac{f(n)}{z-n} = \lim_{p\infty} \sum_{|n|<p}.$$

For this, let us consider the meromorphic function[46] $g(w) = \pi f(w)/\sin \pi w$. Its only singularities are, at worst, simple poles at points $n \in \mathbb{Z}$ with $\text{Res}(g,n) = (-1)^n f(n)$. For given $z \notin \mathbb{Z}$, the function

$$h(w) = g(w)/(w-z) = \pi f(w)/(w-z)\sin \pi w$$

has simple poles at $w \in \mathbb{Z}$, with

---

[46] Recall that the only singularities of a meromorphic function on a domain G (here, $\mathbb{C}$) are necessarily isolated poles. If $f = p/q$ where q has a simple zero at a and p is holomorphic at a, then $\text{Res}(f,a) = p(a)/q'(a)$.

(9.3) $\mathrm{Res}(h, n) = (-1)^n f(n)/(n - z) = -(-1)^n f(n)/(z - n)$,

and, also a simple pole at z, where

(9.4) $$\mathrm{Res}(h, z) = g(z).$$

If, for given $p \in \mathbb{N}$, $h(w)$ is integrated along the rectangle of figure 11, an idea already applied in the previous n°, the result is then equal, up to a factor $2\pi i$, to the difference between $g(z)$ and the partial sum $|n| < p$ of series (2). Hence it all amounts to proving that the integral of h tends to 0 as p increases indefinitely. For this, find upper bounds for the contributions from each of the sides of the rectangle.

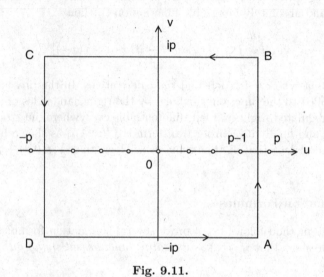

Fig. 9.11.

On the side BC,

$$\left|e^{\pi i w}\right| = e^{-\pi p}, \quad \left|e^{-\pi i w}\right| = e^{\pi p}$$

and as a result $\left|e^{\pi i w} - e^{-\pi i w}\right| \geq e^{\pi p} - e^{-\pi p} \geq e^{\pi p}/2$ for large p, and so, using (1), the following is an upper bound

(9.5) $$|f(w)/\sin \pi w| \leq c_1 e^{\pi(a-1)p}$$

where c_1 does not depend on p. Since also

$$|w - z| \geq |\mathrm{Im}(w - z)| = |p - \mathrm{Im}(z)| \geq p/2 \text{ for large } p,$$

on BC,

(9.6) $$|h(w)| \leq c_2 p e^{\pi(a-1)p} \text{ for large } p.$$

The integral along BC is, therefore, up to a constant, bounded above by $(2p+1)pe^{\pi(a-1)p}$ and tends to 0 since $a < 1$. A similar argument holds for the contribution from side DA.

Next, let us investigate the contribution from AB. As $w = p - 1/2 + iv$,

$$\sin \pi w = \pm \cosh \pi v \,,$$

whence $|\sin \pi w| \geq 1/2e^{-\pi|v|}$ and

$$|h(w)| \leq 2e^{-\pi|v|}|f(p-1/2+iv)|/|p-1/2+iv-z| \leq$$
$$\leq c_3 e^{\pi(a-1)|v|}/|p-1/2-\operatorname{Re}(z)|$$

for all v. Integrating from $v = -p$ to $v = p$, the upper bound

$$c_4 |p-1/2-\operatorname{Re}(z)|^{-1} \int_{-p}^{p} e^{\pi(a-1)|v|}dv$$

is obtained. It tends to 0 like the first factor since the extended integral over $\mathbb{R}$ converges for $a < 1$.

Therefore, the contribution from AB and obviously that from CD too both tend to 0, giving formula (2). The series extended to $\mathbb{Z}$ needs to be interpreted as the limit of its partial symmetric sums since it may very well not converge unconditionally, for instance in the case of the function $f(z) = 1$.

In this case, the partial fraction expansion of the function $1/\sin z$:

(9.7')
$$\frac{\pi}{\sin \pi z} = \sum_{\mathbb{Z}} \frac{(-1)^n}{z-n} = \frac{1}{z} + 2z \sum_{n \geq 1} \frac{(-1)^n}{z^2 - n^2}.$$

Most authors reject the first series or interpret it as the limit of its symmetric sums, but in fact taken separately its "positive" and "negative" parts are convergent though not absolutely. Indeed,

$$1/(z-n) = -1/n + z/n(z-n) = -1/n + O(1/n^2)$$

for large $|n|$, so that the sum extended to $n > 0$ is the sum of an absolutely convergent series and of the alternating series $\sum(-1)^{n+1}/n$, and so converges. (7') can also be written as

$$\frac{\pi}{\sin \pi z} = \frac{1}{z} + \sum_{n \neq 0} (-1)^n \left[\frac{1}{z-n} + \frac{1}{n} \right]$$

with, this time, an absolutely convergent series.

Replacing z by $\frac{1}{2}(1-z)$,

(9.7")
$$\frac{\pi}{2 \cos \pi z/2} = \sum_{\mathbb{Z}} \frac{(-1)^n}{z+2n+1} \,,$$

follows. We will use this formula in n° 15.

Exercise. Deduce from (2) the formulas

$$\pi\frac{\sin az}{\sin \pi z} = 2\sum_{n\geq 1}(-1)^n\frac{n.\sin na}{z^2 - n^2} \quad (-\pi < a < \pi),$$

$$\pi\frac{\cosh az}{\sinh \pi z} = 1/z + 2z\sum_{n\geq 1}(-1)^n\frac{\cos na}{z^2 + n^2}, \quad (-\pi < a < \pi).$$

Show that the convergence of the first amounts to that of the series $\sum(-1)^n\sin(na)/n$. Can these formulas be obtained from the theory of Fourier series?

10 – The Gamma Function, The Fourier Transform of $e^{-x}x_+^{s-1}$ and the Hankel Integral

Formula (8.15) which can also be written

$$\int (y-a)^{-n}e(xy)dy = (2\pi i)^n x_+^{n-1}e(ax)/(n-1)!;$$

was proved above. It assumes that n is an integer ≥ 2, $x \in \mathbb{R}$ and that $\text{Im}(a) > 0$. We intend to show that more generally

$$(10.1) \qquad \int (y-a)^{-s}e(xy)dy = (2\pi i)^s x_+^{s-1}e(ax)/\Gamma(s)$$

for $x \in \mathbb{R}$, $\text{Re}(s) > 1$ and $\text{Im}(a) > 0$, where

$$\Gamma(s) = \int_0^{+\infty} e^{-t}t^s d^*t, \quad \text{Re}(s) > 0$$

is Euler's function (Chapter V, §7, n° 22) and where $d^*t = dt/|t|$. Since, in (1), the function $(y-a)^{-s}$ and that of the right hand side are continuous and integrable for $\text{Re}(s) > 1$, this amounts (inversion formula) to showing that $(y-a)^{-s}$ is the Fourier transform of the function

$$(10.2) \qquad \varphi(x) = e(ax)x_+^{s-1} = \exp(2\pi iax)x_+^{s-1},$$

where $\text{Im}(a) = c > 0$. As $|\exp(2\pi iax)| = \exp(-2\pi cx)$, $\text{Re}(s) > 0$ is sufficient for $\varphi \in L^1(\mathbb{R})$ to hold. Then, setting $w = 2\pi i(y-a)$,

$$(10.3) \qquad \hat{\varphi}(y) = \int x_+^{s-1}e(ax - xy)dx = \int_0^{+\infty} \exp(-wx)x^s d^*x$$

and so $\text{Re}(w) = 2\pi\,\text{Im}(a) > 0$. If w were real > 0, the following formula could be obtained by a change of variable $x \mapsto x/w$:

(10.4)
$$\hat{\varphi}(y) = w^{-s} \int_0^{+\infty} e^{-x} x^s d^* x = \Gamma(s) w^{-s}$$

and so (1) would follow. The analyticity with respect to w remains to be proved.

(i) *The gamma function.* Some of the many properties[47] of this function are already known, to begin with the formulas

(10.5.1) $\Gamma(s+1) = s\Gamma(s), \quad \Gamma(n) = (n-1)!$ for $n \geq 1$.

This relation immediately shows that $\Gamma(s)$ can be extended analytically to all of $\mathbb{C}$, excepting the simple poles at $s = 0, -1, \ldots$, where

(10.5.2) $\mathrm{Res}(\Gamma, -n) = (-1)^n/n!$

(Chapter V, § 7, n° 25, *Example* 5); in fact,

(10.5.3) $\Gamma(s) = \int_0^1 e^{-x} x^s d^* x + \int_1^{+\infty} e^{-x} x^s d^* x =$

$$= \sum_{\mathbb{N}} (-1)^n/n!(s+n) + \Gamma^+(s).$$

The integral over $(1, +\infty)$, $\Gamma^+(s)$, converges for all s and is an entire function. Contrary to the integral over $(0, 1)$, the series obtained by integrating term by term the exponential series is also convergent for all s.

Setting

$$f_n(x) = \begin{array}{ll} (1-x/n)^n x^s & \text{for } x \leq n, \\ 0 & \text{for } x > n, \end{array}$$

we get a sequence of functions converging to $e^{-x} x^s$ while remaining dominated by $e^{-x} x^s$ (exercise!); hence

(10.5.4) $\Gamma(s) = \lim \int f_n(x) d^* x = \lim n! n^s / s(s+1) \ldots (s+n)$

(Chapter V, § 7, n° 23, *Example* 1), a priori for $\mathrm{Re}(s) > 0$. This leads to the expansion

(10.5.5) $1/\Gamma(s) = se^{Cs} \prod (1 + s/n) e^{-s/n}$

---

[47] See for example Dieudonné, *Calcul infinitésimal* (Hermann, 1968), IV.3, IX-4 to IX-8, Remmert, *Funktionentheorie 2*, Chap. 2, § 2, where the function is defined by using its infinite product, Freitag and Busam, *Funktionentheorie*, chap. IV, § 1 and in particular the exercises, not to mention earlier authors. Entire books have been written on it, notably N. Nielsen, *Handbuch der Theorie der Gammafunktion* (Leipzig, 1906, reedit. Chelsea, 1965).

of the function $\Gamma(s)$ into an infinite product convergent everywhere, and so valid everywhere by analytic extension;

$$C = \lim(1 + \ldots + 1/n - \log n) = 0,577215664\ldots$$

is the Euler constant (Chapter VI, § 2, n° 18). The complement formula (same reference)

(10.5.6) $$\Gamma(s)\Gamma(1 - s) = \pi/sin\pi s$$

is, for example, obtained by comparing the infinite product expansions of both sides with that of

$$\sin \pi s = \pi s \prod_{n \geq 1} \left(1 - s^2/n^2\right)$$

(Chap. IV, n° 18). It follows that

(10.5.7) $$\Gamma(1/2) = \pi^{1/2},$$

a result that reduces to the integral of $\exp(-\pi x^2)$, and that

(10.5.8) $$|\Gamma(1/2 + it)|^2 = \pi/\cosh \pi t \text{ for } t \in \mathbb{R}.$$

The *duplication formula*

(10.5.9) $$\Gamma(2s) = \pi^{-1/2}2^{2s-1}\Gamma(s)\Gamma(s + 1/2)$$

which is often used in analytic number theory will be needed. A very (too?) ingenious method[48] for obtaining it consists in using a characterization due to Helmut Wielandt (1939) of the gamma function by the following two properties:

(a) f is holomorphic on a domain G containing the strip $1 \leq \operatorname{Re}(z) \leq 2$ and bounded on it;

(b) $f(s + 1) = sf(s)$ whenever $s, s + 1 \in G$.

Property (b) allows us, as in the case of Euler's function, to first find an analytically extension of f to all of $\mathbb{C}$ with, at worst, simple poles at integers ≤ 0 and

$$\operatorname{Res}(f, -n) = (-1)^n f(1)/n!.$$

As a result, $g(s) = f(s) - f(1)\Gamma(s)$ is an entire function also satisfying $g(s + 1) = sg(s)$. The entire function $h(s) = g(s)g(1 - s)$ then satisfies $h(s + 1) = -h(s)$.

---

[48] I find it in Freitag-Busam, Chap. IV, §1 and in Remmert 2, Chap. 2, §2. A more general, but far less useful, formula can be found in Dieudonné, *Calcul infinitésimal*, IX.4.

As $f(s)$ and $\Gamma(s)$ are bounded on $1 \leq \operatorname{Re}(s) \leq 2$, because of their functional equation, they are bounded at infinity on the entire strip $a \leq \operatorname{Re}(s) \leq a + 1 \leq 2$; the same, therefore, holds for g, and as g does not have poles, it is globally bounded on that strip. Hence $h(s)$ is bounded on the strip $0 \leq \operatorname{Re}(s) \leq 1$.

The pseudo periodicity of h then shows that h is bounded on $\mathbb{C}$, and so is constant, and hence is zero since $g(1) = f(1) - f(1)\Gamma(1) = 0$. Since $g(s)g(1 - s) = 0$ for all s, $g(s) = 0$, and so

$$f(s) = f(1)\Gamma(s),$$

qed.

Having done this, we can return to (10.5.9), which can also be written

$$(10.5.10) \qquad \Gamma(s) = \pi^{-1/2}2^{s-1}\Gamma(s/2)\Gamma\left[(s + 1)/2\right],$$

and observe that the function $f(s) = 2^s\Gamma(s/2)\Gamma\left[(s+1)/2\right]$ satisfies Wielandt's assumptions, which is immediate. Hence $f(s) = f(1)\Gamma(s) = 2\pi^{1/2}\Gamma(s)$, and the formula follows. See Chap. XII, n° 1 also.

(ii) *Fourier transform of* $e^{-x}x_+^{s-1}$. To show that (10.4) remains valid for $\operatorname{Re}(w) > 0$, it suffices – analytic extension – to check that both sides are holomorphic functions of w on this half-plane. Setting

$$(10.6) \qquad w^{-s} = \exp\left[-sL(w)\right] = |w|^{-s}e^{-is\operatorname{Arg}(w)}$$

where $L(w) = \log|w| + i\operatorname{Arg}(w)$ is a uniform branch of the pseudo-function $Log(w)$ on $\operatorname{Re}(w) > 0$ (§ 2, n° 4, (i)), it is the case of the right hand side. As our aim is to find the usual function for w real > 0, we should choose

$$|\operatorname{Arg}(w)| < \pi/2.$$

As for function (3), theorem 9 on integrals depending holomorphically on a parameter can be applied to it. The only non-obvious condition is the existence of a function $p_H(t) \in L^1(\mathbb{R}_+)$ for every compact subset $H \subset \{\operatorname{Re}(w) > 0\}$ such that

$$\left|\exp(-wt)t^{s-1}\right| \leq |p_H(t)|$$

for all $w \in H$ and $t > 0$. But the distance from a compact subset of an open set U to the boundary of U is > 0. H is therefore contained in a half-plane $\operatorname{Re}(w) \geq a$ with $a > 0$; then

$$\left|\exp(-wt)t^{s-1}\right| \leq \exp(-at)t^{\operatorname{Re}(s)-1} = p_H(t).$$

This is an integrable function since $a > 0$ and $\mathrm{Re}(s) > 0$, qed.

(1) can now be justified. Since, by (3) and (4),

$$\hat{\varphi}(y) = \Gamma(s)w^{-s} = \Gamma(s)\left[2\pi i(y-a)\right]^{-s} \quad \text{for} \quad \mathrm{Re}(s) > 0,$$

this function is of the order of magnitude of y^{-s} at infinity. It is, therefore, integrable over $\mathbb{R}$ if $\mathrm{Re}(s) > 1$, and as this is also the case of

$$\varphi(x) = \mathbf{e}(ax)x_+^{s-1}$$

which, moreover, is then continuous everywhere, including at $x = 0$, Fourier's inversion formula shows that

$$(10.7) \qquad \Gamma(s)\int \left[2\pi i(y-a)\right]^{-s}\mathbf{e}(xy)dy = x_+^{s-1}\mathbf{e}(ax)$$

for $\mathrm{Re}(s) > 1$ and $\mathrm{Im}(a) > 0$, provided we set $|\mathrm{Arg}(w)| < \pi/2$. Choosing

$$-\pi < \mathrm{Arg}(y-a) < 0 \quad \text{and} \quad \mathrm{Arg}(2\pi i) = \pi/2,$$

$\mathrm{Arg}(w) = \mathrm{Arg}\left[2\pi i(y-a)\right] = \mathrm{Arg}(2\pi i) + \mathrm{Arg}(y-a)$, and so

$$\left[2\pi i(y-a)\right]^{-s} = (2\pi i)^{-s}(y-a)^{-s} = (y-a)^{-s}/(2\pi i)^s$$

and (7) can be written

$$(10.8) \qquad \int (y-a)^{-s}\mathbf{e}(xy)dy = \frac{(2\pi i)^s}{\Gamma(s)}x_+^{s-1}\mathbf{e}(ax) \quad \text{for} \quad \mathrm{Im}(a) > 0, \ \ \mathrm{Re}(s) > 1;$$

this is the formula generalizing (8.15). Similarly, we get

$$(10.8') \qquad \int (y+a)^{-s}\mathbf{e}(-xy)dy = \frac{(-2\pi i)^s}{\Gamma(s)}x_+^{s-1}\mathbf{e}(ax) \quad \text{for} \quad \mathrm{Im}(a) > 0,$$

where we have to take $0 < \mathrm{Arg}(y+a) < \pi$ and $\mathrm{Arg}(-2\pi i) = -\pi/2$.

Replacing a by z and applying Poisson's summation formula (Chap. VII, n° 23), we get

$$(10.9) \quad \sum \frac{1}{(z+n)^s} = \frac{(-2\pi i)^s}{\Gamma(s)}\sum_{n\geq 1} n^{s-1}\exp(2\pi inz) \quad \text{for} \quad \mathrm{Im}(z) > 0$$

and $\mathrm{Re}(s) > 1$, a result generalizing (8.16). The reader should not forget to check the assumptions allowing the application of the Possion formula to a function f: the latter is continuous and the series $\sum f(x+n)$ and $\sum \hat{f}(y+n)$ converge normally on any compact set.

(iii) *Hankel's integral.* For $x > 0$ and $a = i$, (8) can also be written as

$$2\pi i / \Gamma(s) = \int \frac{e^{2\pi i x(y-i)}}{[2\pi i x(y-i)]^s} 2\pi i x \, dy, \quad \text{Re}(s) > 1.$$

Setting $2\pi i x(y-i) = z$, we get an integral computed over the vertical $\text{Re}(z) = 2\pi x = a > 0$. As $dz = 2\pi i x \, dy$, we finally get that

$$(10.10) \qquad 2\pi i / \Gamma(s) = \int_{\text{Re}(z)=a>0} e^z z^{-s} dz, \quad \text{Re}(s) > 1.$$

Having said that, and the function z^{-s} being defined on $U = \mathbb{C} - \mathbb{R}_-$ by

$$(10.11) \qquad z^{-s} = |z|^{-s} \exp\left[-is \, \text{Arg}(z)\right] \quad \text{with} \quad |\text{Arg}(z)| < \pi,$$

integral (10) is about a holomorphic function on U. We show that a formula holding for all $s \in \mathbb{C}$ can be obtained by deforming the integration contour. This is not the case of (10) since the integral diverges for $\text{Re}(s) \leq 1$. We use

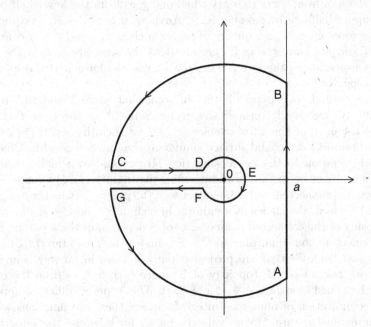

Fig. 10.12.

the above path, along which the integral is zero, and set r and R to be the radii of the two circular arcs. By (11), $|z^{-s}| \asymp |z|^{-\text{Re}(s)}$ on $U = \mathbb{C} - \mathbb{R}_-$ and in particular for small or large $|z|$. Since, moreover,

$$|e^z| = e^{\text{Re}(z)} \leq e^a$$

on the integration contour, the contribution from the large arcs to the integral is $O(R^{1-\mathrm{Re}(s)})$, hence tends to 0 since $\mathrm{Re}(s) > 1$. For the same reason, the contribution from the small circular arc is $O(r^{1-\mathrm{Re}(s)})$, but it does not follow that it approaches 0 as r tends to 0; as always, we have to choose between convergence at infinity and convergence at 0...

As R increases indefinitely, the integral along AB approaches the integral (10) we started with and the integrals over the large circular arcs tend to 0. In view of the directions of integration,

$$(10.12) \quad 2\pi i/\Gamma(s) = \int_{-\infty}^{(0+)} e^z z^{-s} dz \qquad \text{(Hankel's Formula)}$$

where the notation, traditional among specialists of special functions,[49] denotes the path $GFEDC$ expanded at infinity on both sides of the *cut* plane along the real negative axis. The use of this word, also traditional, suggests that if points located over or under $\mathbb{R}_-$ are "infinitely near" in $\mathbb{C}$, from the point of view of the values of z^{-s}, they are not near because the argument of z^{-s} changes from $-\pi i s$ to $+\pi i s$ when one goes from the lower half-plane to the upper half one by crossing $\mathbb{R}_-$, . Anyhow, if $U = \mathbb{C} - \mathbb{R}_-$ is equipped with the topology of $\mathbb{C}$, a sequence of points such as $a_n = -1 + i/n$, converging in $\mathbb{C}$, do not converge *in* U; neither does the sequence $b_n = -1 - i/n$. Despite appearances, the latter is in no way near the former with respect to the topology of U.

If we wanted to explain all this in somewhat more "modern" terms than W&W, we could consider the graph S in $\mathbb{C}^2$ of the $\zeta = \mathcal{L}og\, z = \log|z| + i\, \mathcal{A}rg(z)$, i.e.the set of couples $(z, \zeta) \in \mathbb{C}^2$ such that $\exp(\zeta) = z$. It is for good reason a helicoidal surface similar to the one described in Chapter IV, §4 in relation to the pseudo-function $\mathcal{A}rg(z)$ and on which a genuine function z^{-s} can be defined without ambiguity, namely $(z, \zeta) \mapsto \exp(-s\zeta)$. This graph is connected, but is no longer so if the points for which $z \in \mathbb{R}_-$ are removed. Indeed, the choice of a uniform branch of z^{-s} on $\mathbb{C} - \mathbb{R}_-$ amounts to choosing of the connected components of S deprived of these points. Such a component is homeomorphic to $\mathbb{C} - \mathbb{R}_-$ under the projection $(z, \zeta) \mapsto z$, but its two "sides", that are projected onto $\mathbb{R}_-$, are in no way near each other with respect to the topology of $\mathbb{C}^2$ since each follows from the other under the translation $(z, \zeta) \mapsto (z, \zeta + 2\pi i)$. This type of difficulty appears in the computation of numerous integrals, where there are functions whose analytic extensions are "many valued" on $\mathbb{C}$, for example the integral of $(4x^3 - g_2 x - g_3)^{1/2}$ which occurs in the theory of elliptic functions and is the easiest instance of an integral of an algebraic function, or else integrals involving the function $\log x$, etc.

Returning to Hankel's formula, the exact form of the integration contour is unimportant, and integrating over any path homotopic to the vertical

---

[49] See in particular Whittaker and Watson, *A Course of Modern Analysis*, Cambridge UP, 1902.

$\mathrm{Re}(z) = a > 0$ in $\mathbb{C} - \mathbb{R}_-$ and not only in $\mathbb{C}$ would be sufficient, for example over the curve that a metallurgist would obtain by bending a rectilinear iron

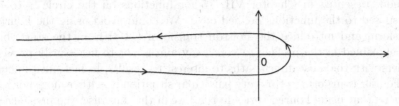

Fig. 10.13.

wire of infinite length around the impassible barrier erected along $\mathbb{R}_-$. The importance of integral (12) lies in the fact that *it is well-defined for all* $s \in \mathbb{C}$, because as $\mathrm{Re}(z)$ tends to $-\infty$, the function e^z tends to 0 quickly enough to neutralize the power functions with which it is multiplied. Using theorem 8, it is not hard to check that the right hand side of (12) is a holomorphic function of s and thus does represent the left hand side on all of $\mathbb{C}$.

$\mathbb{R}_-$ could be replaced by any half-lines with initial point 0 located in the half-plane $\mathrm{Re}(z) < 0$, provided the chosen uniform branch of z^{-s} is in consequence defined. When moving away to infinity along such a cut (and not randomly in $\mathbb{C}$)), $|e^z| = e^{\mathrm{Re}(z)}$ with $\mathrm{Re}(z) \asymp -|z|$, so that the factor e^z tends to 0 sufficiently quickly for the integral to converge for all s.

If we are going to present pre-modern mathematics, let us show how the relation $\Gamma(s)\Gamma(1-s) = \pi/\sin \pi s$ can be recovered by integrating along the path $GFEDC$ extended to infinity. If the ordinates of the lines GF and DC are $-\varepsilon$ and $+\varepsilon$, whence $z = -t \pm i\varepsilon$ with $t > 0$, then the approximations $z^{-s} = t^{-s}\exp(-i\pi s)$ on GF and $z^{-s} = t^{-s}\exp(i\pi s)$ on DC follow. In view of the direction followed,

$$\int_{GF} + \int_{DC} = -e^{-\pi i s} \int_r^{+\infty} t^{-s} e^{-t} dt + e^{\pi i s} \int_r^{+\infty} t^{-s} e^{-t} dt.$$

Relation (10) used above to show that the contribution from the large circular arcs tends to 0 if $\mathrm{Re}(s) > 1$ equally shows that that of the small arc FED tends to 0 if $\mathrm{Re}(s) < 1$; in this case, r can be made to approach 0. At the limit, Hankel's integral becomes

$$2\pi i/\Gamma(s) = 2i \sin \pi s \int_0^{+\infty} e^{-t} t^{-s} dt = 2i \sin \pi s . \Gamma(1-s).$$

This gives the complement formula for $\mathrm{Re}(s) < 1$ and hence, by analytic extension, in all of $\mathbb{C}$.

11 – The Dirichlet problem for the half-plane

The computations of the previous n° allow the Dirichlet problem[50] solution method presented in Chapter VII, § 5 for functions on the circle $\mathbb{T}$ to be transposed to the functions defined on $\mathbb{R}$. We could avoid using the Fourier transform and introduce the Poisson transform (11.3) from the start, but as I explained in the preface to vol. I, my aim is not to necessarily provide readers with the most direct paths to interesting results. In fact, introducing the Fourier transform in this very particular situation is neither more nor less artificial that using Fourier series in the case of the unit disc; the presence of a group is exploited: the group of rotations about 0 in the case of the disc, the group of horizontal translations in the case of the half-plane. The method could be generalized to heat or wave propagation equations

$$(*)\qquad\qquad du/dt = \Delta u\,,\quad d^2u/dt^2 = \Delta u$$

where $u(t, x)$ is a function on $\mathbb{R}_+ \times \mathbb{R}^n$ with (as well as its partial derivative du/dt in the second case) given values for $t = 0$. The reader will already be able to practice on the case $n = 1$, the basic idea being the one applied by Fourier to go from the heat equation on the unit circle to his series: find the "simple" solutions of the form $f(t)g(x)$, then try to express the general solution as a "continuous sum" of such solutions; the calculation is easy for the first equation, but less so for the second one.

Allowing for notation, in the previous n°, the Fourier transform of the function

$$\varphi(t) = t_+^{s-1}\mathbf{e}(zt)\,,\quad \operatorname{Im}(z) > 0\,,\ \operatorname{Re}(s) > 0\,,$$

has been shown to be

$$\hat{\varphi}(u) = \Gamma(s)\left[2\pi i(u - z)\right]^{-s}\,.$$

On the other hand, we know (Chap. VII, § 6, n° 30) that if $f, g \in L^1(\mathbb{R})$, then

$$(*)\qquad\qquad \int f(t)\hat{g}(t)dt = \int \hat{f}(u)g(u)du;$$

this formula is obtained by calculating the double integral

$$\iint f(t)g(u)\mathbf{e}(-tu)dtdu$$

---

[50] Recall that, generally speaking, it consists in constructing a harmonic function on an open set with given values on the boundary. The cases of the unit disc or of the half-plane cannot give any idea of the difficulty involved in the general problem in $\mathbb{C}$, even less so in $\mathbb{R}^n$ – not to mention generalizations to elliptic PDEs.

in two different ways. The easiest proof consists in applying the (genuine) Lebesgue-Fubini theorem; also with a few acrobatics, we could just consider the version proved in Chapter V, n° 33 for lsc functions (separate the real and imaginary parts of the functions in question), but it is not worth it.

Choosing[51] $g = \varphi$, for any function $f \in L^1(\mathbb{R})$,

$$(11.1) \qquad \int_0^{+\infty} \hat{f}(u)u^{s-1}\mathbf{e}(zu)du = \Gamma(s) \int_\mathbb{R} f(t) \left[2\pi i(t - z)\right]^{-s} dt$$

for $\operatorname{Re}(s) > 0$ and $\operatorname{Im}(z) > 0$ since these conditions imply $g \in L^1(\mathbb{R})$. In any event, the case $s = 1$ shows that for every $f \in L^1(\mathbb{R})$,[52]

$$(11.2') \qquad \int_0^{+\infty} \hat{f}(u)\mathbf{e}(zu)du = \frac{1}{2\pi i} \int_\mathbb{R} \frac{f(t)}{t - z}dt = F^+(z) \quad \text{for} \quad \operatorname{Im}(z) > 0,$$

where F^+ is defined and holomorphic for $\operatorname{Im}(z) > 0$. The similarity of this result with Cauchy's integral formula will not escape anyone, but it is misleading: $F^+(z)$ is not a holomorphic function reducing to f on $\mathbb{R}$. If this formula is applied to $f(-t)$, a function whose Fourier transform is $\hat{f}(-u)$, and if z is replaced by $-z$, then similarly

$$(11.2'') \qquad \int_{-\infty}^0 \hat{f}(u)\mathbf{e}(zu)du = -\frac{1}{2\pi i} \int \frac{f(t)}{t - z}dt = -F^-(z) \quad \text{for }, \operatorname{Im}(z) < 0$$

where F^- is defined and holomorphic for $\operatorname{Im}(z) < 0$. This leads us to associate to every *integrable* function f *on* $\mathbb{R}$ the function[53]

$$(11.3) \quad F(z) = \frac{1}{2\pi i} \int \frac{f(t)dt}{t - z} = \begin{array}{l} \int_0^{+\infty} \hat{f}(u)\mathbf{e}(zu)du \quad \text{if} \quad \operatorname{Im}(z) > 0 \\[2mm] -\int_{-\infty}^0 \hat{f}(u)\mathbf{e}(zu)du \quad \text{if} \quad \operatorname{Im}(z) < 0 \end{array}$$

analogous to the Poisson transform P_f introduced in Chapter V, §5 in the case of the unit disc [see in particular formula (21.7)]. It is holomorphic

---

[51] The reader will probably observe that φ is a regulated function on $\mathbb{R}$ only if $\operatorname{Re}(s) > 1$ or $s = 1$ since the factor x_+^{s-1} is not bounded in the neighbourhood of 0 for $\operatorname{Re}(s) < 1$ or, for $\operatorname{Re}(s) = 1$, $s \neq 1$, does not approach a limit as $x \longrightarrow +0$. All these difficulties disappear in Lebesgue theory.

[52] (2') could be directly obtained by computing the Fourier transform of the function equal to $\mathbf{e}(\zeta y)$ for $y > 0$ and to 0 otherwise, and by applying the general formula (*).

[53] $f(t)dt$ could be replaced by a measure $d\mu(t)$ with finite total mass, and it is precisely by studying such functions that Stieltjes was led to define his integrals. Example: write the set of rational numbers as a sequence (u_n) and choose the measure μ given by $\int f(t)d\mu(t) = \sum f(u_n)/n^2$ for continuous f with compact support. The behaviour of the corresponding function

$$F(z) = \sum 1/n^2(u_n - z)$$

in the neighbourhood of the real axis is not obvious.

outside the real axis and decomposes into the two functions F^+ and F^- on the upper and lower half-planes H^+ and H^-. These functions being given by the Fourier integrals occurring in (3), they extend by continuity to the closed half-planes $\mathrm{Im}(z) \geq 0$ and $\mathrm{Im}(z) \leq 0$, provided these integrals converge for real z, which assumes that $f \in F^1(\mathbb{R})$ since f has already been assumed to be integrable. It is not in general possible to go from one to the other by analytic extension since, as we will see, their limit values on the real axis do not coincide. Nonetheless, it is necessary to observe that the left hand side of (3) continues to be well-defined in a neighbourhood of z if z does not belong to the support S of f, so that (3) defines a holomorphic function on the *connected* open set $H^+ \cup H^- \cup (\mathbb{R} - S)$ of $\mathbb{C}$ if the open subset $\mathbb{R} - S$ of $\mathbb{R}$ is not empty; under these circumstances, the analytic extension across $\mathbb{R} - S$ is possible, which does not prevent the function obtained to suffer a discontinuities across S.

Set

$$(11.4) \quad u_f(x,y) = F^+(z) - F^-(\bar{z}) = \frac{1}{2\pi i} \int \left(\frac{1}{t-z} - \frac{1}{t-\bar{z}} \right) f(t)dt =$$

$$= \frac{1}{\pi} \int \frac{y}{|t-z|^2} f(t)dt ,$$

where integration is as always over $\mathbb{R}$ when it is not specified. (2') and (2") show that[54]

$$(11.5) \quad u_f(x,y) = \int_0^{+\infty} \hat{f}(u) \exp(-2\pi uy)e(ux)du +$$

$$+ \int_{-\infty}^0 \hat{f}(u) \exp(+2\pi uy)e(ux)du =$$

$$= \int e(xu) \exp(-2\pi y|u|)\hat{f}(u)du .$$

As y tends to $+0$, the function $\exp(-2\pi y|u|)$ converges uniformly to 1 on every compact set by remaining ≤ 1. Therefore, the most elementary version of the dominated convergence theorem shows that

$$(11.6) \quad \lim_{y \to +0} \left[F^+(x+iy) - F^-(x-iy) \right] =$$

$$= \lim_{y \to +0} u_f = \int e(xu)\hat{f}(u)du = f(x)$$

if $f \in F^1(\mathbb{R})$, which explains the impossibility of "gluing back" F^+ and F^- into a single holomorphic function on $\mathbb{C}$.

---

[54] These computations have already been used in Chapter VII, §6, n° 30 to prove Fourier's inversion formula (Theorem 26).

Contrary to (3), definition

$$(11.7) \qquad u_f(x,y) = \frac{1}{\pi} \int \frac{y}{(t-x)^2 + y^2} f(t) dt$$

continues to be well-defined for $y > 0$, provided f is *bounded on* $\mathbb{R}$. Hence if it is not possible to define the Poisson transform $F = P_f$, u_f can then be defined on the upper half-plane; this function is harmonic since the relation

$$u_f(x,y) = \lim_{n=+\infty} \frac{1}{2\pi i} \int_{-n}^{n} \left(\frac{1}{t-z} - \frac{1}{t-\bar{z}} \right) f(t) dt = \lim_{n=+\infty} (F_n(z) + F_n(\bar{z}))$$

shows that, on the upper half-plane H^+, harmonic functions converge uniformly to u_f on all compact subsets (exercise), and that u_f is therefore harmonic[55] (Chapter VII, §5, n° 25, Theorem 21). Relation (6), i.e.

$$(11.6') \qquad\qquad \lim u_f(x,y) = f(x),$$

obtained by supposing $f \in F^1(\mathbb{R})$, applies in fact to all *continuous and bounded* functions f on $\mathbb{R}$. Indeed, set

$$(11.8) \quad P_y(t) = \frac{y}{\pi (t^2 + y^2)} = y^{-1} P(t/y) \text{ where } P(t) = \frac{1}{\pi (t^2 + 1)}$$

for $t \in \mathbb{R}$ and $y > 0$. When $y \longrightarrow +0$, these functions form, up to countability, a Dirac sequence on $\mathbb{R}$ in the sense of Chapter V, §8, n° 27: they are continuous, positive with total integral 1 and, for any $r > 0$, as y tends to 0, so does the integral

$$\int_{|t|>r} P_y(t) dt = 2 \int_{r}^{+\infty} y^{-1} P(t/y) dt = \frac{2}{\pi} \int_{r/y}^{+\infty} \frac{dt}{t^2 + 1}.$$

The arguments of Chapter V immediately imply (6') for continuous and bounded f and even

$$(11.6'') \qquad\qquad \lim u_f(x,y) = \frac{1}{2} [f(x+0) + f(x-0)]$$

for f regulated and bounded on $\mathbb{R}$.

The function u_f is itself continuous and bounded on the closed half-plane $\text{Im}(z) > 0$ if f is continuous and bounded on $\mathbb{R}$. Since $P_y(t)$ is an even function of t,

$$(11.9) \quad u_f(x,y) = \int P(x-t,y) f(t) dt = \int P_y(u) f(x-u) du,$$

---

[55] It may also be observed that, up to a factor i, $1/(t-z) - 1/(t-\bar{z})$ is the imaginary part of the holomorphic function $1/(t-z)$, and so is harmonic for any $t \in \mathbb{R}$, and the Laplacian of u_f may be calculated by differentiating under the $\int$ sign.

indeed follows. This is a product of convolutions on $\mathbb{R}$, and since $\int P_y(u)du = 1$, it proves that

(11.10) $$|u_f(x,y)| \leq \|f\|,$$

where the norm is a uniform one on $\mathbb{R}$. On the other hand, for all $a \in \mathbb{R}$,

$$|u_f(x,y) - f(a)| \leq \int P_y(u)\,|f(x-u) - f(a)|\,du.$$

For all $r > 0$, the extended integral over $\{|u| \geq r\}$ is, up to a factor $2\|f\|$, bounded above by the extended integral of P_y over the same set, est hence, as seen above, is $\leq \varepsilon$ for all x given any sufficiently small y . In the interval $|u| \leq r$, $|x - u - a| \leq |x - a| + r$, and so $|f(x - u) - f(a)| \leq \varepsilon$ for any u in this interval provided $|x - a|$ and r are sufficiently small. This integral is, therefore, $\leq \varepsilon$ since the value of the total integral of P_y is 1. Finally, $|u_f(x,y) - f(a)| \leq 2\varepsilon$ for sufficiently small y and $|x - a|$, qed.

Exercise. If $f(x)$ is uniformly continuous on $\mathbb{R}$, the function $x \mapsto u_f(x,y)$ converges uniformly to $f(x)$ on $\mathbb{R}$ as $y \longrightarrow +0$.

The function u_f, therefore, solves the Dirichlet problem for the half-plane and for continuous and bounded functions on $\mathbb{R}$: find a harmonic function on an open set G with given values on the boundary. More precisely, it is one of the possible solutions since any function of the form $u_f(x,y) + ay$, where a is a constant, is also a solution of the problem. Uniqueness holds in the case of the unit disc D considered in Chapter VII, §5 (Theorem 22) because, if a continuous function on the *closed* disc and harmonic on the interior is zero on the boundary, then a compactness argument and the maximum principle show that it is identically zero. But the half-plane is not compact. In fact, the map

$$z \longmapsto \zeta = (z - i)/(z + i),$$

is a conformal representation of the half-plane $\mathrm{Im}(z) > 0$ on the unit disc D : $|\zeta| < 1$; it, therefore, transforms every holomorphic (resp. harmonic) function on the former to a holomorphic (resp. harmonic) function on the latter. The real axis $y = 0$ is homeomorphically mapped onto the boundary $|\zeta| = 1$ *with the point 1 removed*, which could be obtained by making z approach infinity. Hence, for continuous functions $f(x)$ on $\mathbb{R}$, the Dirichlet problem in the half-plane, translated into the language of the unit disc amounts to finding a continuous function on $\overline{D} - \{1\}$, harmonic on D and with given values on $\mathbb{T} - \{1\}$, which allows every kind of behaviour in the neighbourhood of 1. In fact,

$$y = \mathrm{Im}(z) = (1 - |\zeta|^2)/|1 - \zeta|^2$$

is easily computed to be the Poisson disc functional; it is harmonic on D, continuous on $\overline{D} - \{1\}$ and zero on $\mathbb{T} - \{1\}$, but takes every positive value in all neighbourhoods of 1, so that it does not contradict Dirichlet's principle for the unit disc...

For f integrable, apart from the function,

$$u_f(x, y) = F^+(z) - F^-(\bar{z}),$$

the function

(11.11) $$v_f(x, y) = -i\left[F^+(z) + F^-(\bar{z})\right]$$

may be introduced; if f is real, clearly $F^-(\bar{z}) = -\overline{F^+(z)}$, so that u_f and v_f are, up to a factor of 2, the real and imaginary parts of $F^+(z)$ on the upper half-plane. Changing the computations leading to (4) and (5),

$$(11.12) \quad v_f(x, y) = \frac{1}{\pi}\int \frac{t-x}{(t-x)^2 + y^2} f(t)dt = \frac{1}{\pi}\int \frac{t}{t^2 + y^2} f(x+t)dt$$

$$= -i\int e(xu)\text{sgn}(u)\exp(-2\pi y|u|)\hat{f}(u)du$$

follows. Let us investigate the limit behaviour of this function when $y > 0$ approaches 0, but by supposing now that $f \in F^1(\mathbb{R})$. As done above for u_f, it is possible to pass to the limit under the $\int$ sign in the Fourier integral, and so

(11.13) $$\lim_{y=0+} v_f(x, y) = -i\int e(xu)\text{sgn}(u)\hat{f}(u)du.$$

Let us now consider the first integral (12), which can also be written

$$\pi v_f(0, y) = \int \frac{t}{t^2 + y^2} f(t)dt = \int_0^{+\infty} \frac{t^2}{t^2 + y^2} g(t)dt/t$$

for $x = 0$, where $g(t) = f(t) - f(-t)$. If f is differentiable at the origin, then

$$f(t) = f(0) + f'(0)t + o(t), \quad f(-t) = f(0) - f'(0)t + o(t)$$

and, therefore, $g(t) = 2f'(0)t + o(t)$; hence the function $g(t)/t$ is integrable in the neighbourhood of 0, as well as at infinity like f. As y approaches 0, the function $t^2/(t^2 + y^2)$ approaches 1 while always remaining ≤ 1. The elementary version of the dominated convergence theorem can, therefore, be applied again, which shows that

$$\lim \pi v_f(0, y) = \int_{t>0} g(t)dt/t = \lim_{r=0}\int_{|t|>r} f(t)dt/t.$$

Following traditions, set

(11.14) $$\text{p.v}\int_{-\infty}^{+\infty} f(t)dt/t = \lim_{r=0}\int_{|t|>r} f(t)dt/t.$$

This is the Cauchy principal value for integrable functions on $\mathbb{R}$ and differentiable at the origin, which an essential condition. We thus finally get the formula

(11.15) $\text{p.v.} \int f(t)dt/t = -\pi i \int \text{sgn}(u)\hat{f}(u)du\,,$

which at least holds under the following conditions: f is differentiable at the origin and belongs to $F^1(\mathbb{R})$.

It is in particular the case if $f \in \mathcal{S}(\mathbb{R})$, the Schwartz space. As a function of f, the left hand side $T(f)$ can immediately be checked to be a tempered distribution (Chapter VII, § 6, n° 32), denoted v.p.1/t by some; indeed,

$$|T(f)| \leq \pi \int \left|\hat{f}(u)\right| du = \pi \int \left|\hat{f}(u)\left(1 + 4\pi^2 u^2\right)\right| \cdot \left(1 + 4\pi^2 u^2\right)^{-1} du \leq$$
$$\leq \sup \left|\hat{f}(u)\left(1 + 4\pi^2 u^2\right)\right|\,.$$

Now, $(1 + 4\pi^2 u^2)\hat{f}(u)$ is the Fourier transform of $f(t) - f''(t)$, and hence for all u is bounded above by the integral

$$\int |f(t) - f''(t)| \, dt = \int \left(1 + t^2\right)|f(t) - f''(t)| \cdot \left(1 + t^2\right)^{-1} dt \leq$$
$$\leq \pi . \sup \left(1 + t^2\right)|f(t) - f''(t)| \leq$$
$$\leq \pi . \sup \left(1 + t^2\right)|f(t)| + \pi . \sup \left(1 + t^2\right)|f''(t)| \leq N_2(f)$$

up to a constant factor, and where, as in Chapter VII, § 6, eq. (32.3), the topology of $\mathcal{S}(\mathbb{R})$ is defined by the seminorms

$$N_r(f) = \sum_{p,q \leq r} \sup \left|t^p f^{(q)}(t)\right|\,.$$

This proves that $f \mapsto T(f)$ is continuous with respect to the topology of $\mathcal{S}(\mathbb{R})$.

The Fourier transform of the tempered distribution T is, by definition, the distribution $\hat{T}(f) = T(\hat{f})$; therefore, the significance of the formula obtained is that *the Fourier transform of the distribution* v.p. 1/t *is the distribution* $-\pi i.\text{sgn}(u)$, which is in fact a function since, in distribution theory, a function $\varphi(u)$ is always identified with the distribution $f \mapsto \int f(u)\varphi(u)du$. Conversely, the Fourier transform of the function $-\pi i.\text{sgn}(u)$, a Fourier transform that is not well-defined in classical theory as the function $\text{sgn}(u)$ is not integrable, is $v.p. 1/t$. This theory has given rise to generalizations in several dimensions that play an important role in some aspects of the theory of partial differential equations.

When f is bounded on $\mathbb{R}$ without being integrable, the function

$$F(z) = \frac{1}{2\pi i} \int \frac{f(t)dt}{t-z} = \lim F_n(z)$$

is not well-defined in general. $u_f(x,y)$ can indeed be defined: if f is real, it is, up to a factor of 2, the real part of the non-existent function $F(z)$; but its imaginary part is defined by a divergent integral (12). However, if the integral that should define F is integrated term by term with respect to z, we now get a function

$$(11.16) \quad G(z) = \frac{1}{2\pi i} \int \frac{f(t)dt}{(t-z)^2} = \frac{1}{2\pi i} \int \frac{d}{dz}\left(\frac{1}{t-z}\right) f(t)dt, \quad \mathrm{Im}(z) \neq 0$$

defined and holomorphic for $\mathrm{Im}(z) \neq 0$ and satisfying $G(z) = F'(z)$ when F exists. Failing to use (3') in order to define the function F that we are looking for, a primitive for G on $\mathrm{Im}(z) > 0$ could perhaps instead be used as a substitute. To simplify, f will be assumed to be real in the rest of this n°.

Let us first prove an important result which could have been another corollary of Theorem 3 in n° 3 related to the existence of primitives of holomorphic functions:

Theorem 10. *On a* simply connected *domain, a real harmonic function is the real part of a holomorphic function and is unique up to the addition of a pure imaginary constant.*

Uniqueness is obvious. Let u be a harmonic function; suppose there is a holomorphic function $f = u + iv$ such that $u = \mathrm{Re}\, f$. Then

$$f' = D_1 u + i D_1 v \quad \text{and} \quad D_1 v = -D_2 u,$$

and so $f' = D_1 u - i D_2 u$. Conversely, starting with a harmonic function u, Laplace's equation says precisely that $D_1 u - i D_2 u$ is holomorphic. Now, on a simply connected domain, this function has a primitive

$$(**) \qquad f(z) = \int_a^z [D_1 u(\zeta) - i D_2 u(\zeta)]\, d\zeta,$$

where integration is along an arbitrary path connecting a fixed point a to the point z in G. Setting $f = p + iq$,

$$f' = D_1 p - i D_2 p = D_1 u - i D_2 u,$$

so that the derivatives of u and p are identical. Hence $u = p + c$, where c is a real constant and the function $f(z) + c$ solves the problem, qed.

Coming back to the construction of a primitive for G, we see that a holomorphic function F_1 exists on the half-plane $H^+ : \mathrm{Im}(z) > 0$ such that

$$(11.17) \qquad u_f(x,y) = F_1(z) + \overline{F_1(z)}.$$

Differentiate (17) with respect to x. For holomorphic functions, differentiating with respect to x is the same as differentiating with respect to z. Thus

$$(11.18) \quad F_1'(z) + \overline{F_1'(z)} = \frac{d}{dx} u_f(x,y) = \frac{1}{2\pi i} \frac{d}{dx} \int \left(\frac{1}{t-z} - \frac{1}{t-\bar{z}} \right) f(t) dt$$

$$= \frac{1}{2\pi i} \int \left(\frac{1}{(t-z)^2} - \frac{1}{(t-z)^2} \right) f(t) dt = G(z) + \overline{G(z)}$$

provided – an easy application of Theorem 9 of n° 7 – differentiation under the $\int$ sign with respect to x in (7) is justified. As holomorphic functions F_1' and G have the same real parts,

$$G(z) = F_1'(z) + ia,$$

where a is a real constant; the function

$$F^+(z) = F_1(z) + iaz$$

is, therefore, a primitive for G on $\mathrm{Im}(z) > 0$. Since a is real,

$$F^+(z) + \overline{F^+(z)} = u_f(x,y) - 2ay.$$

Denoting by $F(z)$ the function equal to $F^+(z)$ for $\mathrm{Im}(z) > 0$ and to $-\overline{F^+(z)}$ for $\mathrm{Im}(z) < 0$, the following result is finally obtained:

Theorem 11. *For any continuous and bounded function f on $\mathbb{R}$, there is a function F defined and holomorphic on $\mathbb{C} - \mathbb{R}$ such that*

$$\lim_{y=0+} [F(x+iy) - F(x-iy)] = f(x) \quad \text{for all } x \in \mathbb{R}.$$

12 – The Complex Fourier Transform

(i) *Generalities*. Given a function on $\mathbb{R}$ that will be denoted $\hat{f}(t)$ for reasons specified later, the function

$$(12.1) \qquad f(z) = \int \mathbf{e}(tz) \hat{f}(t) dt = \int \mathbf{e}(tx) \exp(-2\pi ty) \hat{f}(t) dt$$

is called the (inverse...) *complex Fourier transform* of $\hat{f}$. This assumes that the integral converges absolutely for a non-empty set of values of z, which for example excludes the function $\exp(\pi t^2)$. It also excludes functions for which (1) only converges for real z, for example rational functions, since in this context, $f(z)$ is expected to be defined and holomorphic on an open subset of $\mathbb{C}$.

Exercise 1. Show that for $\hat{f}(t) = \exp(-\pi t^2)$ the integral converges for all z and is equal to $\exp(-\pi z^2)$.

Formula

$$\int_0^{+\infty} t^{s-1}\mathbf{e}(zt)dt = \Gamma(s)(-2\pi i z)^{-s}, \quad \mathrm{Re}(s) > 0, \ \mathrm{Im}(z) > 0,$$

which amounts to re-writing relations (10.2) to (10.4) differently, is another example; here the function $\hat{f}$ is t_+^{s-1}.

As $|\mathbf{e}(tz)| = \exp(-2\pi ty)$ and

$$\exp(y) \leq \exp(a') + \exp(b') \quad \text{if} \ a' \leq y \leq b'.$$

If (1) converges absolutely for $y = a'$ and $y = b' > a'$, it converges for $a' \leq y \leq b'$. The set of values of y such that

(12.2)
$$\int \left| \hat{f}(t) \right| \exp(-2\pi ty)dt < +\infty$$

is, therefore, an interval $I = (a, b)$ of a priori arbitrary nature; $f(z)$ is defined on the horizontal strip $B: y \in I$. For any *compact* interval $I' = [a', b'] \subset I$, the function being integrated is dominated by the integrable function[56] $|f(t)|(e^{-2\pi a't} + e^{-2\pi b't})$ on the closed strip $\mathrm{Im}(z) \in I'$ since $a', b' \in I$. In other words, integral (1) converges *normally*[57] on the closed strip $B' : y \in I'$. As a result (theorem 9 of n° 7),

(1) f is continuous and bounded on B', and so is continuous but not neces- sarily bounded on B since I is the union of the intervals I',
(2) f is holomorphic on the *open* strip $y \in]a', b'[$, hence on the interior $a < y < b$ of the strip B which is the union of these open strips,
(3) the derivatives of f can be computed by differentiating under the $\int$ sign.

To make sure that condition (2) is satisfied in a given *open* interval $I =]a, b[$, it *suffices* to suppose that

(12.3)
$$a < y < b \Longrightarrow \sup_{t \in \mathbb{R}} \left| \hat{f}(t) \right| \exp(-2\pi ty) < +\infty.$$

Indeed, if this condition holds, for given y, a' and b' can be chosen so that $a < a' < y < b' < b$. As (3) is satisfied for $y = a'$ and $y = b'$, for large $|t|$,

$$\hat{f}(t) = O\left[\exp(2\pi a't)\right], \quad \hat{f}(t) = O\left[\exp(2\pi b't)\right],$$

and so

---

[56] Recall that "integrable" means "absolutely integrable", except in very rare cases when it is specified to be otherwise.

[57] An integral $\int f(x, y)d\mu(x)$, defined for $y \in E$, converges normally on $A \subset E$ if there is a μ-integrable positive function $p_A(x)$ such that $|f(x, y)| \leq p_A(x)$ for all $y \in A$ and all x. This is clearly analogous to the normal convergence of a sequence of functions.

$$\left|\hat{f}(t)\right|\exp(-2\pi ty) \le M\exp\left[2\pi\left(a'-y\right)t\right],$$

$$\left|\hat{f}(t)\right|\exp(-2\pi ty) \le M\exp\left[2\pi\left(b'-y\right)t\right].$$

As $a'-y < 0$, the first relation shows that $\hat{f}(t)\exp(-2\pi ty)$ approaches 0 exponentially as t tends to $+\infty$; since $b'-y > 0$, the second shows that the same is true as t tends to $-\infty$; this is more than needed to ensure (2). This argument also shows that if $\hat{f}$ satisfies (3), the same holds for $t^n\hat{f}(t)$ for all $n \in \mathbb{N}$.

Setting $z = x+iy$, the function $x \mapsto f(x+iy)$ is the usual inverse Fourier transform of $\hat{f}(t)\mathbf{e}(ity)$. If

(12.4) $$\int |f(x+iy)|\,dx < +\infty \quad \text{for all } y \in I$$

and if $\hat{f}$ is continuous,[58] the Fourier inversion formula applies (Chap. VII, n° 30, theorem 26) and

$$\hat{f}(t)\mathbf{e}(ity) = \int f(x+iy)\mathbf{e}(-tx)dx,$$

in other words,

(12.5) $$\hat{f}(t) = \int_{\mathrm{Re}(z)=y} f(z)\mathbf{e}(-tz)dz \quad \text{for all } y \in I,$$

which is an integral à la Cauchy along the unbounded path $t \mapsto t+iy$.

Finding conditions ensuring (4) and hence (5) is easy. Indeed, if the derivatives of order $\le p$ of a function φ of class C^p are integrable over $\mathbb{R}$, then $\hat{\varphi}(v) = o(|v|^{-p})$ is known to hold at infinity (Chap. VII, n° 31, lemma 2); $\hat{\varphi}$ is, therefore, integrable, if $p \ge 2$. The method used then (integration by parts) applies here: as $2\pi iz\mathbf{e}(tz)$ is the derivative of $\mathbf{e}(tz)$,

$$2\pi izf(z) = \hat{f}(t)\mathbf{e}(tz)\Big|_{-\infty}^{+\infty} - \int \hat{f}'(t)\mathbf{e}(tz)dt = \int \hat{f}'(t)\mathbf{e}(tz)dt$$

if the function $\hat{f}'(t)\mathbf{e}(tz)$ is integrable. Iterating the computation, we conclude that if $\hat{f}^{(r)}$ exists and if $\hat{f}^{(r)}(t)\mathbf{e}(tz)$ is integrable for $y \in I$ and all $r \le p$ – in other words if the $\hat{f}^{(r)}$s satisfy the same assumption as $\hat{f}$ –, then the function $z^p f(z)$ is, like $f(z)$, bounded on every strip of finite width contained in B. When $p \ge 2$, this is enough to prove (4) since in this case, $f(x+iy) = 0(x^{-2})$ at infinity.

---

[58] When Lebesgue theory is available, this assumption is superfluous: if the Fourier transform of an integrable function is integrable, then the given function is almost everywhere equal to a continuous function for which the inversion formula holds everywhere.

Exercise 3. Supposing that $\hat{f}$ satisfies the conditions just proved, show directly that integral (5) is independent of $y \in B$. (Integrate along a horizontal rectangle in B).

We could also use Dirichlet's theorem (Chap. VII, n° 30, theorem 27) and show that

$$\lim \int_{-N}^{N} f(z)\mathbf{e}(-tz)dz = \frac{1}{2}\left[\hat{f}(t+0) + \hat{f}(t-0)\right]$$

for all $y \in I$ if $\hat{f}$ is right and left differentiable everywhere; the extended integral is over the interval $|x| \leq N$ of the horizontal $\operatorname{Im}(z) = y$.

(ii) *A Paley-Wiener theorem.* One of the problems of the theory is characterizing the complex Fourier transforms of functions $\hat{f}$ of a given category. The simplest result is related to the space $\mathcal{D}(\mathbb{R})$ of C^∞ functions with compact support in $\mathbb{R}$. In this case, f is defined for all $z \in \mathbb{C}$, and hence is an entire function, and if $\hat{f}$ vanishes outside a compact interval $[a, -a]$, then

$$|f(z)| \leq \int_{-a}^{a} \left|\hat{f}(t)\right| \exp(-2\pi ty)dt \, .$$

The exponential is bounded above for all t by $\exp(2\pi a|y|)$ on the integration interval, and so

$$f(z) = O\left(e^{2\pi a|y|}\right) \quad \text{on } \mathbb{C}\, .$$

We saw that if $\hat{f}$ is replaced by its derivative of order n, then the function $f(z)$ is replaced by $(-2\pi iz)^n f(z)$. Thus

(PW)
$$z^n f(z) = O\left(e^{2\pi a|y|}\right) \quad \text{on } \mathbb{C}$$

for all $n \in \mathbb{N}$. Conversely:

Theorem 12 (Paley-Wiener). *Let φ be an entire function. The following two conditions are equivalent :*

(i)   *There is a number $a > 0$ such that φ satisfies (PW) for all n;*
(ii) *φ is the complex Fourier transform of a C^∞ function vanishing outside $[-a, a]$.*

It suffices to prove that (i)$\Longrightarrow$(ii). The function $z^n f(z)$ being bounded on every horizontal and in particular on $\mathbb{R}$, the Fourier transform

$$\hat{f}(t) = \int f(x)\mathbf{e}(-tx)dx$$

of f on $\mathbb{R}$ is well-defined and is C^∞: this can be seen by differentiating under the $\int$ sign. To show that it is zero for $t \geq a$, integrate $f(z)\mathbf{e}(tz)$ along the path consisting of the interval $(-R, R)$ followed by the upper half-circle; the result is zero since the function integrated is holomorphic everywhere. By (PW), for all n, there is an upper bound

$$|f(z)\mathbf{e}(tz)| = |f(z)| \exp(-2\pi ty) \leq c_n R^{-n} \exp\left[2\pi(a-t)y\right)]$$

over the half-circle for $y > 0$. For $t \geq a$, the exponential is ≤ 1; the integral along the half-circle is then $O(R^{1-n})$ for all n and tends to 0. For $t \leq -a$, replace the upper half-circle by the lower one. Then $\hat{f}(t) = 0$ for $|t| \geq a$.

It remains to check that $f(z) = \int \hat{f}(t)\mathbf{e}(-tz)dt$. As the Fourier transform of $f(x)$ is in $\mathcal{D}(\mathbb{R})$, the inversion formula shows that f is the inverse Fourier transform of $\hat{f}$ on $\mathbb{R}$. The complex Fourier transform of $\hat{f}$ being holomorphic everywhere and equal to F on $\mathbb{R}$, the analytic extension principle (Chap. II, n° 20) shows they are both identical on $\mathbb{C}$, qed.

Exercise 4. Let $\mathcal{S}$ be the Schwartz space, i.e. the set of C^∞ functions $\varphi(t)$, $t \in \mathbb{R}$, such that all functions $t^p \varphi^{(q)}(t)$ are bounded on $\mathbb{R}$ (Chap. VII, §6, n° 31). Let $\mathcal{S}_+$ be the set of the $\varphi \in \mathcal{S}$ that are zero for $t \leq 0$, so that $\varphi^{(n)}(0) = 0$ for all n. (i) Show that the complex Fourier transform f of any $\varphi \in \mathcal{S}_+$ is defined and holomorphic for $y > 0$ and that, for all $p, p \in \mathbb{N}$, the function $z^p f^{(q)}(z)$ is bounded on the half-plane $\mathrm{Im}(z) \geq 0$. (ii) Conversely, let f be a holomorphic function on $y > 0$ satisfying this condition. Show that, for any $y > 0$, the function $x \mapsto f(x + iy)$ is in the space $\mathcal{S}$ and the the integral of $f(z)\mathbf{e}(-tz)$ along the horizontal $\mathrm{Im}(z) = y$ does not depend on y. Denoting this integral $\varphi(t)$, show that $\varphi \in \mathcal{S}_+$ and that f is its complex Fourier transform.

(iii) *Holomorphic functions integrable over a strip.* Let

$$I =]a, b[\subset \mathbb{R}$$

be an *open* interval, $\rho(y)$ a continuous function with values > 0 on I, and B the open horizontal strip $\mathrm{Im}(z) \in I$. Let f be a holomorphic function on B; suppose that

$$(12.6) \qquad \iint_{B'} |f(z)| \rho(y)dm(z) < +\infty$$

for any *closed* horizontal strip *of finite width* $B' \subset B$.[59] We show that, under these conditions, f is a complex Fourier transform. The proof uses the relation shown in n° 4, (iv) between compact convergence and mean convergence, i.e in the sense of the L^1 norm. In what follows, we will use the notation

$$\rho(y)dm(z) = d\mu(z).$$

---

[59] As inequality $0 < m \leq \rho(y) \leq M < +\infty$ holds in every compact set contained in I, condition (6) does not really involve ρ. We will see in the next section that this is no longer the case if B' is replaced in (6) by the strip B. This is why I introduce a seemingly superfluous function $\rho(y)$ here.

Lemma. *Condition (6) holds for any closed horizontal strip $B' \subset B$ of finite width if and only if the series $\sum f(z+n)$ converges normally on every compact subset of B.*

Suppose that (6) holds and that $K \subset B$ is a compact set; normal convergence on a compact set being a local property (BL), K may be assumed to be contained in the interior G of a compact rectangle $K' \subset B$ defined by

$$K' : m \le x \le m+1, \quad a' \le y \le b'$$

where $[a', b'] \subset I$ is compact and where $m \in \mathbb{R}$. (4.16) can then be applied to K and G, giving an upper bound

(12.7) $$|f(z)| \le M \iint_{K'} |f(w)|\, d\mu(w) \quad \text{for all } z \in K,$$

with a constant M independent of f. Replacing $z \mapsto f(z)$ by $z \mapsto f(z+n)$, we get

(12.8) $$|f(z+n)| \le M \iint_{K'} |f(w+n)|\, d\mu(w)$$

$$= M \iint_{K'+n} |f(w)|\, d\mu(w),$$

where $K'+n$ is the image of K' under the translation $w \mapsto w+n$, a translation that leaves the measure μ invariant. For any $z \in K$, the series $\sum f(z+n)$ is, therefore, up to a factor M, dominated by the numerical series

(12.9) $$\sum_{\mathbb{Z}} \iint_{K'+n} |f(w)|\, d\mu(w) = \iint_{B'} |f(w)|\, d\mu(w) \le +\infty.$$

Hence the series $\sum f(z+n)$ is normally convergent on K if integral (6) is finite. Conversely, this condition implies (6) since if there is a convergent numerical series u_n such that $|f(z+n)| \le u_n$ for all $z \in K'$ and all n, then

$$\iint_{B'} |f(z)|\, d\mu(z) = \sum \iint_{K'+n}$$

$$= \sum \iint_{K'} |f(z+n)|\, d\mu(z) \le \mu(K') \sum u_n,$$

qed.

Exercise 5. We replace $|f(z)|$ by $|f(z)|^p$ in condition (6) for a given real number $p > 1$. Using (4.16) for the exponent p, show that the series

$$\sum \left| f^{(r)}(z+n) \right|^p$$

converge normally on all compact subsets and that

$$\iint \left| f^{(r)}(x+iy) \right|^p dx < +\infty$$

for all $y \in]a, b[$ and all $r \in \mathbb{N}$. Converse?

Returning to the case $p = 1$, this result sends us back to Chap. VII, n° 29 where it was shown how to deduce Fourier's inversion formula from Poisson's summation formula. Consider the function $x \mapsto f(x+iy)$ for given $y \in I$. As the series $\sum f(z+n)$ converges normally on every compact subset,[60]

$$(12.10) \qquad \int |f(x+iy)| \, dx < +\infty$$

for all $y \in]a, b[$, which allows the Fourier transform

$$(12.11) \qquad \int f(x+iy)\mathbf{e}(-tx)dx = F(t,y)$$

to be defined; it is a continuous function of t.

Let us show that there is a function $\hat{f}(t)$ on $\mathbb{R}$ such that

$$(12.12) \qquad F(t,y) = \hat{f}(t)\mathbf{e}(ity).$$

Formula (11), multiplied by $\mathbf{e}(-ity) = \exp(2\pi ty)$, can also be written

$$F(t,y)\mathbf{e}(-ity) = \int f(z)\mathbf{e}(-tz)dz.$$

This is the Cauchy integral along the horizontal $\text{Im}(z) = y$, and it all amounts to showing that it is independent of y. To compare its values at y' and y'', let us integrate over the rectangle $ABCD$ bounded by the horizontals AB : $\text{Im}(z) = y'$ and $DC : \text{Im}(z) = y''$ and the verticals $DA : \text{Re}(z) = -n$ and BC : $\text{Re}(z) = n$; by Cauchy, the result is zero. $|f(z)\mathbf{e}(-tz)| = |f(n+iy)| \exp(2\pi ty)$ on BC; as the series $\sum f(z+n)$ converges normally on the compact interval $[y', y'']$ of the imaginary axis, the function $y \mapsto f(n+iy)$ converges uniformly to 0 on $[y', y'']$ as n increases. At the limit, the contributions from the vertical sides are, therefore, zero. This gives the expected result. Hence, there is a relation

$$(12.13) \qquad \int f(x+iy)\mathbf{e}(-tx)dx = \hat{f}(t)\mathbf{e}(ity)$$

---

[60] Applied to the double integral (6), the version of the Lebesgue-Fubini theorem for lsc positive functions (Chap. V, n° 33, theorem 31), only states that the function $y \mapsto \int |f(x+iy)|dx$, with values $\leq +\infty$ (large inequality), is integrable as a lsc function, and so is finite "almost everywhere"; but it could very well be infinite for some values of y. The fact that it is finite everywhere, with no zero measure exception, is one of the many miracles of the theory of holomorphic functions: all the seemingly pathological phenomena of the theory of integration disappear. This meta-theorem useful for making conjectures on what is going on, does not exempt from giving proper proofs.

with a function

$$(12.14) \qquad \hat{f}(t) = \int f(z)\mathbf{e}(-tz)dx$$

independent of y and that can serve as the Fourier transform of f.

Let us show that the Fourier inversion formula can be applied to (11). Since the series $\sum f^{(r)}(z+n)$ converges, $f^{(r)}(x+iy)$ is integrable at x and approaches 0 as $|x|$ increases indefinitely; (11) can, therefore, be computed by integrating by parts. f being holomorphic, $D_1 f = f'$ and the usual computation shows that

$$(12.15) \qquad \int f^{(r)}(x+iy)\mathbf{e}(-tx)dx = (-2\pi i t)^r F(t, y).$$

It follows that for all $y \in I$ and all r,

$$(12.16) \qquad F(t, y) = O\left(|t|^{-r}\right) \quad \text{as } |t| \longrightarrow +\infty,$$

a condition more than sufficient to justify the use of the inversion formula. In view of (12), it can be written

$$f(z) = \int \hat{f}(t)\mathbf{e}(tz)dt$$

and proves that f is the complex Fourier transform of $\hat{f}$.

Besides, (16) shows that $\sum |F(n, y)| < +\infty$ for all y, which allows Poisson's summation formula to be applied (Chap. V, n° 27, theorem 24) to $x \mapsto f(x+iy)$; in view of (12), it can be written

$$\sum f(z+n) = \sum \hat{f}(n)\mathbf{e}(nz).$$

Ultimately, the following formulas are obtained:

$$(12.17) \qquad \hat{f}(t) = \int_{\mathrm{Re}(z)=y} f(z)\mathbf{e}(-tz)dz,$$

$$(12.18) \qquad f(z) = \int \hat{f}(t)\mathbf{e}(tz)dt,$$

$$(12.19) \qquad \sum f(z+n) = \sum \hat{f}(n)\mathbf{e}(nz).$$

They obviously assume that $a < y < b$, a condition that, as seen earlier, makes the integrals absolutely convergent.

(iv) *Holomorphic functions integrable over a half-plane.* Let P be the half-plane $\mathrm{Im}(z) > 0$. In what precedes, take $I =]0, +\infty[$ and $B = P$ and instead of (6), impose the stronger condition

$$(12.20) \qquad \iint_P |f(z)|\, \rho(y)dxdy < +\infty$$

on the functions $f(z)$. The results of section (iii) apply. On the other hand, (13) and (16) show that, for any $y \in I$ and any $r \in \mathbb{N}$,

(12.21) $\hat{f}(t) = O\left[|t|^{-r} \exp(2\pi t y)\right]$ as $|t| \longrightarrow +\infty$;

this relation seemingly better than (3) suffices to ensure convergence of (18) for $y > 0$. Relation (15) also shows that

$$\left|\hat{f}(t)\right| \exp(-2\pi t y) \leq \int |f(x + iy)| \, dx$$

and so

$$\int_0^{+\infty} \left|\hat{f}(t)\right| \exp(-2\pi t y) \rho(y) \, dy \leq \iint_P |f(x + iy)| \, \rho(y) \, dx \, dy < +\infty$$

for all t. The function $\hat{f}$ is, therefore, zero for the values of t for which the integral $\int \exp(-2\pi t y) \rho(y) \, dy$ is divergent. For example, if

$$\rho(y) = y^{k-2}$$

with k real but not necessarily an integer as will now be supposed, which is an important case in the theory of automorphic functions, then

(12.22) $\hat{f}(t) \neq 0 \Longrightarrow \int_0^{+\infty} \exp(-2\pi t y) y^{k-2} \, dy < +\infty$.

The convergence of the integral requires $k > 1$ and $t > 0$, which shows in particular that, for $k \leq 1$, the only holomorphic solution of (20) is $f(z) = 0$.

There still remains to be shown that non-zero solutions of (20) effectively exist for $k > 1$. To obtain a holomorphic function on P, let us try

(12.23) $g(z) = (z - \bar{w})^{-p}$

with $\text{Im}(w) > 0$. Setting $w = u + iv$, the change of variable $x = u + \xi(y + v)$ shows that

$$\int |g(x + iy)| \, dx = \int \left[(x - u)^2 + (y + v)^2\right]^{-p/2} dx =$$

$$= \int (y + v)^{1-p} \left(\xi^2 + 1\right)^{-p/2} d\xi \, .$$

This result is finite if $p > 1$. It remains to check that the integral

$$\int_0^{+\infty} (y + v)^{1-p} y^{k-2} dy \, , \quad v > 0 \, ,$$

converges. This suppose that at $y = 0$, $k > 1$ and that at infinity, $p > k$, which implies the condition $p > 1$ that has already been found. Hence (20) has non-zero solutions for $k > 1$, but none for $k \leq 1$. Summarizing:

Theorem 13. *Let k be a real number. The space $\mathcal{H}_k^1(P)$ of holomorphic functions such that*

$$(12.24) \qquad \iint_{\text{Im}(z)>0} |f(z)| y^{k-2} dx dy < +\infty$$

does not reduce to $\{0\}$ if and only if $k > 1$. Any holomorphic solution of (24) is the complex Fourier transform of a continuous function $\hat{f}(t)$ which is zero for $t < 0$. Formulas (17), (18) and (19) hold.

Exercise 6. Calculate the function $\hat{f}(t)$ corresponding to (23). p need not be assumed to be an integer nor a real since function (24) has uniform branches on $y > 0$.

To conclude this section, note that, while we have obtained important properties of the functions $\hat{f}$, we have not *characterized* them as we did in the Paley-Wiener theorem; as far as I know – given the flood of publications since the last fifty years, it is necessary to be prudent –, the answer to this question will never be known. It is, however, fully known if the functions $f(z)$ are taken to be *square* integrable on the half-plane: these are complex Fourier transforms of the functions $\hat{f}(t)$ which are zero for $t \leq 0$ and for which

$$\int_0^{+\infty} \left|\hat{f}(t)\right|^2 t^{1-k} dt < +\infty;$$

but these are now functions on the Lebesgue L^2 space and the result cannot be obtained without the help of the whole theory of Fourier transforms on L^2. This topic will be presented in Chap. XI.

13 – The Mellin Transform

(i) *Questions of convergence.* To obtain Paley-Wiener type theorems that hold for *meromorphic* functions rather than holomorphic ones as in the previous n°– an important problem in analytic number theory for example –, it is useful to reformulate the complex Fourier transform. If the change of variable $\exp(2\pi t) = u$ is carried out in integral (12.1) which it is defined by, then $u > 0$ and

$$2\pi dt = du/u = d^*u.$$

Setting $iz = s$, we get $\mathbf{e}(tz) = u^s$ and $\hat{f}(t) = F(e^{2\pi t})$, and so

$$2\pi f(z) = \int_0^{+\infty} F(u) u^s d^*u.$$

We often come across integrals of this type; they fall within the general framework of the Mellin transform, which associates the function

$$(13.1) \qquad \Gamma_f(s) = \int_0^{+\infty} f(x) x^s d^* x \,,$$

where $x^s = \exp(s \log x)$ is the real, positive function of Chap. IV for real s, to "any" function[61] $f(x)$ defined for $x > 0$. The notation recalls the fact that

$$(13.2) \qquad \Gamma_f(s) = \Gamma(s) \quad \text{if} \ \ f(x) = e^{-x} \,.$$

As $\Gamma_f(is)$ is the complex Fourier transform of $t \mapsto 2\pi f(e^{2\pi t})$, the general statements of n° 12, (i) are easily translated.

In conformity to a long tradition, setting

$$s = \sigma + it \,,$$

absolute convergence of (1) only depends on $\mathrm{Re}(s) = \sigma$. Clearly, Γ_f is defined on a strip of the plane of the form $\mathrm{Re}(s) \in I$, where I is a priori an arbitrary interval; it is holomorphic on the interior of this strip and bounded on any *closed* vertical strip *of finite width* where it is defined. Theorem 9 of n° 7 shows that its derivatives are given by

$$(13.3) \qquad \Gamma_f^{(n)}(s) = \int_0^{+\infty} f(x) \log^n x . x^s d^* x$$

for all $n \in \mathbb{N}$. The logarithmic factor does not destroy convergence: since we are in the interior of the convergence strip, the integral $\int f(x) x^s d^* x$ indeed converges at $s = a$ and $s = b$, where $a < \mathrm{Re}(s) < b$, and it is sufficient to observe that

$$(\log x)^n x^s = o(x^a) \quad \text{as } x \text{ tends to } 0,$$

$$(\log x)^n x^s = o(x^b) \quad \text{as } x \text{ tends to } +\infty$$

to justify the result, which is anyhow guaranteed by theorem 9 of n° 7.

Convergence of the Mellin integral in the neighbourhood of 0 is ensured for $\mathrm{Re}(s) > 0$ if f is bounded in the neighbourhood of 0, though it converges at infinity if f is integrable at infinity with respect to dx and if the function x^{s-1} is bounded at infinity, i.e. for $\mathrm{Re}(s) < 1$. The strip where the function Γ_f is defined is, therefore, at least $0 < \mathrm{Re}(s) < 1$ if f is *bounded and integrable* with respect to x over $\mathbb{R}_+$.

As $\Gamma_f(is)$ is the value of the complex Fourier transform $g(z)$ of $2\pi f(e^{2\pi u}) = \hat{g}(u)$ at $z = is$, and as there is sometimes an inversion formula

$$\hat{g}(u) = \int_{\mathrm{Re}(z)=y} g(z) \mathbf{e}(-uz) dz \,,$$

in horizontal strip where g is defined, a similar result on Mellin transforms will presumably follow by making suitable assumptions.; variable changes

---

[61] In practice, $f(x)$ is regulated and almost always continuous for $x > 0$.

$z = is$ (and so $dz = ids$) and $e^{2\pi u} = x$ as above are sufficient to obtain it. The corresponding formula can be written

$$(13.4) \qquad 2\pi i f(x) = \int_{\mathrm{Re}(s)=\sigma} \Gamma_f(s) x^{-s} ds,$$

where integration is over a vertical $t \mapsto \sigma + it$ *located in the convergence strip of the integral defining* Γ_f. This is the Mellin inversion formula.. Like the Fourier inversion formula it is equivalent to, it holds under the following assumptions, which are merely translations of theorem 26 of Chap. VII, n° 30:

(a)   the function f is continuous for $x > 0$,
(b) $\int |f(x)x^s| d^*x < +\infty$ for $\mathrm{Re}(s) = \sigma$,
(c)   Integral (4) is absolutely convergent.

For the Mellin version of Paley-Wiener, a given function $\varphi(s)$ has to be shown to be a Mellin transform. The problem is the same: suppose φ to be holomorphic on a strip $a < \mathrm{Re}(s) < b$ and define a function f using (4), where Γ_f is replaced by φ and where $a < \sigma < b$. If for a given σ, f and φ satisfy conditions (a), (b) and (c), then conversely $\varphi(s) = \int f(x)x^s d^*x$ for $\mathrm{Re}(s) = \sigma$. Again, this is only a translation of Fourier's inversion formula. In practice, the given function φ decreases at a sufficiently rapidly at infinity for integral (4) to be independent of $\sigma \in]a, b[$.

The most frequent tendency is to move the vertical over which integration is performed to regions where, like Euler's function, Γ_f is only defined by analytic extension; condition (b) is no longer satisfied and formula (4) becomes false in general. To rectify it, we take into account of the residues of Γ_f at poles encountered while moving the integration vertical, initially located in the region where condition (b) holds, to a vertical where it no longer is. Section (iv) of this n°, and more so chapter XII, will explain this fundamental point.

(ii) *Analytic extension of a Mellin transform.* In practice, often the aim is to show that the function Γ_f can, like Euler's function, be analytically extended beyond the integral's domain of convergence. This question is determined by the behaviour of f at x in the neighbourhood of 0 or very large because functions the transformation is applied to, in practice are, as will be assumed, at least continuous for $x > 0$. Powerful results can be obtained from simple assumptions on the asymptotic behaviour of f in the neighbourhood of 0 and infinity.

First consider the integral

$$(13.5') \qquad \Gamma_f^-(s) = \int_0^1 f(x)x^s d^*x$$

(the choice of the limit 1 is convenient but not essential) and suppose that the function f has a bounded expansion

(13.6') $$f(x) = a_1 x^{u_1} + \ldots + a_n x^{u_n} + O\left(x^{u_{n+1}}\right)$$

in the neighbourhood of 0, with real exponents $u_1 < \ldots < u_n < u_{n+1}$. As

$$\int_0^1 x^s d^* x = 1/s \quad \text{if} \quad \mathrm{Re}(s) > 0,$$

integral (5') converges for $\mathrm{Re}(s) > -u_1$ and, is equal to

(13.7) $$\sum_{1 \leq k \leq n} \frac{a_k}{s + u_k} + \int_0^1 O\left(x^{u_{n+1}}\right) x^s d^* x$$

on this strip. The $\sum$ is a rational function whose poles and residues are prominently displayed, though the integral in the last term converges and is holomorphic for $\mathrm{Re}(s) > -u_{n+1}$. (5') can, therefore, be defined by analytic extension on this half-plane. If f admits an unbounded asymptotic expansion[62] of the form

(13.8')

$$f(x) \approx \sum a_n x^{u_n}, \quad x \longrightarrow 0 \quad \text{with} \quad u_n < u_{n+1} \quad \text{and} \quad \lim u_n = +\infty,$$

function (5') can be analytically extended to all of $\mathbb{C}$, with simple poles and residues equal to a_n at all points $-u_n$. If for example $f(x)$ is C^∞ on $\mathbb{R}_+$, with the origin included, then by MacLaurin's formula [Chap. V, n° 18, eq. (18.11)], there is an unbounded asymptotic expansion

$$f(x) \approx \sum f^{(n)}(0) x^n / n!.$$

Function (5'), a priori defined for $\mathrm{Re}(s) > 0$, can, therefore, be extended to $\mathbb{C}$ with the points $0, -1, -2, \ldots$ removed, points where it has simple poles and residues equal to the corresponding numbers $f^{(n)}(0)/n!$. This has already been seen for $f(x) = e^{-x}$ in relation to Euler's Γ function; in this case, the MacLaurin series can even be integrated term by term over $]0, 1]$, which gives an expansion

$$\Gamma_f^-(s) = \sum f^{(n)}(0)/n!(s + n)$$

with a convergent series for all s. The same holds for any *analytic* function f in the neighbourhood of 0 provided the radius of convergence R of its MacLaurin series is > 1. If $R \leq 1$, decompose $\mathbb{R}_+^*$ at $(0, a)$ and $(a, +\infty)$ with

---

[62] As a general rule, formula $f(x) \approx \sum u_n(x)$ means that (i) $u_{n+1}(x) = o\left[u_n(x)\right]$ in the neighbourhood of the point considered, (ii) $f(x) = u_1(x) + \ldots + u_n(x) + O\left[u_{n+1}(x)\right]$ for any n. This does not in any way imply that the series $\sum u_n(x)$ converges to $f(x)$; it can in fact diverge and that is often the case in practice, in particular for the Taylor series of a non-analytic C^∞ function. See Chap. VI, n° 10.

$a < R$ in order to integrate the power series term by term over $(0, a)$; the contribution from $]0, a]$ is then equal to $\sum f^{(n)}(0)a^{n+s}/n!(s+n)$, a series that converges even better than the power series of f at $x = a$ since $|s + n|$ increases indefinitely. The residue of the function at $s = -n$ can be immediately calculated since $a^{n+s} = 1$ at this point; Hence we again get $f^{(n)}(0)/n!$. This result is, as it should be, independent of the chosen point a.

Through a change of of variable $x \mapsto 1/x$, which leaves the measure d^*x invariant, the integral

$$(13.5") \qquad \qquad \Gamma_f^+(s) = \int_1^{+\infty} f(x)x^s d^*x$$

reduces to the preceding case. However,

$$\int_1^{+\infty} x^s d^*x = -1/s \quad \text{if} \ \ \mathrm{Re}(s) < 0.$$

Hence if there is a bounded expansion

$$(13.6") \qquad f(x) = b_1 x^{-v_1} + \ldots + b_n x^{-v_n} + O\left(x^{-v_{n+1}}\right)$$

at infinity, with $v_1 < \ldots < v_n < v_{n+1}$, integral (5"), a priori defined and holomorphic on $\mathrm{Re}(s) < v_1$, can be analytically extended to the half-plane $\mathrm{Re}(s) < \mathrm{Re}(v_{n+1})$, with simple poles at the points v_k and residues equal to the coefficients b_k. In particular, if there is an unbounded asymptotic expansion

$$(13.8") \qquad f(x) \approx \sum b_n x^{-u_n}, \quad x \longrightarrow +\infty, \quad \text{with} \ \lim v_n = +\infty,$$

it can be extended to all of $\mathbb{C}$, excepting at some simple poles.

For example, choose the function $f(x) = x/(1+x^2) = f(x^{-1}); f(x) \sim x$ in the neighbourhood of 0. This gives the convergence condition $\mathrm{Re}(s) + 1 > 0$, and $f(x) \sim 1/x$ at infinity, which in turn gives the convergence condition $\mathrm{Re}(s) - 1 < 0$; the Mellin integral, therefore, converges on the strip $|\mathrm{Re}(s)| < 1$. For $x \leq a < 1$, $f(x) = \sum(-1)^n x^{2n+1}$, which is much better than an asymptotic expansion. $\Gamma_f^-(s)$, a priori defined for $\mathrm{Re}(s) > -1$, can be analytically extended to all of $\mathbb{C}$, with simple poles at the points $-2n - 1$ and residues equal to $(-1)^n$. Similarly,

$$f(x) = \sum(-1)^n x^{-2n-1}$$

at infinity. So $\Gamma_f^+(s)$, a priori defined for $\mathrm{Re}(s) < 1$, can be extended to $\mathbb{C}$, with simple poles at the points $2n+1$. However, $\Gamma_f(s) = \Gamma_f^-(s) + \Gamma_f^+(s)$ on the strip $|\mathrm{Re}(s)| < 1$ where these three functions are defined and holomorphic. Since the right hand term is meromorphic on $\mathbb{C}$, it follows that Γ_f can be analytically extended to all of $\mathbb{C}$, its singularities being simple poles at the points $2n + 1$, $n \in \mathbb{Z}$, with residues equal to $(-1)^n$. In fact, we will show in n° 15 that

$$\Gamma_f(s) = \pi/2 \cos(\pi s/2),$$

which is much more precise, but we rarely have the chance of being able to calculate everything explicitly.

If f decreases rapidly at infinity, i.e. if $f(x) = O(x^{-N})$ for all N, there is no need to think: integral (5") converges on all of $\mathbb{C}$ and is an entire function of s.

As in the previous example, these two types of results apply if f admits unbounded asymptotic expansions in the neighbourhood of 0 or for large x, obviously provided that the integral defining Γ_f is to start with convergent on a strip of non-zero width. Otherwise "gluing" the meromorphic functions defined by (5') and (5") would be impossible: the function 1 has the nicest asymptotic behaviour in the world, but its Mellin transform is not defined; in this case, $\Gamma_f^-(s) = 1/s$ and $\Gamma_f^+(s) = -1/s$, and so $\Gamma_f(s) = 0$ if Γ_f were well-defined: absurd assumptions lead to absurd conclusions.

The fact that the Mellin transform does not have simple poles is due to the nature of the asymptotic expansions that we have admitted. In more complicated cases, there may be multiple poles. Suppose for example that, in the neighbourhood of 0, f is the sum of a function of type (6') and of a finite number of terms $x^p \log^q x$, with $p > 0$. The contribution made by these terms to (5') converges for $\mathrm{Re}(s) > -p$ since $\log x = O(x^{-r})$ for all $r > 0$; it is equal to $c_q(s+p)$, where

$$(13.9) \qquad c_q(s) = \int_0^1 \log^q x . x^{s-1} dx, \quad \mathrm{Re}(s) > 0.$$

Integrating by parts, we get

$$sc_q(s) = [1 - qc_{q-1}(s)].$$

As $c_0(s) = 1/s$, $c_1(s) = 1/s - 1/s^2$, it follows that $c_2(s) = 1/s - 2/s^2 + 2s^3$ and more generally

$$(13.10) \quad c_q(s) = 1/s - q/s^2 + q(q-1)/s^3 - \ldots + (-1)^q q!/s^{q+1}.$$

Replacing s by $s + p$, we see that the presence of a term in $x^p \log^q x$ in the asymptotic expansion introduces a pole of order $q + 1$ at the point $-p$.

(iii) *Example: the Riemann zeta function.* The function $\exp(-\pi u^2)$ being equal to its Fourier transform, that of $u \mapsto \exp(-\pi x u^2)$, where $x > 0$, is $v \mapsto x^{-1/2} \exp(-\pi v^2/x)$; at infinity, these functions approach 0 sufficiently rapidly for Poisson's formula to be written

$$\sum \exp(-\pi n^2 x) = x^{-1/2} \sum \exp(-\pi n^2/x),$$

where summation is over $\mathbb{Z}$ (Chap. VII, n° 28). As will be seen below, the results of section (ii) apply to the Jacobi function

$$\theta(x) = \sum \exp(-\pi n^2 x) \quad (x > 0).$$

First, it is C^∞; differentiating the series term by term p times, up to a constant factor, we indeed get $\sum n^{2p} \exp(-\pi n^2 x)$. Since, for $t > 0$, $t^N e^{-t}$ is bounded by a constant M_N for all $N > 0$, for any $p \neq 0$ there is an upper bound of the form $n^{2p} \exp(-\pi n^2 x) \leq M_N n^{2p}/(n^2 x)^N$. Choosing $N \geq 2p+2$, it follows that the derived series are all normally convergent in $x \geq c$ for any $c > 0$, where this is a strict inequality. The result follows. This calculation also shows that

$$\theta(x) = 1 + O\left(x^{-N}\right) \qquad \text{at infinity}$$

for all N, the term 1 coming from the term $n = 0$ of the series, while the derivatives satisfy

$$\theta^{(p)}(x) = O\left(x^{-N}\right) \qquad \text{at infinity}.$$

The functional equation then shows that

$$\theta(x) = x^{-1/2}\left[1 + O\left(x^N\right)\right] \quad \text{for } x \longrightarrow 0.$$

The results of section (ii) can, therefore, be applied to

$$f(x) = \theta\left(x^2\right) - 1,$$

a function for which

$$f(x) = O\left(x^{-N}\right) \text{ at infinity}, \quad f(x) = 1/x - 1 + O\left(x^N\right) \text{ at } 0.$$

Convergence at infinity of the integral defining $\Gamma_f(s)$ does not need to satisfy any conditions, but convergence at 0 supposes that $\mathrm{Re}(s) > 1$. A formal calculation shows that $\Gamma_f(s)$ is equal to

$$\int_0^{+\infty} dx \sum_{n \neq 0} \exp\left(-\pi n^2 x^2\right) x^{s-1} = \sum \int_0^{+\infty} \exp\left(-\pi n^2 x^2\right) x^s d^* x$$

$$= \sum_{n \neq 0} \frac{1}{2}\left(\pi n^2\right)^{-s/2} \int_0^{+\infty} \exp\left(-y\right) y^{s/2} d^* y = \pi^{-s/2}\Gamma(s/2)\zeta(s),$$

where $\zeta(s) = \sum_{n>0} 1/n^s$ is the Riemann series converging for $\mathrm{Re}(s) > 1$. To justify the permutation of the signs $\int$ and $\sum$, it suffices (Chap. V, n° 23, theorem 21) to show that (1) the series being integrated converges normally on every compact set $K \subset]0, +\infty[$, which is obvious since this is the case of the theta series and since x^{s-1} is bounded on K, (2) the series

$$\sum_{n \neq 0} \int \left|\exp\left(-\pi n^2 x^2\right) x^s\right| d^* x$$

converges, which is obvious by our formal calculation since it is proportional to that of Riemann.

Hence the function

$$\Gamma_f(s) = \pi^{-s/2}\Gamma(s/2)\zeta(s) = \xi(s)$$

can be analytically extended to all of $\mathbb{C}$, with simple poles at $s = 0$ and $s = 1$ coming from the first two terms of the asymptotic expansion of f at 0; the residues are equal to 1 at $s = 1$ and to -1 at $s = 0$. As the function $1/\Gamma(s/2)$ is entire and has simples zeros at $0, -2, -4, \ldots$, the function $\zeta(s) = \pi^{s/2}\Gamma_f(s)/\Gamma(s/2)$ is meromorphic on $\mathbb{C}$. As $\Gamma(1/2) = \pi^{1/2} \neq 0$, the simple pole of Γ_f at $s = 1$ spreads to $\zeta(s)$, with a residue equal to $\pi^{-1/2}\pi^{1/2} = 1$. As $1/\Gamma(s/2)$ vanishes at $s = 0$, the pole of Γ_f at this point is neutralized by the zero of the function $1/\Gamma$. Therefore, the function $\zeta(s)$ has a unique singularity in $\mathbb{C}$: a simple pole at $s = 1$. It vanishes at $s = -2, -4, \ldots$, as well as at several other points less obvious at first.

It also satisfies a functional equation which can be deduced from that of the Jacobi function. Indeed,

$$f(1/x) = \theta\left(1/x^2\right) - 1 = x\theta\left(x^2\right) - 1 = x\left[f(x) + 1\right] - 1,$$

i.e.

$$f(1/x) = xf(x) + x - 1.$$

Hence, for $\mathrm{Re}(s) > 1$,

$$\Gamma_f^-(s) = \int_0^1 f(x)x^s d^*x = \int_1^{+\infty} f(1/x)x^{-s}d^*x$$

$$= \int_1^{+\infty} \left[f(x)x^{1-s} + x^{1-s} - x^{-s}\right] d^*x.$$

Each of these three function occurring in the last integral is integrable over $[1, +\infty[$ with respect to d^*x, where $\mathrm{Re}(s) > 0$: this is obvious for the last two, and the the first one is integrable for all s since it decreases rapidly at infinity. In conclusion,

$$\Gamma_f^-(s) = \Gamma_f^+(1-s) + 1/(s-1) - 1/s$$

for $\mathrm{Re}(s) > 1$, and so, by analytic extension, for all $s \in \mathbb{C}$. As $\Gamma_f = \Gamma_f^+ + \Gamma_f^-$,

$$\Gamma_f(s) = \Gamma_f^+(s) + \Gamma_f^+(1-s) + 1/(s-1) - 1/s,$$

which proves that

$$\xi(s) = \xi(1-s).$$

As pages and pages have been written and continue to be written on the Riemann function, we will not pursue these investigations here; for the sequel see Chap. XII.

(iv) *A Paley-Wiener type theorem.* There are Paley-Wiener type results for the Mellin transform. For example, the next fairly long application of Cauchy theory; the method and even the result are used to study the relations between Dirichlet series and modular functions.

Theorem 14. *Let $S_+ = S(\mathbb{R}_+)$ be the set of functions defined and infinitely differentiable for $x \geq 0$ and that are together with their derivatives rapidly decreasing at infinity . For any $f \in S(\mathbb{R}_+)$, the Mellin transform*

$$\Gamma_f(s) = \int_0^{+\infty} f(s)x^s d^* x = \varphi(s)$$

has the following properties:

(i) Γ_f *is defined and holomorphic for $\mathrm{Re}(s) > 0$ and can be extended analytically to a meromorphic function on all of the plane whose only singularities are at most simple poles at $s = 0, -1, -2, \ldots;$*

(ii) *for any $n \in \mathbb{N}$, the function $s^n \Gamma_f(s)$ is bounded at infinity on every vertical strip of finite width.*[63]

Moreover,

$$(13.11) \qquad 2\pi i f(x) = \int_{\mathrm{Re}(s)=\sigma} \Gamma_f(s)x^{-s} ds \quad \text{if } \sigma > 0$$

and for all $p \in \mathbb{N}$,

$$(13.12) \quad 2\pi i f(x) = \int_{\mathrm{Re}(s)=\sigma} \Gamma_f(s)x^{-s} ds + \sum_{0 \leq k < p} a_k x^k \text{ if } -p-1 < \sigma < -p,$$

where $a_k = \mathrm{Res}(\Gamma_f, -k) = f^{(k)}(0)/k!$.

Conversely, any function φ satisfying conditions (i) and (ii) is the Mellin transform of a unique $f \in S_+$, given by (11); then for all $m, n \in \mathbb{N}$ $s^m \varphi^{(n)}(s)$ are rapidly decreasing functions at infinity on every vertical strip of finite width.

The proof can be split up into several parts.

(a) *Assertions (i) and (ii) for $\varphi = \Gamma_f$, where $f \in S$.* Assertion (i) was proved before the theorem. So was the formula

$$(13.13) \qquad\qquad \mathrm{Res}(\varphi, -k) = a_k = f^{(k)}(0)/k! .$$

---

[63] it is in fact bounded on every subset of $\mathbb{C}$ defined by inequalities of the form $a \leq \sigma \leq b$, $|t| \geq c$ where $a, b, c \in \mathbb{R}$ and $c > 0$ (set $s = \sigma + it$). Getting near the poles of the function must clearly be avoided.

To prove (ii), first note that $\Gamma_f(s)$ is bounded on any strip $0 < a \le \operatorname{Re}(s) \le b < +\infty$ due to the simple fact that the integral converges for $s = a$ and $s = b$. Having said this, let us integrate by parts for $\operatorname{Re}(s) > 0$:

$$s\Gamma_f(s) = \int_0^{+\infty} f(x)sx^{s-1}dx = f(x)x^s \Big|_0^{+\infty} - \int_0^{+\infty} f'(x)x^{s+1}d^*x =$$
$$= -\Gamma_f'(s+1);$$

the part fully integrated is zero because (1) f is continuous for $x \ge 0$, where this is a large inequality, and x^s approaches 0 as x tends to 0, (2) f is a rapidly decreasing function at infinity. The previous relation generalizing the formula $s\Gamma(s) = \Gamma(s+1)$ can be integrated since $f \in \mathcal{S}_+$ implies $f' \in \mathcal{S}_+$ and leads to

$$(13.14) \qquad s(s+1)\dots(s+n-1)\Gamma_f(s) = (-1)^n\Gamma_{f^{(n)}}(s+n).$$

By analytic extension, this result holds for all s. If s remains in a vertical strip of finite width and if n is chosen to be sufficiently large, then $s + n$ remains in a strip $0 < a \le \operatorname{Re}(s) \le b < +\infty$, in which the right hand side is bounded. As

$$s(s+1)\dots(s+n-1) \sim s^n \qquad \text{for large } |s|$$

and given n, assertion (ii) follows.

(b) *Inversion formula.* To prove this, it suffices to check conditions (a), (b) and (c) of section (i). The continuity of f and the convergence of $\int f(x)x^s d^*x$ for $\operatorname{Re}(s) > 0$ are obvious since $f \in \mathcal{S}(\mathbb{R}_+)$; integral (4) involved in the inversion formula converges for all non-integral negative σ because of assertion (ii) of the theorem.

(c) *The function f associated to a function φ satisfying (i) and (ii).* By (ii), $t \mapsto \varphi(\sigma + it)$ is a rapidly decreasing function at infinity for all $\sigma \in \mathbb{R}$; it can, therefore, be integrated over any vertical $\operatorname{Re}(s) = \sigma \ne 0, -1, \dots$ and integral (4) can be computed. We show that it is independent of σ on any interval $[a, b]$ which does not contain an integer ≤ 0.

Let us integrate $\varphi(s)x^{-s}$ over a rectangle bounded by the verticals $\operatorname{Re}(s) = a$ and $\operatorname{Re}(s) = b > a$ and by the horizontals $\operatorname{Im}(s) = \pm T$.

$$\left|x^{-s}\right| \le x^{-a} + x^{-b},$$

on the horizontal sides. This constant is independent of T. As for the factor $\varphi(s)$, it is $O(T^{-n})$ for all n since the function $s^n\varphi(s)$ is bounded at infinity on the vertical strip $a \le \operatorname{Re}(s) \le b$. The contributions from these sides, therefore, clearly tend to 0 as $T \longrightarrow +\infty$. Setting $\psi(s) = \varphi(s)x^{-s}$, at the limit,

(13.15) $\displaystyle\int_{\mathrm{Re}(s)=b}\psi(s)ds-\int_{\mathrm{Re}(s)=a}\psi(s)ds=2\pi i\sum_{a<-k<b}\mathrm{Res}(\psi,-k)\,.$

So the integrals over the verticals a and b are indeed equal if ψ, i.e. φ, has no poles between a and b.

Let us calculate the residues of ψ. By assumption,

$$\varphi(s)=a_k/(s+k)+\dots$$

in the neighbourhood of the simple pole $s=-k$, where the unwritten terms represent a power series in $s+k$. Besides,

$$x^{-s}=x^k x^{-(s+k)}=x^k\exp\left[-(s+k)\log x\right]=x^k\left[1-(s+k)\log x+\dots\right]$$

is a power series in $s+k$ whose first term is x^k. Hence

(13.16) $\displaystyle\mathrm{Res}(\psi,-k)=a_k x^k\,.$

Next, (15) applied to $0<a<b$ shows that, setting

(13.17) $\displaystyle 2\pi i f(x)=\int_{\mathrm{Re}(s)=\sigma>0}\varphi(s)x^{-s}ds=ix^{-\sigma}\int\varphi(\sigma+it)x^{-it}dt$

for $x>0$, where this is a strict inequality, defines a function over $\mathbb{R}_+^*$ without any ambiguity. As $|x^{-it}|=1$,

$$2\pi x^\sigma|f(x)|\le\int|\varphi(\sigma+it)|dt$$

for any $\sigma>0$; the second expression being independent of x, f is a rapidly decreasing function at infinity.

(d) *Differentiability of f for $x>0$.* To show that f is C^∞ for *strictly* positive x, it must first be shown that (17) can be differentiated with respect to x. Thanks to Theorem 24 of Chapter V, § 7, n° 25, this amounts to checking that:

(a) with respect to x, the function under the $\int$ sign has as derivative a continuous function of the couple (x,t), which is obvious,
(b) for any compact subset $H\subset\mathbb{R}_+^*$, this derivative, namely $-s\varphi(s)x^{-s-1}$, is dominated by a function $p_H(t)$ integrable over $\mathbb{R}$ and not depending on the parameter $x\in H$ (normal convergence).

But if x remains in an interval $H=[a,b]$ with $0<a<b<+\infty$, then

$$\left|s\varphi(s)x^{-s-1}\right|\le|s\varphi(s)|\left(a^{-\sigma-1}+b^{-\sigma-1}\right)=p_H(t)\,;$$

$s\varphi(s)$ being a rapidly decreasing as a function of t, the function p_H is integrable, giving (b).

This allows us to write that

(13.18) $$2\pi i f'(x) = -\int_{\mathrm{Re}(s)=\sigma>0} s\varphi(s)x^{-s-1}ds\,.$$

However, the function $s\varphi(s)$, or more generally the product of $\varphi(s)$ with a polynomial in s, visibly satisfies conditions (i) and (ii). Therefore, the arguments used for f show that f' is a rapidly decreasing function at infinity with derivative given by

$$2\pi i f''(x) = +\int s(s+1)\varphi(s)x^{-s-2}ds\,,$$

where integration is over a vertical $\mathrm{Re}(s) = \sigma > 0$. Iterating the process, f is seen to be infinitely differentiable for $x \geq 0$, where this is a strict inequality, and all its derivatives

(13.19) $$2\pi i f^{(n)}(x) = (-1)^n \int s(s+1)\ldots(s+n-1)\varphi(s)x^{-s-n}ds$$

are seen to be rapidly decreasing functions at infinity like f itself and for the same reason. Integration is obviously over a vertical $\mathrm{Re}(s) = \sigma > 0$.

(e) *Behaviour of f in the neighbourhood of 0.* The aim is to show that f, for the moment defined for $x > 0$, can be extended to a C^∞ function for $x \geq 0$.

Integral (17) is extended to a vertical $\mathrm{Re}(s) = \sigma > 0$, but it can be moved to the left, provided the poles of φ are taken into account: indeed, the argument that has led to (15) relies only on φ decreasing at infinity. Hence, in view of calculation (16) of these residues,

(13.20) $$2\pi i f(x) = 2\pi i(a_0 + a_1 x + \ldots + a_n x^n) + \int_{\mathrm{Re}(s)=-n-1/2} \varphi(s)x^{-s}ds$$

for all $n \in N$; the point $-n-1/2$ is only noteworthy for being located between $-n-1$ and $-n$. For $\mathrm{Re}(s) = -n-1/2$,

$$\int |\varphi(s)x^{-s}|\,dt = x^{n+1/2}\int |\varphi(s)|\,dt\,.$$

The additional integral in (20) is, therefore, $O(x^{n+1/2})$, so that (20) is a bounded expansion of f in the neighbourhood of 0. Since $n \in \mathbb{N}$ is arbitrary, this gives an asymptotic expansion

(13.20') $$f(x) \approx \sum a_k x^k\,, \quad x \longrightarrow 0.$$

Let us show that it can be differentiated term by term, i.e. that

(13.20") $$f'(x) \approx \sum k a_k x^{k-1}\,, \quad x \longrightarrow 0.$$

Indeed, formula (18) can also be written as

$$2\pi i x f'(x) = -\int s\varphi(s)x^{-s}ds\,.$$

Passing from $f(x)$ to $xf'(x)$ is done by replacing $\varphi(s)$ by $-s\varphi(s)$, a function still satisfying conditions (i) and (ii) in the statement. Hence (20') can again be applied in this case provided that the residue a_k of $\varphi(s)$ at $s = -k$ is replaced by that of $-s\varphi(s)$, namely ka_k since the point $s = -k$ is a *simple* pole of φ; so

$$xf'(x) \approx \sum ka_kx^k\,,$$

which gives (20").

Iterating the argument, for all n, the derivative $f^{(n)}(x)$ is, therefore, seen to have an asymptotic expansion in the neighbourhood of 0, obtained by differentiating term by term that of f n times. To deduce that f can be extended to a C^∞ function on $x \geq 0$, where this is a large inequality, it remains to prove a rather easy general result:

Lemma 1. *Let f be a function defined and infinitely differentiable on an open interval $0 < x < b$. f can be extended to an infinitely differentiable function on the interval $0 \leq x < b$ if and only if the following conditions hold:*

(a) *f has an asymptotic expansion*

$$f(x) \approx \sum_{k \in \mathbb{N}} a_k x^k\,, \quad x \longrightarrow 0;$$

(b) *for all $n \in \mathbb{N}$, the derivative $f^{(n)}(x)$ has an asymptotic expansion obtained by differentiating term by term that of f n times.*

First of all, the relation $f(x) = a_0 + a_1 x + o(x)$ shows both that $f(x)$ tends to a_0 as x approaches 0 and that if we define $f(0) = a_0$, then the function f thus extended has a derivative equal to a_1 at the origin. Since, by (ii), $f'(x) \approx a_1 + 2a_2 x + \dots$ and so $\lim f'(x) = a_1$, the extension of f at $0 \leq x < b$ is C^1. Applying these arguments to f' instead of f, f' is seen to have as extension a C^1 function, so that f is C^2, etc.

A variation of lemma 2: suppose that

$$\lim_{x=0+} f^{(n)}(x) = f^{(n)}(0+)$$

exists for all n. Indeed, for $0 < x < x + h$,

$$|f(x + h) - f(x) - f'(x)h| \leq h.\sup|f'(x + k) - f'(x)|\,,$$

where the sup is extended to $k \in [0, h]$; since $f'(0+)$ exists, this sup is $\leq r$ for sufficiently small h, and so as x tends to 0,

$$|f(h) - f(0+) - f'(0+)h| \leq rh \, ;$$

the function f, extended at $x = 0$, therefore, has a derivative $f'(0) = f'(0+)$. A recurrence argument generalizes the result to successive derivatives.

As shown above, f and its successive derivatives are rapidly decreasing functions at infinity. So $f \in \mathcal{S}(\mathbb{R}_+)$.

(f) *The Mellin inversion formula for φ.* We now show that φ is indeed the Mellin transform of f, i.e. that formula (17) defining f can be inverted. Since it is merely a Fourier transform in disguise, it amounts to checking the conditions used in section (i) for showing that the inversion formula applies *in both directions* :

(a)   the function φ is continuous on the vertical $\text{Re}(s) = \sigma > 0$,
(b)   integral (17) is (absolutely) convergent,
(c) $\int |f(x)x^s| d^*x < +\infty$ for $\text{Re}(s) > 0$;

the order of the conditions to be checked is changed as here we need to calculate φ in terms of f and not f in terms of φ.

Checking (a) is trivial (φ is holomorphic), we would never have thought of writing (17) if condition (b) was not satisfied, finally (c) is obvious since, as seen above, f is continuous at $x = 0$ and, as noticed at the end of part (c) of the proof, is a rapidly decreasing function at infinity.

The last assertion of the statement is totally unrelated to the Mellin transform; apply the following result:

Lemma 2. *Let φ be a function defined and holomorphic on an open set*

$$U : a < \text{Re}(s) < b, \quad \text{Im}(s) > c$$

and m be a real number. If the function $s^m \varphi(s)$ is bounded at infinity on the closed vertical strip of finite width contained in U, then the same holds for all of φ's derivatives.

To see this, argue as in n° 4, (iv). Take a closed strip

$$B : \text{Im}(s) \geq c' > c, \ a' \leq \text{Re}(s) \leq b' \ \text{with} \ a < a' < b' < b$$

contained in U and choose some $r > 0$ such that

$$c < c' - r, \quad a < a' - r, \quad b' + r < b.$$

The closed strip

$$B' : \text{Im}(s) \geq c' - r, \quad a - r \leq \text{Re}(s) \leq b + r$$

is contained in U and, for all $s \in B$, the disc centered at s and of radius r is contained in B'. For $s \in B$, Cauchy's formula (not quite correct, but the forgotten numerical factor has no influence on the orders of magnitude)

$$r^n \varphi^{(n)}(s) = \varphi \oint [s + re(t)] \, \mathbf{e} \, [-(n+1)t] \, dt$$

shows that (same remark)

$$\left| \varphi^{(n)}(s) \right| \le r^{-n} \cdot \sup |\varphi(w)| ,$$

where the sup is extended to the points w of the circle $|w - s| = r$. By assumption, there is a constant c_m such that $|w^m f(w)| \le c_m$ at infinity in B'. $|s| - r \le |w| \le |s| + r$ on the circle. Hence $|w| \asymp |s|$ for large $|s|$, and

$$|\varphi(w)| = O\left(|w|^{-m} \right) = O\left(|s|^{-m} \right) .$$

So

$$\left| s^m \varphi^{(n)}(s) \right| \le |s^m| r^{-n} O\left(|s|^{-m} \right) = O(1) \quad \text{in } B ,$$

and the lemma follows. It comes under the same philosophy as that of Weierstrass' convergence theorem (Chap. VII, n° 19).

Theorem 12 characterizes Mellin transforms of functions belonging to $\mathcal{S}(\mathbb{R}_+)$, but the method cannot obviously be applied to other cases. For example, let us try to characterize the Mellin transforms of function f that have the following properties on $\mathbb{R}_+^*$:

(a) f and its successive derivatives are C^∞ and rapidly decreasing functions at infinity;

(b) f has an unbounded asymptotic expansion

$$f(x) \approx \sum_N a_n x^{u_n}$$

in the neighbourhood of 0, with real exponents réels $u_0 < u_1 < \dots$ such that $\lim u_n = +\infty$;

(c)   for all $k \in \mathbb{N}$, in the neighbourhood of 0, the derivative $f^{(k)}(x)$ has an unbounded asymptotic expansion obtained by formally differentiating that of f.

As seen at the start of this n°, the Mellin transform

$$\varphi(s) = \int f(x) x^s d^* x = \Gamma_f(s) ,$$

a priori defined for $\operatorname{Re}(s) > -u_0$, can be extended to a meromorphic function on all of $\mathbb{C}$ whose poles, all simple, are the points $-u_n$. The formula $s\Gamma_f(s) = -\Gamma_{f'}(s+1)$ obviously continues to hold for $\operatorname{Re}(s) > -u_0$. The proof is the same as before. As thanks to (c), the successive derivatives of f also clearly satisfy above conditions (a) and (b), it can be iterated and as in the proof of theorem 14, it follows that the function $s^n \varphi(s)$ is bounded at infinity (hence so are its successive derivatives) on any vertical strip of finite width.

We leave it to the reader to show that conversely any meromorphic function $\varphi(s)$ on $\mathbb{C}$ with these two properties (its only singularities are simple poles at points $-u_0 > -u_1 > \ldots$ on the real axis and it is rapidly decreasing at infinity on any vertical strip of finite width) is the Mellin transform of a function of the previous type.

14 – Stirling's Formula for the Gamma Function

The simplest example of an application of the inversion formula can be obtained by choosing $f(x) = e^{-x}$, obviously in $\mathcal{S}(\mathbb{R}_+)$. Its Mellin transform is, by definition, $\Gamma(s)$. Hence

$$(14.1) \quad e^{-x} = \frac{1}{2\pi i} \int_{\mathrm{Re}(s)=\sigma} \Gamma(s) x^{-s} ds = \frac{1}{2\pi} \int \Gamma(\sigma + it) x^{-\sigma - it} dt, \quad \sigma > 0,$$

and a slightly less simple formula for $-p - 1 < \sigma < -p$. This result is due to Mellin himself (1910), but it can be easily obtained without invoking the general theorem: it suffices to reconstruct the proof in this particular case...

Theorem 14 also shows that, on any vertical not passing through a pole, the function $t \mapsto \Gamma(\sigma + it)$ is in the Schwartz space. This is a weak result: indeed, formula (27) that will be proved at the end of this section shows that the Γ function decreases exponentially on every vertical.

This result is based on an evaluation (Stieltjes) which for $s \in \mathbb{N}$ reduces to Stirling's formula

$$(14.2) \qquad n! \sim (2\pi)^{1/2} n^{n+1/2} e^{-n} \quad \text{as } n \longrightarrow +\infty$$

of Chapter VI, n° 18, namely

$$(14.3) \qquad \Gamma(s) \sim (2\pi)^{1/2} s^{s-1/2} e^{-s}$$

as s tends to infinity in an angle $|\mathrm{Arg}(s)| \geq \pi - \delta$ with $\delta > 0$; for integer s, the equivalence with (2) follows from the relation

$$(n - 1)! = n!/n \sim (2\pi)^{1/2} n^{n-1/2} e^{-n}.$$

We first prove[64] relation (3), then we show how to deduced the behaviour of the Γ function on the verticals. These results arise in fields such as analytic number theory and in the study of the asymptotic behaviour of important

---

[64] The rest of this n° is essentially a fairly concise, detailed presentation of Remmert 2, Chap. 2, §4. Dieudonné, *Calcul infinitésimal*, IX.7.6, gives a genuine asymptotic expansion by using in full the Euler-MacLaurin formula. N. Bourbaki, *Fonctions d'une variable réelle* is another reference.

special functions. Proving them is a complex logarithm calculation exercise involving all the pitfalls of the topic.[65] In what follows, set

$$\mathbb{C} - \mathbb{R}_- = \mathbb{C}_+ .$$

The idea behind the proof is simple. A naive calculation shows (3) to be seemingly equivalent to

(14.4) $$\log \Gamma(s) = \frac{1}{2} \log 2\pi + (s - 1/2) \log s - s + o(1) .$$

However, the formula

(14.5) $$\Gamma(s) = \lim n! n^s / s(s+1) \ldots (s+n) ,$$

which holds for all non-integral $s \leq 0$, seems to show that

(14.6) $$\log \Gamma(s) = \lim \left\{ \log(n!) + s \log n - \sum_0^n \log(s+p) \right\}$$

and (2) shows that

(14.7) $$\log(n!) = 1/2 \log(2\pi) + (n + 1/2) \log n - n + o(1) .$$

The problem, therefore, appears to lie in the evaluation of the sum of the $\log(s+p)$, which, as we shall see, is made possible by the Euler-MacLaurin summation formula (Chapter VI, § 2, n° 16). Combining these results, (4) and hence (3) can be expected to be justified. But complex logarithms, not to speak of their limits, cannot be used like those of Neper; hence their meaning will first need to be specified.

Let us start by specifying the meaning of the expression $s^{s-1/2}$ occurring in (3), namely

(14.8) $$s^{s-1/2} = \exp\left[(s - 1/2) \operatorname{Log} s\right] \quad \text{for } s \in \mathbb{C}_+ ,$$

where, on $\mathbb{C}_+$, the Log function is the uniform branch which reduces to the Neper function on the positive real axis:

(14.9) $$\operatorname{Log} z = \log |z| + i \operatorname{Arg}(z) \quad \text{with } |\mathcal{A}rg(z)| < \pi .$$

This function allows the following more general definition

$$z^w = \exp(w \operatorname{Log} z) \quad \text{for } z \in \mathbb{C}_+ , \ w \in \mathbb{C} .$$

For technical reasons, the Log function needs to be extended to all of $\mathbb{C}^*$ by setting

---

[65] See Serge Lang, *Complex Analysis* (Springer-New York, 4th. ed., 1999), pp. 422–428 for an example of a proof where complex logs are used without precaution.

(14.9') $$L(z) = \log|z| + i\,\mathrm{Arg}(z) \quad \text{with} \quad -\pi < \mathrm{Arg}(z) \leq +\pi$$

on $\mathbb{C}^*$, and so $z = \exp[L(z)]$ for all $z \in \mathbb{C}^*$. The L function is not holomorphic on all of $\mathbb{C}^*$; it is discontinuous at every point of $\mathbb{R}^*_-$. To be precise, let us consider a sequence of points $z_n \in \mathbb{C}^*$ converging to some $z \in \mathbb{C}^*$. If $z \in \mathbb{C}_+$, clearly $\mathrm{Arg}(z) = \lim \mathrm{Arg}(z_n)$, and as a result $L(z) = \lim L(z_n)$. But if $z \in \mathbb{R}_-$, then $\mathrm{Arg}(z) = +\pi$ by convention whereas the arguments of the z_n are, for large n, near $+\pi$ or to $-\pi$ according to the values of n; hence there are integers $k_n \in \{-1, 0\}$ such that

$$\mathrm{Arg}(z) = \lim[\mathrm{Arg}(z_n) + 2k_n\pi i] .$$

In conclusion, in all cases, there exist k_n such that such that

(14.10) $$L(\lim z_n) = \lim[L(z_n) + 2k_n\pi i] .$$

In such a case, $L(z_n)$ may approach a limit. The same then holds for $2k_n\pi i$, which is, therefore, constant for large n; as a result,

(14.10') $$L(\lim z_n) = \lim L(z_n) \mod 2\pi i \quad \text{if} \quad \lim L(z_n) \text{ exists.}$$

The functional equation of the logarithm can be generalized with some precaution to the function $L(s)$. With definition (9') for the argument, clearly[66]

$$z = z_1 \ldots z_n \implies \mathrm{Arg}(z) = \sum \mathrm{Arg}(z_p) \mod 2\pi ,$$

so that

(14.11) $$L(z_1 \ldots z_n) = \sum L(z_p) \mod 2\pi i ,$$

and

(14.12) $$L(z_1 \ldots z_n) = \sum L(z_p) \iff -\pi < \sum \mathrm{Arg}(z_p) \leq +\pi .$$

For real $x > 0$ and $s = \sigma + it \in \mathbb{C}$, $x^s = x^\sigma \exp(it \log x)$; Since the argument of $x \in \mathbb{R}^*_+$ is zero, (12) shows that

$$L(x^s) = L(x^\sigma) + L[\exp(it \log x)]$$

and as the argument of $\exp(it \log x)$ is equal to $t \log x \mod 2\pi$,

(14.13) $$L(x^s) = sL(x) \mod 2\pi i \quad \text{for } x \in \mathbb{R}^*_+ , \; s \in \mathbb{C};$$

the formula $L(x^s) = sL(x)$ holds if $s \in \mathbb{R}$ since calculations are then in $\mathbb{R}^*_+$.

---

[66] In what follows, write $a = b \mod 2\pi$ or $\mod 2\pi i$ to mean that $a - b$ is a multiple of 2π or of $2\pi i$ as the case may be. The traditional notation is the sign $\equiv$, which is unnecessary if it is followed by an indication such as "$\mod 25$".

We will also need the relation

(14.14) $L(1 + z) = z - z^2/2 + z^3/3 \ldots$ for $|z| < 1$;

it is true for real z since $L(1 + z) = \log(1 + z)$ then, and so by analytic extension also on the unit disc; besides, in effect (14) only deals with the Log function.

Let us now specify what should be understood by $\operatorname{Log} \Gamma(s)$. The Γ function is holomorphic and never zero (infinite product expansion) on the simply connected domain $\mathbb{C}_+$. Hence the equation $\exp[f(s)] = \Gamma(s)$ has holomorphic solutions in $\mathbb{C}_+$, namely (§ 1, n° 3, Corollary 2 of Theorem 3) some primitives of $\Gamma'(s)/\Gamma(s)$. Choose the function

(14.15) $\operatorname{Log} \Gamma(s) = \displaystyle\int_1^s \frac{\Gamma'(z)}{\Gamma(z)} dz,$

where integration is along a path connecting 1 to s in $\mathbb{C}_+$, the simplest being the line segment. With this definition,

(14.16) $\operatorname{Log} \Gamma(s) = \log \Gamma(s)$ for $s \in \mathbb{R}_+^*$

since then the real function $\Gamma'(x)/\Gamma(x)$ is integrated over the interval $(1, s)$.

It may be though that $\operatorname{Log} \Gamma(s) = L[\Gamma(s)]$. This would be the case if $s \in \mathbb{C}_+ \implies \Gamma(s) \in \mathbb{C}_+$ was known to hold, for then the right hand side, consisting of two holomorphic functions on $\mathbb{C}_+$, would, like the left hand one, also be so. The obviously exact formula on $\mathbb{R}_+^*$, would then hold in all of $\mathbb{C}_+$. But the assumption on which this argument is based does not seem to be exact.[67] Since, by definition,

$$\exp[\operatorname{Log} \Gamma(s)] = \Gamma(s) = \exp\{L[\Gamma(s)]\}$$

the relation

(14.17) $\operatorname{Log} \Gamma(s) \doteq L[\Gamma(s)] \mod 2\pi i$

is exact and for us, this suffices.

Exercise. Using the infinite product, show that

(14.18) $-\Gamma'(s)/\Gamma(s) = C + 1/s + \displaystyle\sum_{n \geq 1} \left(\frac{1}{s+n} - \frac{1}{n} \right)$

if s is not an integer ≤ 0, then that

(14.19) $\operatorname{Log} \Gamma(s) = -Cs - \operatorname{Log} s + \displaystyle\sum [s/n - \operatorname{Log}(1 + s/n)]$ for $s \in \mathbb{C}_+$.

---

[67] Remmert 2 alludes discretely to it on p. 42 but, unfortunately, does not prove it, and I do not see how to do so.

These preliminary explanations allow us to return to

$$\Gamma(s) = \lim \left[n! n^s / s(s+1) \ldots (s+n) \right].$$

First of all, by (10), (11) and (17),

$$(14.20) \quad \text{Log}\, \Gamma(s) = \lim \left\{ \log(n!) + s \log n - \sum_{0}^{n} L(s+p) + 2k_n \pi i \right\}$$

for properly chosen $k_n \in \mathbb{Z}$, and by Stirling,

$$(14.21) \qquad \log(n!) = \frac{1}{2} \log(2\pi) + (n+1/2) \log n - n + o(1).$$

To evaluate the sum of the $L(s+p)$ for given $s \in \mathbb{C}_+$, set $f(x) = L(s+x) = \text{Log}(s+x)$ for $x > 0$; we get a C^∞ function such that $f'(x) = (s+x)^{-1}$, $f''(x) = -(s+x)^{-2}$ since $\text{Log}\, z$ is holomorphic on $\mathbb{C}_+$ and has $1/z$ as derivative. Instead of referring the reader to Chapter VI, § 2, n° 16 for the general Euler-MacLaurin formula, let us introduce the functions

$$(14.22) \qquad P_1(x) = x - 1/2, \quad P_2(x) = \frac{1}{2}\left(x^2 - x\right).$$

So $P_1' = 1$ and $P_2' = P_1$. As $P_2(0) = P_2(1) = 0$, integrating twice by parts immediately show that

$$\int_p^{p+1} f(x)dx = \int_0^1 f(x+p)dx =$$

$$= \frac{1}{2}\left[f(x+p) + f(x+p+1)\right] + \int_0^1 f''(x+p)P_2(x)dx.$$

Setting

$$(14.23) \qquad P_2^*(x) = P_2\left(x - [x]\right)$$

to be a function with period 1 equal to P_2 on $(0,1)$, transform the last integral into that of the function $f''(x)P_2^*(x)$ on $(p, p+1)$. Summing from $p = 0$ to $p = n - 1$,

$$\int_0^n f(x)dx = -\frac{1}{2}\left[f(0) + f(n)\right] + \sum_0^n f(x+p) + \int_0^n f''(x)P_2^*(x)dx$$

follows. Integrating by parts $f(x) = \text{Log}(s+x)$, we get

$$\int_0^n f(x)dx = (s+n)\,\text{Log}(s+n) - s\,\text{Log}\, s - n,$$

and so

$$\sum_{0}^{n} \mathrm{Log}(s + p) = (s + n)\,\mathrm{Log}(s + n) -$$

$$-s\,\mathrm{Log}\,s - n + \frac{1}{2}\left[\mathrm{Log}\,s + \mathrm{Log}(s + n)\right] -$$

$$-\int_{0}^{n}(s + x)^{-2}P_2^*(x)dx\,.$$

Formula (20) then shows that, modulo some small calculations, $\mathrm{Log}\,\Gamma(s)$ is the limit of a sequence whose general term is equal to

$$z_n = \frac{1}{2}\log(2\pi) - (n + s + 1/2)\left[\mathrm{Log}(s + n) - \log n\right] + (s - 1/2)\,\mathrm{Log}\,s +$$

$$+ \int_{0}^{n}(s + x)^{-2}P_2^*(x)dx$$

up to a multiple of $2\pi i$. If $\lim z_n = z$ is shown to exist, then relation (10') will show that $\mathrm{Log}\,\Gamma(s) = 2k\pi i + z$, and hence that $\Gamma(s) = e^z$.

First, the function $P_2^*(x)$ being bounded, the integral over $(0, n)$ converges to what Remmert denoted by

$$(14.24) \quad \mu(s) = \int_{0}^{+\infty}(s + x)^{-2}P_2^*(x)dx = -\int_{0}^{+\infty}(s + x)^{-1}P_1^*(x)dx\,.$$

On the other hand, $s + n = (1 + s/n)n$, and as $\mathrm{Arg}(n) = 0$, this leads to (12) and $\mathrm{Log}(s + n) = \mathrm{Log}(1 + s/n) + \log n$. So, by (14),

$$\mathrm{Log}(s + n) - \log n = \mathrm{Log}(1 + s/n) = s/n + O\left(1/n^2\right)\,.$$

As a result,

$$\lim(n + s + 1/2)\left[\mathrm{Log}(s + n) - \log n\right] = s\,.$$

Thus

$$\lim z_n = \frac{1}{2}\log(2\pi) - s + (s - 1/2)\,\mathrm{Log}\,s + \mu(s)$$

exists and the following formula holds:

$$(14.25) \qquad \Gamma(s) = (2\pi)^{1/2}s^{s-1/2}e^{-s}e^{\mu(s)} \quad \text{for all } s \in \mathbb{C}_+\,.$$

Hence it remains to show that $e^{\mu(s)}$ approaches 1 as s tends to infinity in $\mathbb{C}_+$ in not too arbitrarily, and that because of this $\mu(s)$ tends to 0. Set $s = r.\exp(i\varphi)$. For $x > 0$,

$$|s + x|^2 = (x + r\cos\varphi)^2 + r^2\sin^2\varphi = r^2 + 2xr\cos\varphi + x^2 =$$
$$= (r + x)^2 - 4xr\sin^2\varphi/2\,;$$

as $4xr \le (r + x)^2$ – calculate the difference –,

$$|s + x|^2 \geq (r + x)^2 \cos^2 \varphi/2 \,.$$

Since the periodic function $P_2(x)$ is contained between 0 and 1/8 everywhere,

$$8|\mu(s)| \cos^2 \varphi/2 \leq \int_0^{+\infty} (r + x)^{-2} dx = 1/r$$

or (Remmert, p. 52)

$$(14.26) \qquad\qquad |\mu(s)| \leq 1/8|s| \cos^2 \varphi/2 \quad \text{in } \mathbb{C}_+ \,.$$

Hence $\lim \mu(s) = 0$ if s tends to infinity in such a way that the product $|s| \cos^2 \varphi/2$ does so as well, for example if we consider a subset of $\mathbb{C}$ defined by

$$|\operatorname{Arg}(s)| \leq \pi - \delta \quad \text{with } 0 < \delta < \pi \,.$$

The Stieltjes formula holds under this condition, for example if s remains in the half-plane $\operatorname{Re}(s) > c$ and hence in a vertical strip of finite width.

As for Remmert, he avoids all the limit calculations that have been detailed here. He a priori introduces the function $\mu(s)$ and using the second integral, he notices through an elementary calculation that

$$\mu(s) - \mu(s + 1) = (s + 1/2) \operatorname{Log}(1 + 1/s) - 1 \,,$$

then that the function $f(s) = s^{s-1/2} e^{-s} e^{\mu(s)}$, holomorphic on $\mathbb{C}_+$, satisfies Wielandt's assumptions [n° 10, (i)]. Hence $f(s) = f(1)\Gamma(s)$, and in particular $f(n) = f(1)(n - 1)!$; as $\mu(n)$ obviously tends to 0, comparison with Stirling's formula shows that $f(1) = (2\pi)^{1/2}$. An excellent example of *Blitzbeweis*!

To show that the gamma function decreases exponentially on the verticals,

$$|s^{s-1/2}| = e^{\operatorname{Re}[(s-1/2) \operatorname{Log} s]}$$

still needs to be evaluated For $s = \sigma + it$,

$$\operatorname{Re}\left[(s - 1/2) \operatorname{Log} s\right] = (\sigma - 1/2) \log|s| - t \operatorname{Arg}(s) \,.$$

As s tends to infinity in the vertical strip B of finite width, the argument of s tends to $\pi/2$ if t tends to $+\infty$, and to $-\pi/2$ if t tends to $-\infty$; in the first case, using the power series for Arctg,

$$\pi/2 - \operatorname{Arg}(s) = \operatorname{Arctg}(\sigma/t) = \sigma/t + O(t^{-3})$$

since σ remains in a compact set, and so

$$-t \operatorname{Arg}(s) = -\pi|t|/2 + \sigma + O(t^{-2}) \,,$$

a result which also holds in the second case. For the same reason,

$$|s| = \left(\sigma^2 + t^2\right)^{1/2} = |t| \left(1 + O\left(t^{-2}\right)\right),$$
$$\log|s| = \log|t| + \log\left(1 + O\left(t^{-2}\right)\right) = \log|t| + O\left(t^{-2}\right).$$

So finally,

$$\mathrm{Re}\left[(s - 1/2)\,\mathrm{Log}\,s\right] = -\pi|t|/2 + (\sigma - 1/2)\log|t| + \sigma + O\left(t^{-2}\right),$$

and so

$$\left|s^{s-1/2}\right| \sim e^{-\pi|t|/2}|t|^{\sigma-1/2}e^{\sigma} \quad \text{at infinity in } B$$

since the factor $\exp\left[O(|t|^{-2})\right]$ tends to $\exp(0) = 1$. Returning to (29), we finally get the evaluation sought (Remmert tells us it is due to the Italian Salvatore Pincherle, 1889), namely

$$(14.27) \qquad |\Gamma(\sigma + it)| \sim (2\pi)^{1/2}|t|^{\sigma-1/2}e^{-\pi|t|/2}.$$

It is reassuring to see that this result is compatible with formula (10.5.8)

$$|\Gamma(1/2 + it)|^2 = \pi/\cosh \pi t.$$

15 – The Fourier Transform of $1/\cosh \pi x$

Like $\Gamma(1/2 + it)$, the function $1/\cosh \pi x$ is in the space $\mathcal{S}(\mathbb{R})$. This can be directly verified since its n-th derivative is obtained by dividing by $\cosh^{2n} \pi x$ a polynomial in $\sinh \pi x$ and $\cosh \pi x$ all of whose monomials are of total degree $< 2^n$; now, $\cosh \pi x \sim \frac{1}{2}\exp(\pi|x|)$ at infinity.

We show that it is identical to its Fourier transform. This result will later give rise to the strange identity (17) that can be found at the end of this n°.

The Fourier transform of $1/\cosh \pi x$ is the integral

$$(15.1) \qquad 2\int \frac{\exp(-2\pi i y t)}{e^{\pi t} + e^{-\pi t}}\,dt = \frac{2}{\pi}\int_0^{+\infty} \frac{x^{2iy}}{1 + x^2}\,dx$$

as shown by the change of variable $e^{-\pi t} = x$. More generally, the integral

$$(15.2) \qquad \varphi(s) = \int_0^{+\infty} \frac{x^s}{1 + x^2}\,dx, \quad |\mathrm{Re}(s)| < 1,$$

therefore, remains to be computed. It is the Mellin transform of

$$f(x) = x/(1 + x^2) = f(1/x).$$

If

$$\varphi(s) = \pi/2\cos(\pi s/2)$$

is shown to hold, the Fourier transform sought will then be equal to

$$2\varphi(2iy)/\pi = 1/\cosh \pi y$$

as expected. We give three methods for this calculation.

First proof. It is the shortest. Start with the formula

$$\Gamma(s)\Gamma(1-s) = \pi/\sin \pi s$$

and, for $0 < \mathrm{Re}(s) < 1$, write

$$\Gamma(s)\Gamma(1-s) = \int e^{-x}x^s d^*x \int e^{-y}y^{-s}dy = \iint e^{-x-y}(x/y)^s d^*x dy =$$

$$= \int dy \int e^{-x-y}(x/y)^s d^*x = \int dy \int e^{-xy-y}x^s d^*x =$$

$$= \int x^s d^*x \int e^{-(x+1)y}dy = \int \frac{x^s}{x+1}d^*x \,,$$

where all integrals are over $]0,+\infty[$. These transformations are justified by the Lebesgue-Fubini Theorem (theorem 25 of Chap. V, n° 26 would suffice) since all functions considered are integrable over $(0,+\infty)$ for $0 < \mathrm{Re}(s) < 1$. The change of variable $x \mapsto x^2$ in the last integral, which transforms d^*x into $2d^*x$, then shows that

$$\Gamma(s)\Gamma(1-s) = 2\varphi(2s-1)\,,$$

and so, replacing s by $(s+1)/2$,

$$\varphi(s) = \pi/2 \cos(\pi s/2)$$

for $|\mathrm{Re}(s)| < 1$, qed.

Second Proof.

(15.3') $$f(x) = x - x^3 + x^5 + \ldots \qquad \text{for } |x| < 1\,,$$

(15.3") $$f(x) = x^{-1} - x^{-3} + \ldots \qquad \text{for } |x| > 1.$$

A simple idea consists in multiplying series (3') and (3") by x^s and in integrating them term by term over $(0,1)$ and $(1,+\infty)$ with respect to d^*x taking into account the fact that the integral $\int x^s d^*x$ extended to $(0,1)$ or to $(1,+\infty)$ is equal to $1/s$ or $-1/s$ *when it converges*. A formal calculation thus gives

$$\varphi(s) = [1/(s+1) - 1/(s+3) + \ldots] - [1/(s-1) - 1/(s-3) + \ldots]$$

for $|\mathrm{Re}(s)| < 1$ since all integrals in question are then convergent. But the two series obtained, though semi-convergent (n° 9), are not absolutely convergent.

Thus a priori the permutation of the signs $\sum$ and $\int$ seems suspect. To justify it, replace (3') by the identity

$$f(x) = x - x^3 + \ldots + (-1)^n x^{2n+1} + (-1)^{n+1} x^{2n+2} f(x)$$

which is, up to a factor x, just the relation

$$1/(1-q) = 1 + q + \ldots + q^n + q^{n+1}/(1-q)$$

for $q = -x^2$. Hence the contribution $\varphi^-(s)$ from the interval $(0,1)$ is equal to

$$\sum_{0 \leq p \leq n} \frac{(-1)^p}{s + 2p + 1} + (-1)^{n+1} \varphi^-(s + 2n + 2)$$

for all $n > 0$. This presupposes that $\mathrm{Re}(s) > -1$ in order for the integral at $p = 0$ and hence for the following ones to be convergent, but in fact holds for all $s \in \mathbb{C}$ by analytic extension since $\varphi^-(s)$ is obviously meromorphic on all of the plane. However, for any $s \in \mathbb{C}$,

$$\varphi^-(s + 2n + 2) = \int_0^1 \frac{x^{s+2n+2}}{1 + x^2} d^* x \quad \text{for} \quad \mathrm{Re}(s) + 2n + 2 > 0,$$

hence for large n. As n increases, the function $x^{s+2n+2}/(1+x^2)$ converges to 0 everywhere on $]0,1[$ while remaining, in modulus, ≤ 1 for large n. Therefore, the integral tends to 0 (dominated convergence with respect to the measure $d^* x$), and once again the series is convergent and the relation

$$\varphi^-(s) = \sum_{p \geq 0} \frac{(-1)^p}{s + 2p + 1}$$

holds for all s.

To deal with the contribution $\varphi^+(s)$ from the interval $(1, +\infty)$ to the calculation of $\varphi(s)$, note that $\varphi^+(s) = \varphi^-(-s)$, and so

$$\varphi^+(s) = \sum_{p \leq -1} \frac{(-1)^p}{s + 2p + 1}.$$

For $|\mathrm{Re}(s)| < 1$,

$$\varphi(s) = \int_0^{+\infty} \frac{x^s}{1 + x^2} dx = \sum_{\mathbb{Z}} \frac{(-1)^p}{s + 2p + 1} = \pi/2 \cos(\pi s/2)$$

again holds thanks to (9.7").

Third Proof. The reader may be happy with these proofs, but this § is supposed to present applications of Cauchy's formula. So here is another way of calculating $\varphi(s)$. It is far less simple and miraculous, but can be generalized to all rational fractions.

It consists in integrating the function

$$g(z) = z^s/(1+z^2)$$

along the contour μ (see figure below) for $|\operatorname{Re}(s)| < 1$. In the above, $z^s = \exp(s\operatorname{Log} z)$ in $\mathbb{C} - \mathbb{R}_+$, where

(15.4) $\operatorname{Log} z = \log|z| + i\operatorname{Arg} z$, $0 < \operatorname{Arg} z < 2\pi$.

We get $2\pi i(\rho_i + \rho_{-i})$, which involves the residues of the function at the simple poles i and $-i$. Now, by (4),

$$\rho_i = \lim_{z=i}(z - i)z^s/(1+z^s) = i^s/2i = e^{\pi i s/2}/2i.$$

$\rho_{-i} = -e^{3\pi i s/2}/2i$ follows similarly. As a result,

(15.5) $$\int_\mu z^s\,dz/(1+z^2) = \pi e^{\pi i s/2}\left(1 - e^{\pi i s}\right).$$

The integral $\varphi(s)$ remains to be deduced from all this.

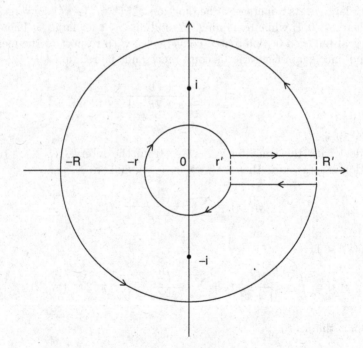

Fig. 15.14.

Let r and R be the radii of the two circles and $\pm\delta$ be the ordinates of the two segments of the horizontals that μ consists of. The segment with ordinate $+\delta$ contributes

$$(15.6) \qquad \int_{r'}^{R'} (x + i\delta)^s dx \big/ \left[1 + (x + i\delta)^2\right]$$

to the integral, where the limits $r' < r$ and $R' < R$ tend to r and R as δ tends to 0. Since the argument of $x + i\delta$ approaches 0 as δ tends to 0, for all real $x > 0$, the function

$$F(x, \delta) = (x + i\delta)^s \big/ \left[1 + (x + i\delta)^2\right]$$

tends to $F(x, 0+) = x^s(1 + x^2)^{-1}$, where $x^s = e^{s \log x}$ takes its usual value for real $x > 0$ (Chap. IV). The limit of integral (6) can therefore be assumed to be the extended integral over all of the interval $[r, R]$ of $F(x, 0+)$, but this requires justification, which will be provided by the theorem of dominated convergence.

First, it is clear that $r' > r/2$ for δ sufficiently small. Integral (6) is, therefore, the one over the fixed interval $(r/2, R)$ of the function equal to $F(x, \delta)$ between r' and R' and 0 elsewhere. This new function tends to $F(x, 0+)$ in $]r, R[$ and to 0 elsewhere in the interval $(r/2, R)$. On the other hand, the formula defining $F(x, \delta)$ continues to be well-defined for $x > 0$ and $\delta = 0$ since the result is obviously a continuous function on the product set $\mathbb{R}_+^* \times \mathbb{R}_+$. It follows that f is bounded on the compact set $\{x \in [r/2, R] \ \& \ 0 \le \delta \le 1\}$. As δ tends to 0, the modified function $F(x, \delta)$ integrated over $[r/2, R]$ tends to the function equal to $F(x, 0+)$ on $]r, R[$ and zero elsewhere, while remaining dominated by a fixed constant. As integration is over a compact set, passing to the limit is justified and finally

$$(15.7) \quad \lim \int_{r'}^{R'} F(x, \delta)dx = \int_{r}^{R} F(x, 0+)dx = \int_{r}^{R} x^s dx/(1 + x^2)$$

as expected.

The integral along the segment with ordinate $-\delta$ can be dealt with in a similar way; it is necessary to change the direction followed and to take into account that the argument of $x - i\delta$ tends to 2π, which introduces a factor $e^{2\pi i s} = \mathbf{e}(s)$ in the calculation. The limit value is, therefore, integral (7) multiplied by $-\mathbf{e}(s)$.

Hence, for given r and R and δ tending to 0, the total contribution from the segments of horizontals is equal to

$$(15.8) \qquad\qquad (1 - \mathbf{e}(s)) \int_{r}^{R} x^s dx \big/ \left(1 + x^2\right) .$$

By (2), this expression tends to $(1 - \mathbf{e}(s))\varphi(s)$ as r and R tend to 0 and $+\infty$.

To show that the contributions from the circular arcs tend to 0, use the lemma from the introduction to this §: in $U = C - \mathbb{R}_+$ and for all $s \in \mathbb{C}$,

$$(15.9) \qquad |z^s| \asymp |z|^{\mathrm{Re}(s)} \quad \text{in } U.$$

As $|1 + z^2|^{-1} = O(R^{-2})$ over the large circle, the integral is $O\left(R^{\mathrm{Re}(s)-1}\right)$ and tends to 0 since $\mathrm{Re}(s) < 1$.

Over the small circle, $|z^s| = O\left(r^{\mathrm{Re}(s)}\right)$ and $(1 + z^2)^{-1} \sim 1$. Therefore, the integral is $O\left(r^{1+\mathrm{Re}(s)}\right)$, and so we reach the same conclusion since $\mathrm{Re}(s) > -1$.

Ultimately, taking (4) into account, we get

$$(1 - e^{2\pi i s})\varphi(s) = \int_\mu = \pi e^{\pi i s/2}(1 - e^{\pi i s})$$

for $|\mathrm{Re}(s)| < 1$, and so

$$(15.10) \qquad \varphi(s) = \pi/2 \cos(\pi s/2),$$

which ends the third proof.

This method shows how to calculate the Mellin transform of a rational function $f(x) = p(x)/q(x)$ without poles in $\mathbb{R}_+$. As f is finite at $x = 0$, the integral converges in the neighbourhood of 0, at least for $\mathrm{Re}(s) > 0$. If $d°(q) - d°(p) = n$, then at infinity, $f(x) \asymp x^{-n}$ and hence $f(x)x^{s-1} \asymp x^{s-n-1}$, so that the integral converges for $\mathrm{Re}(s) < n$. The Mellin transform is, therefore, a priori defined on the vertical strip $0 < \mathrm{Re}(s) < n$. As $\Gamma_f(s)$ is obtained by integrating $f(x)x^{s-1}$ with respect to the measure dx,

$$(15.11) \qquad [1 - \exp(2\pi i s)]\, \Gamma_f(s) = 2\pi i \sum \mathrm{Res}\left[z^{s-1} f(z), a\right],$$

the sum being extended to all the poles of f. If

$$f(z) = \sum A_k / (z - a_k)$$

only has simple poles, then $\mathrm{Res}\left[z^{s-1} f(z), a_k\right] = A_k a_k^{s-1}$.

For example, for $f(z) = (1 + z)^{-1}$ the Mellin integral converges for $0 < \mathrm{Re}(s) < 1$ and

$$(-1)^{s-1} = \exp[\pi i(s - 1)] = -\exp(\pi i s)$$

needs to be calculated.

$$(15.12) \qquad \int_0^{+\infty} \frac{x^s}{1 + x} d^* x = \pi/\sin \pi s \quad \text{for } 0 < \mathrm{Re}(s) < 1$$

immediately follows. For $f(z) = (z - a)^{-k}$ with $a \notin \mathbb{R}_+$ and $k > 1$, the residue of $z^s f(z)$ is the coefficient of $(z - a)^{k-1}$ in the Taylor series for z^s at a. Now, by Newton, for $|z - a| < |a|$,

$$z^s = [a + (z - a)]^s = a^s [1 + (z - a)/a]^s =$$
$$= a^s \sum s(s - 1) \dots (s - p + 1) [(z - a)/a]^p /p! .$$

The residue at a is, therefore, $\begin{pmatrix} s \\ k-1 \end{pmatrix} a^{s-k+1}$, and so

$$\left(1 - e^{2\pi i s}\right) \int_0^{+\infty} \frac{x^s}{(x - a)^k} d^*x = 2\pi i \begin{pmatrix} s \\ k-1 \end{pmatrix} a^{s-k+1} \quad \text{for } 0 < \mathrm{Re}(s) < k .$$

The fact that the function $1/\cosh \pi x$ is identical to its Fourier transform, can at first seem a mere curiosity only interesting because it gives rise to exercises. The Fourier transform of the function $1/\cosh \pi t x$ is $t^{-1}/\cosh(\pi x/t)$ for $t = 0$, and as these are functions in $\mathcal{S}(\mathbb{R})$, Poisson's summation formula applies:

$$(15.13) \qquad \sum 1/\cosh(\pi n/t) = t \sum 1/\cosh \pi n t .$$

Let us then consider the similar series

$$(15.14) \qquad f(z) = \sum 1/\cos(\pi n z), \quad z \notin \mathbb{R},$$

and show first that it converges normally on all of the half-plane of the form $\mathrm{Im}(z) \geq r > 0$. Indeed, in this half-plane,

$$2 |\cos(\pi \hbar z)| = \left| e^{\pi(ny - inx)} + e^{\pi(-ny + inx)} \right| \geq$$
$$\geq \left| e^{\pi|n|r} - e^{-\pi|n|r} \right| \geq e^{\pi|n|r} - 1 .$$

Therefore, the convergent series $\sum 1/(e^{\pi|n|r} - 1)$ dominates series (14) in the half-plane considered.

Let us now show that f satisfies two simple functional equations. First,

$$(15.15) \qquad f(z + 2) = f(z) .$$

On the other hand, relation (13) means that

$$(15.16) \qquad f(-1/z) = (z/i).f(z)$$

holds for purely imaginary $z = it$. The two sides being analytic on the half-plane $\mathrm{Im}(z) > 0$, (16) holds in it.

Relations (15) and (16) resemble those proved in Chap. VII, n° 28 for the Jacobi function

$$\theta(z) = \sum \exp\left(\pi i n^2 z\right) = 1 + 2\left(q + q^4 + q^9 + q^{16} + \dots\right),$$

where $q = \exp(\pi i z)$. As an aside, note that the $\theta(x)$ series used to obtain the functional equation of the zeta function is in fact the value of $\theta(z)$ for $z = ix$. Here too, the series converges for $\mathrm{Im}(z) > 0$. Thanks to Poisson's summation formula and to the fact that the function $x \mapsto \exp(-\pi x^2)$ is equal to its Fourier transform, we showed that

$$\theta(-1/z) = (z/i)^{1/2}\theta(z).$$

This is why Riemann used it to prove the functional equation for his series $\zeta(s)$. We also have

$$\theta(z+2) = \theta(z).$$

Therefore, the function $\theta(z)^2$ satisfies (15) and (16). In fact,

(15.17) $$f(z) = \theta(z)^2.$$

In Chap. XII, the proof of (17) will lead us directly to the theory of modular functions and to the classical formula giving the number of ways an integer can be represented as the sum of two squares.

IX – Multivariate Differential and Integral Calculus

§ 1. Classical Differential Calculus – § 2. Differential Forms of Degree 1 – § 3. Integration of Differential Forms – § 4. Differential Manifolds

§ 1. Classical Differential Calculus

The aim of this § is to present the differential calculus of multivariate functions in the framework of finite dimensional real vector spaces; the case of a 2-dimensional space having been dealt with in Chap. III, § 5, we will mostly generalize the results, the proofs being the same as in dimension 2. Nonetheless, this § also contains considerations about tensors that are not found everywhere.

1 – Linear Algebra and Tensors

(i) *Finite-dimensional vector spaces* .[1] The elements of such a space E are, depending on the context, called "points" or "vectors", the numbers – real or complex depending on needs of the analysis – by which vectors are multiplied being "scalars"; in what follows, the letter K will indifferently denote $\mathbb{R}$, $\mathbb{C}$ or any other field in which the scalars vary. A *basis* for an n-dimensional vector space E is a family (a_i) of n linearly independent vectors, i.e. such that any $h \in E$ can be written in a unique way as $h = \sum h_i a_i$, the scalars h_i being the "components" or "coordinates" of h with respect to the basis considered.[2]

---

[1] For details and proofs, see for example sections §§ 10 to 24 in *Cours d'algèbre* (Hermann, 1966 or 1997) by the author, or else the somewhat condensed thirteen pages of the Annex to *Eléments d'analyse*, vol. 1, by Dieudonné, or Serge Lang, *Linear Algebra* (Springer, 1987), etc.

[2] The use of the letter h instead of x to denote vectors is due to the fact that in analysis, vectors occur mostly as increases of a variable, like in the notion of a differential defined later.

If E and F are vector spaces over the same field K, their Cartesian product $E \times F$ can be regarded as a vector space over K by defining the fundamental operations by

$$(x', y') + (x'', y'') = (x' + x'', y' + y'') , \quad t(x, y) = (tx, ty) .$$

With the same assumptions, a map $u : E \longrightarrow F$ is said to be *linear* if $u(x + y) = u(x) + u(y)$ and $u(tx) = tu(x)$ for any vector x and y and any scalar t; more generally,

$$u\left(\sum t_i x_i\right) = \sum t_i u(x_i)$$

for all vectors x_i and scalars t_i. If (a_i) forms a basis for the initial space E, then for any vector $h = \sum h^i a_i$,

(1.1) $$u(h) = \sum h^i u_i \quad \text{where} \quad u_i = u(a_i) \in F .$$

If (b_j) is a basis for F, setting $u(a_i) = \sum u_i^j b_j$, the table of coefficients u_i^j is the *matrix* of u with respect to the chosen bases of E and F.

For $F = K$, the term *linear functionals* or, sometimes that of *covectors* is used. The $u_i \in K$ are the *coefficients* of u. The set of these forms, equipped with the obvious algebraic operations (sum of two forms, scalar product), is the *dual space* of E, written E^*; the following notation is often used:

$$\langle h, u \rangle = u(h) \quad \text{for} \quad h \in E, \ u \in E^* ,$$

similarly to a scalar product. In analysis, the case $K = \mathbb{R}$ and $F = \mathbb{C}$ is constantly needed; the term *complex linear functionals* is then used; they are given by (1) with coefficients $u_i \in \mathbb{C}$ and their set, the *dual complex space* $E_{\mathbb{C}}^*$, is a complex vector space whose dimension over $\mathbb{C}$ is equal to that of E over $\mathbb{R}$. Each basis (a_i) of E has an associated *dual basis* (a^i) of E^* over $\mathbb{R}$ or of $E_{\mathbb{C}}^*$ over $\mathbb{C}$ consisting of linear functionals

(1.2) $$a^i : h \longmapsto h^i$$

on E.

Each linear map $A : E \longrightarrow F$ is associated to its *transpose* ^tA or A' : $F^* \longrightarrow E^*$, given by the relation

$$\langle A(h), u \rangle = \langle h, {}^tA(u) \rangle ; \quad \cdot$$

this definition is justified by the fact that for given u, the left hand side is a linear function of $h \in E$. $^t(BA) = {}^tA \, {}^tB$ holds for all linear maps $A : E \longrightarrow F$ and $B : F \longrightarrow G$.

Classical results on determinants will also be needed. If $\dim(E) = n$, then, up to a constant factor, there is a unique function $D(h_1, \ldots, h_n) \in K$ of n variables $h_i \in E$ satisfying the following two properties: it is *multilinear*, i.e. a linear function of h_i when the other variables are held fixed, (ii) it is *alternating*, i.e. changes sign when two of the variables h_i and h_j are permuted. Choosing a basis (a_i) for E and assuming $D(a_1, \ldots, a_n) = 1$, which determines D, the number $D(h_1, \ldots, h_n)$ is the *determinant of the h_i with respect to the* given *basis*. It is non-zero if and only if the h_i are linearly independent, i.e form a basis of E. The formula for calculating $D(h_1, \ldots, h_n)$ explicitly from the coordinates of the h_i can be found everywhere.

Given a linear map $u : E \longrightarrow E$, the *determinant of u* is the determinant of the vectors $u(a_i)$, where (a_i) is an arbitrary basis for E; it does not depend on it. Its basic properties are (i)

$$(1.3') \qquad D\left[u\left(h_1\right), \ldots, u\left(h_n\right)\right] = \det(u) D\left(h_1, \ldots, h_n\right)$$

for all $h_i \in E$, and so

$$(1.3'') \qquad \det(u \circ v) = \det(u) \det(v)$$

for all $u, v : E \longrightarrow E$, (ii) u is injective (or what amounts to the same in finite dimension, surjective or bijective) if and only if $\det(u) \neq 0$.

More generally, consider a linear map $u : E \longrightarrow F$, where E and F may have different dimensions, and let r be its *rank*, i.e. the dimension of the subspace $u(E)$ of F. Let $A = (u_i^j)$ be the matrix of u with respect to two arbitrary bases of E and F. Square matrices of order $\leq \min[\dim(E), \dim(F)]$ can be extracted from it by arbitrarily choosing the same number of rows and columns of A. Having said that, r is the largest integer for which a square matrix of order r and non-zero determinant can be extracted from A.

Finally, note that instead of "finite-dimensional vector space", we will mostly use the expression real or complex *Cartesian space* when $K = \mathbb{R}$ or $\mathbb{C}$. These are the only cases occurring in classical analysis.

(ii) *Tensor notation.* When Albert Einstein began his work in general relativity, he learnt the hard way, with the help of his mathematician friends who had read the Italian literature on differential geometry, not to mix up vectors and linear functionals (and, more generally, to distinguish what are called covariant tensors from contravariant ones – see below), despite the fact that a vector has as many components as a linear functional has coefficients. Indeed, if there is basis change, by (1), the coefficients of a linear functional undergo the same linear transformation as the basis vectors, whereas the coordinates of a vector undergo a "contragredient" one. This term is here used in in sense in which it is in the context of square matrices: the inverse of the transpose. For example in the simplest case, where the basis $(a_1, \ldots, a_n)$ is replaced by the basis $(t_1 a_1, \ldots, t_n a_n)$, with scalars $t_i \neq 0$, the u_i are multiplied and the h^i divided by the t_i. Hence, if, with respect to a particular basis, a vector and a linear functional appear to coincide because they have the same coordinates, this is not the case with respect to others; equating them has no physical and mathematical meaning. Anyhow, they are not objects of the same nature: a function defined on a set is not an element of this set.

Specialists of classical tensor calculus had adopted a system of notation, now mostly obsolete (including in my *Cours d'algèbre*), that, nevertheless, had some advantages; it was based on inferior and superior indices and on the summation convention attributed to Einstein.

In a space E with a basis denoted by (a_i), the first convention consists, at the simplest level, in writing vectors and linear functionals as follows:

$$(1.4) \qquad h = \sum h^i a_i, \quad u(h) = \sum u_i h^i$$

where the h^i are the components of h and the $u_i = u(a_i)$ the coefficients of u with respect to the basis considered. By (2), the second relation (4) can also be written $u(h) = \sum u_i a^i(h)$, i.e.

$$(1.5) \qquad u = \sum u_i a^i,$$

so that the u_i are the coordinates of $u \in E^*$ with respect to the dual basis (a^i) of (a_i) given by $a^i(h) = h^i$. In calculations involving both vectors and linear functionals, this notation, with its inferior and superior indices, makes it possible to immediately detect the nature of the objects discussed;[3] this is its first advantage.

We generalize this notation to more complex objects, namely *tensors*. In a finite-dimensional vector space E, a tensor is a function T of several variables, some with values in E, the others in E^*, and satisfying the same multilinearity property as a determinant: if all variables except one are held fixed, we get a linear function of the remaining variable. This property generalizes constantly used calculation rules in elementary algebra:

$$(x + y)z = xz + yz, \quad (tx)z = t(xz).$$

Calculation rules for tensors are, therefore, the same as those for products, excepting commutativity.

The function T is generally real or complex valued. T is said to be *of type* (p, q), or p times *covariant* and q times *contravariant*, if it depends on p variables in E and q variables in E^*. A scalar is a tensor of type $(0, 0)$. A linear functional is a tensor of type $(1, 0)$. A vector $h \in E$ identified with a linear functional $u \mapsto u(h)$ on E^*, becomes a tensor of type $(0, 1)$. An euclidean scalar product $(h|k)$ is a tensor of type $(2, 0)$. A linear map $T : E \longrightarrow E$ becomes a tensor of type $(1, 1)$ if the function $T(h, u) = u[T(h)]$ is associated to it, and conversely. Given a tensor $S(x, u)$ of type $(1, 1)$ and a tensor $T(x, y, u)$ of type $(2, 1)$, the function

$$(x, y, z, u, v) \longmapsto S(x, u)T(y, z, v)$$

is a tensor of type $(3, 2)$, the *tensor product* $S \otimes T$ of S and T. This can be generalized in an obvious way to other types, provided the variables involved in both tensors are fully separated.

---

[3] Purists will reply that coordinates can be dispensed with. It does not seem to be the opinion of physicists.

The function

$$(x, y, u) \longmapsto S(x, u)T(y, x, u)$$

is obviously not a tensor: the function x^2 is not linear on $\mathbb{R}$ or any other field.

Determining the way in which a tensor depends on the coordinates of its variables with respect to a given basis (a_i) is easy. For example, if $T(h, k, u)$ is a tensor of type $(2, 1)$, then leaving out the summation signs with respect to i, j and p – this is *Einstein's summation convention*, which I will mostly use –,

$$T(h, k, u) = T\left(h^i a_i, k, u\right) = h^i T\left(a_i, k, u\right) = h^i T\left(a_i, k^j a_j, u\right) =$$
$$= h^i k^j T\left(a_i, a_j, u\right) = h^i k^j T\left(a_i, a_j, u_p a^p\right) =$$
$$= h^i k^j u_p T\left(a_i, a_j, a^p\right),$$

and so

(1.6) $$T(h, k, u) = T_{ij}^p h^i k^j u_p,$$

where the

$$T_{ij}^p = T\left(a_i, a_j, a^p\right)$$

play the role of coefficients, components or coordinates – the terminology matters little – of T with respect to the basis considered. Conversely, any function given by a formula of type (6) is clearly trilinear in h, k, u.

Relation (6) easily gives the transformation undergone by the coefficients T_{ij}^k under a change of basis from (a_i) to (b_p) given by

(1.7) $$b_p = \rho_p^i a_i,$$

where (ρ_p^i) is the transition matrix from the first basis to the second one. It suffices to note that for the corresponding dual bases,

(1.8) $$b^q = \theta_j^q a^j \quad \text{with} \quad \rho_p^j \theta_j^q = \delta_p^q.$$

The well-known "Kronecker delta" equals 1 or 0 according to whether p and q are equal or not; indeed, by definition of a dual basis

$$\delta_p^q = b^q\left(b_p\right) = \theta_j^q a^j\left(\rho_p^i a_i\right) = \theta_j^q \rho_p^i \delta_i^j = \rho_p^j \theta_j^q.$$

Formula (6) then shows that

(1.9) $$T\left(b_p, b_q, b^r\right) = \rho_p^i \rho_q^j \theta_k^r T\left(a_i, a_j, a^k\right),$$

where the summation is over i, j and k. Conversely, it is easy to see that if we associated numbers T_{ij}^k transformed according to (9) under basis change to each basis (a_i) for E, then the trilinear functional defined by (6) is independent of the basis (a_i).

The convention of placing some indices in an inferior position and others in a superior one can then be explained as follows. Given a basis (a_i), choose two linear functionals $f = f_i a^i$ and $g = g_i a^i$, a vector $h = h^i a_i$ and consider their tensor product $f \otimes g \otimes h$, i.e. the tensor

$$S(x, y, u) = f(x)g(y)u(h) \quad (x, y \in E, u \in E^*)$$

of type $(2, 1)$ like T. Its coefficients in the given base are the numbers

$$S_{ij}^k = f(a_i) g(a_j) a^k(h) = f_i g_j h^k.$$

Formula (9), therefore, expresses that the coefficients of T transform like those of $f \otimes g \otimes h$; besides, by (6), a tensor of type $(2, 1)$ is clearly a linear combination of such products. Hence, the position of the indices immediately leads to formula (9).

There are more complicated formulas, but in all cases within the ambit of tensor calculus, both sides are sums of monomials with indices, for example

$$(*) \qquad A_{kh}^{ij} = B^{pqi} C_{pq}^j D_{kh} + M_{pkh} N^{ijp}.$$

A formula of this type is a relation between the tensors A, B, C, D, M and N with coordinates or components in each basis, written A_{kh}^{ij}, etc.; it has relevance only if it holds for any basis. For this and for the formula to have a chance of being correct, it must satisfy the following conditions:

(a)   An index appearing only once in a monomial is a free variable on which the monomial considered depends; it must occur once and only once in the inferior or superior position in all the monomials of the relation considered in order to ensure that with respect to this index, all the monomials and hence their sum are transformed likewise;

(b)   unless otherwise indicated, an index occurring twice in a monomial is a summation index, hence a bound or ghost variable on which the monomial does not depend; it must occur once in an inferior position and once in a superior position in order to ensure that, under basis change, linear transformations undergone by the monomials relative to these two indices cancel out;

(c)   a summation index cannot occur more than twice in a given monomial and cannot occur as a free variable in any other monomials since then, by rule (a), it would also be occurring as a free variable in all other monomials.

Exercise. Using formulas such as (9), show that the A_{kh}^{ij} given in each basis by (*) are indeed the components of a tensor.

Exercise. Let T be a tensor of type $(5, 3)$. Show that the numbers

$$T_{ijkpq}^{jqr} = S_{ikp}^r$$

are the components of a tensor of type $(3, 1)$.

These conventions hold in (*); p and q are summation indices whose name matter little, which makes it possible to use the same summation index p in many different monomials, but does not allow the first term of the right hand side to be written as $B^{ppi}C^j_{pp}D_{kh}$ since it is a sum over all couples (p, q), and not only over couples such that $p = q$. Similarly, if we multiply two sums, then we face serious problems if we write that

$$\sum a_i b^i . \sum c_i d^i = \sum a_i b^i c_i d^i ,$$

which is a higher level version of the immortal identity

$$(a + b)(c + d) = ac + bd .$$

The correct way to write this, especially if the $\sum$ are omitted, is

$$a_i b^i . c_j d^j = a_i b^i c_j d^j$$

in accordance with the distributivity rule of multiplicity with respect to addition: the ith term of the first sum is multiplies by the jth term of the second one and we sum over all couples (i, j). As a basic precaution, this amounts to denoting the free or bound variables with different meanings by different letters. Similarly, a double integral is not written as $\iint f(x, x)dxdx$; it is written $\iint f(x, y)dxdy$; if the function f depends on an additional variable z, so that its integral with respect to x and y depends on z, it is written $\iint f(x, y, z)dxdy$; calling z the integration variable would lead to a completely different result, namely $\iint f(z, y, z)dydz$.

Einstein's convention aims at simplifying typography; for example

$$c^j_i h^i u_j \quad \text{instead of} \quad \sum_{i,j} c^j_i h^i u_j \quad \text{or of} \quad \sum_{i=1}^{i=p} \sum_{j=1}^{j=q} c^j_i h^i u_j$$

if the indices i and j vary within the indicated limits; besides, in general there is no ambiguity about this matter. As mentioned above, several mathematicians now censor this way of writing on grounds that « this deluge of indices gives me seasickness », as used to say Dieudonné who was very sensitive to the latter (confirmed during a three day tempest in September 1950) and that considering mathematical objects themselves is anyhow better than considering their coordinates or components. This is undeniable, but forces to thinks and is often far less quick.

In fact, tensor notation only applies in some very particular circumstances – multilinear algebra and differential geometry – where it can be very convenient. I will, therefore, use them systematically in this chapter whenever general theoretical calculations will be involved, while giving the intrinsic formulas that allows coordinate calculations to be avoided; the reader will thus be able to compare both viewpoints. Moreover, if the deluge of indices which

I found amusing when eighteen still does not terrify me sixty years later, I do not see why I should deprive my readers of the pleasure of mastering them before casting them out into outer darkness instead of comparing the Γ^i_{jkh} to an "unpleasant insect" like Serge Lang, apparently impressed by Kafka, does in the preface of his *Fundamentals of Differential Geometry*...

As mentioned earlier about Baire, before the war, the municipal library of Le Havre used to propose to its readers all the usual French textbooks and treatises of the time. It also had the complete collection of the *Mémorial des sciences mathématiques*, a short monograph series on the most varied topics, excepting almost all "modern" ones that were then being elaborated outside France. It was in general too difficult or too little interesting for me and I was anyhow sufficiently busy learning more directly useful mathematics. Nonetheless, one day I was stunned by the booklet on *Absolute Differential Calculus* by René Lagrange, professor in Dijon; the word "absolute" that had intrigued me had been introduced by the Italians who had perhaps read Balzac. It was a presentation of tensor analysis in Riemann spaces, a vaguely defined notion: it was more or less possible to understand that it entailed n-dimensional curved spaces and used curvilinear coordinate systems that could be changed at will by formulas involving only functions that were as differentiable as necessary; the square of the distance ds from a point x with coordinates (x^i) to an "infinitely near" point $(x^i + dx^i)$ could be calculated by a formula of type

$$ds^2 = g_{ij}(x)dx^i dx^j \; ;$$

these spaces contained strange objects having an "absolute" meaning – what meaning? a mystery –, tensors represented in each coordinate system by functions of a variable point x in the space assigned inferior and superior indices; these were supposed to be transformed, under any change of coordinates, by formulas provided in advance involving the first derivatives of the coordinates with respect to the old ones; finally, height of virtuosity, the signs $\sum$ were always omitted. All this was presented without any allusion as to what a curved space, a vectorial space, a linear functional or multilinear form, etc. were; inventors and users of tensor calculus were still in the position of a physicist engaged in vector analysis calculations (gradient, rotational, divergence, etc.) without knowing what a vector is. As will be explained below (n° 12 and 14), "tensors" of the time are just tensor *fields*, i.e. functions T that, independently from any coordinate system, associate to each point x of the "curved space" X a tensor $T(x)$ of type (p, q) in the vector space $X'(x)$ depending on x and having the same dimension as X – the "tangent" vector space of X at x, whose somewhat abstract definition will be given in n° 12 –, in the purely algebraic sense given to this notion above. It is, therefore, a generalization of vector fields of physicists and mathematicians, that are just vector valued functions, i.e. tensor fields of type $(0, 1)$. But it was marvelous; only the machinery of traditional differential calculus needed to be known – essentially, the chain rule –, no other idea than that of constructing valid formulas for all systems of curvilinear coordinates was involved. This only implied compliance to Einstein's conventions. The "curvature" and "geodesics" of the space were calculated and finally, the tensor notation presented above was systematically used and led almost automatic calculations. In barely a month, I became an expert of absolute differential calculus without understanding anything. I gained nothing at the time, excepting gymnastics, but it was not worse than spending two hours per day in front of the television, *Flight Simulator* or *Tomb Raider* being as yet unknown.

There was also another exception in this collection: a far more modern and difficult treatise on *La géométrie des espaces de Riemann*, by Élie Cartan. It contained little calculation and indices, but mainly abstract and somewhat vague ides, for example the distinction between "closed" (i.e. compact) spaces and "open" (i.e. non-compact) ones. Formulations invented after 1945 thanks to the clarification of the notion of a topological space, to the definitive crystallization of the theory of abstract differential manifolds,[4] in particular by Claude Chevalley,[5] were unknown at the time, and to the invention of fiber spaces by algebraic topologists partly inspired by Élie Cartan who, during the war, invented a broad generalization of tensors connected to group theory. In the 1950s, the theory acquire the perfect and abstract form that can be found in all presentations of the subject, starting with N. Bourbaki's *Fascicule de résultats*. Naturally all this only concerns its most basic aspects and did not prevent its development in often unexpected directions too hard to present in the Bourbaki style – a maximum of abstractions and generalities.

In the special issue on Bourbaki in the magazine *Pour la Science*, the French version, but not a translation of *Scientific American*, Benoît Mandelbrot is alleged to have said, p. 82, that he left the École Normale Supérieure for the École Polytechnique because *thanks to my uncle,[6] I knew they were a militant gang, that they were strongly prejudiced against geometry and science, and that they tended to despise and even humiliate those who did not follow them*; and the author of this declaration apparently left France for the United States (and the IBM) in 1958 because of their *stifling influence*.

I am in a good position to appreciate the influence of Bourbaki on the École normale in 1944. Between 1940 and 1953, the one and only member of

---

[4] i.e. that are no longer considered subsets of Cartesian spaces, as used to be the case. The most popular example is the general relativity space; most people find it hard to understand precisely because of this and the more so when it appeared that it could be "closed" or bounded, i.e. compact. Mystics, a species that is not endangered, wondered what there could be outside: dread is the feeling of nothingness (Heidegger).

[5] *Theory of Lie Groups*, Princeton UP, 1946.

[6] Szolem Mandelbrojt, professor at the Collège de France and a specialist of quasi-analytic functions of one variable, a difficult subject that did not acquire the importance of the great fields developed after the war; see chapter 19 of Rudin's book. Szolem Mandelbrojt belonged to the initial Bourbaki group, but quickly left due to this very different conception of mathematics. This proves nothing against Bourbaki nor against Mandelbrojt. Everyone is free.

the "militant gang" then in Paris was Henri Cartan, the others being in the provinces or in the United States. He was the only one to look after students who were then supposed to follow the classes of the Sorbonne and to prepare for the advanced degree known as *agrégation*. From this strategic position, he obvious and inevitably influenced them for about twenty years starting from 1940, the year when I went to the École. Like everyone, he had his own conception of mathematics and propagated it, though in a less flamboyant style than M. Mandelbrot.

Anyhow, before the 1950s, there were almost no other mathematician in Paris likely to generate enthusiasm among students from the École seriously attracted to mathematics, and even less to explain to them any thing else apart from pre-1914 stuff.[7] Élie Cartan being too old did not teach anymore. The Lebesgue integral was sometimes taught by Arnaud Denjoy but his lectures were incomprehensible. As for Lebesgue, he preferred to teach elementary geometry at the Collège de France, an institution where, if I am not mistaken, professors are supposed to present recent subjects likely to be further developed; so it was Cartan who taught us in a very concise manner what a (Radon) measure and an integrable function were. Gaston Julia, a specialist of analytic functions reconverted to Hilbert spaces, was certainly around; provided one knew German, we could have learnt much more and much more quickly by reading some sixty pages by von Neumann in *Mathematische Annalen* of 1928–29 than by following his classes; even Henri Cartan, a non-specialist, used to teach us almost as much in a few lectures as Julia whose classes did not lead to any aspects of the subject subsequently developed. Jacques Dixmier experienced this before converting with great success to von Neumann's rings of operators,[8] a theory dating from the 1930s and seemingly unknown to Julia. I also followed lectures by Paul Montel, where he presented his theory of normal (i.e. compact) families of analytic functions, a 1910 model. We could also learn fluid mechanics from Henri Villat and Joseph Pérès, but the subject did not attract very many students from École Normale.

The only major exception was, shortly after 1945, Jean Leray, a very temporary member of the initial Bourbaki group which he used to criticize virulently on a personal plane, as I witnessed during a private conversation with him in 1950 in Cambridge, Mass, A specialist of partial differential equations in fluid dynamics before the war, he was made a prisoner in June 1940; detained for almost five years in an officers' camp in Austria where he had organized some sort of university and not wishing Germans to benefit from his

---

[7] And that is saying a lot since all that the German school had invented, from Gauss to Hilbert, in algebraico-analytic fields had been forgotten in France since at least fifty years. See the chapter on "Jeunes Turcs contre pontifes sclérosés" in the *Pour la Science* issue on Bourbaki.

[8] Dixmier claimed this was due to me. It is true I had discovered them thanks to my habit, acquired in Le Havre, of opening books randomly, for example the volumes of *Annals of Mathematics*. Von Neumann's papers had a great advantage for ignorant youngsters like us: they could be read almost without knowing anything.

expertise, he had converted to algebraic topology. Between 1940 and 1945, he invented the basic ideas of sheaf theory, which he expounded to Andre Weil and to Cartan shortly after his release. The theory was immediately adopted and greatly improved first by the former and then by very young people – my contemporary Jean-Louis Koszul, Jean-Pierre Serre and the Swiss Armand Borel, all members of Bourbaki before long. Sorry! –, and was the subject of a famous Cartan seminar talk in 1950-51. I wrote a book on it some years later[9] when Grothendieck, a member of the group and converted by Serre, was starting to revolutionize the subject and algebraic geometry. Elected to the College de France in 1947, Leray presented in 1947–48 and 1949–50 what was to become his major article on sheaves in the *Journal de Liouville* of 1950 and as such certainly contributed to the education of a few people. In 1950, Leray returned for good to partial differential equations and obviously influenced some of the young people who chose this promising subject, though in the traditional meaning of the word, he had very few students. In any event, it is clear that had in 1944 Benoit Mandelbrot opted to come to the ENS instead of Polytechnique and followed Leray's lectures from 1947 onwards in order not to fall into the hands of the *militant gang*, he would have learnt one of the most abstract and "modern" subjects. Anyhow, no one would have prevented him from choosing what suited him.

In 1949, Gustave Choquet arrived in Paris. Whilst promoting in his undergraduate lectures, when he had the opportunity to do so from 1954–55 onwards, a version of general topology that most members of the group never dared to diffuse at this level of abstraction, he was a major expert of fine structure theory, in line with Lebesgue, Baire and Denjoy and of potential theory together with Jacques Deny and, briefly, Henri Cartan. The inventor of fractals would have got on well with him had he not chosen Polytechnique, an institution where he probably learnt little else than traditional mathematics and the obligation of standing to attention when professors entered the lecture hall. The only advantage of Polytechnique being that one learnt other subjects, in particular physics, slightly more modern on some points – which was not hard – than at the Sorbonne. Besides, Polytechnique offered its best students much better career prospects and influence networks than

---

[9] It was far removed from the mathematics I was involved in at the time. But a Bourbaki member is supposed to write down everything in the program of the groups, and, moreover, I was somewhat upset of constantly hearing discussions on algebraic topology, which I did not understand at all. When very temporarily, Bourbaki decided to prepare a book on the subject, I volunteered to write a first account of sheaf theory. Once the chapter was written and discussed together, it became clear that Bourbaki could not publish this type of mathematics before a long time, and I was advised to turn it into a book. It is quite different from my original report and still sells, Hermann publishers having recently reedited it in French without consulting me, when the necessity of an English version has been long obvious (the first one was quickly translated into Russian)... The moral of the story and of many others of this type is that the group represented for its members, in particular its youngest ones, a fantastic opportunity for learning mathematics.

what rue d'Ulm could propose to its scientists in those days. But no one would have prevented students in mathematics who so desired to for example take an interest in theoretical physics if only its mathematical aspects had been taught in Paris. During the war, in one semester of lectures at the École – I have forgotten the year –, Louis de Broglie had not even gone as far as writing the Schrödinger equation on the blackboard. Physicists themselves learnt quantum mechanics on their own from German, American, English or Soviet writers while waiting for Messiah's classes at the CEA (*Centre d'étude atomique*).

During the era mentioned by M. Mandelbrot, within a few years, Henri Cartan had launched his students from the ENS in fields as diverse as potential theory, topology of Lie groups, Lie algebras, homotopy groups, differential topology, functions of several complex variables, etc. Some of his student from the École normale chose subjects he knew little about, such as noncommutative harmonic analysis in my case (discovered from André Weil's book and my lectures in the library of the École or of the Henri Poincaré Institute), or nothing about, such as algebraic geometry in the case of Pierre Samuel, influenced by Chevalley in Princeton, not to mention those who chose the theory of trigonometric series like Jean-Pierre Kahane, probability theory like Gerard Debreux, etc. Generally speaking, everyone was perfectly free, on the understanding that, like everyone, Cartan did not take charge of those who chose fields he was completely ignorant about or that did not find interesting at all, or that were totally outdated: he directed them to others if they could be found. If four students graduating from the ENS deserved a CNRS (*Centre nationale de recherches scientifiques*) grant and if the CNRS only offered two, he obviously had to choose those he would support.

By 1955 at the latest, students supervised by Cartan had many opportunities to learn from other "pure" or applied mathematicians. Serre's lectures and talks at the College de France and those of Laurent Schwartz at the Henri Poincaré Institute, to mention only these two members of the Bourbaki group, met with prodigious success for several decades. The same is true for those of Jacques-Louis Lions a few years later. As a student of Schwartz, he could not initially avoid Bourbaki's influence.

Contrary to what Benoit Mandelbrot and numerous other critics seem to think, no one in the group was unaware of the existence of other important fields aside from "fundamental structures" or despised them if they were not outdated or too light; most of the original work of its members goes beyond these. The idea that we held *strong prejudices against geometry* is odd from the part of people for whom Élie Cartan was the only French master and when Andre Weil was giving a solid foundation to the Italians' algebraic geometry, with their "generic points"! Bourbaki took no interest in it as such precisely because they were not part of these structures, which kept it sufficiently busy: given our program, we would have aroused much mirth if, instead of starting the *Eléments de Mathématique* with set theory, algebra and general topology, we had directly embarked on PDEs, stochastic processes, operational research, turbulent flows or the mathematics of quantum mechanics, all of which are subjects that we are told today , half a century later, suppose a new conception of mathematics, at the opposite end to ours. Incidentally, reading Dautray-Lions's volumes or some recent talks at the Bourbaki Seminar is far from confirming this point of view. Around 1948 together with Dieudonné and Schwartz, I attended some superb lectures by Jean Delsarte, a founding member of Bourbaki, on analytic number theory, following Hardy-Littlewood-Rademacher-Winogradov's version; though at the opposite of the Bourbaki spirit, the subject did not exactly arouse reactions of scorn from the audience. Dieudonné's superb article on analytic number theory in the *Encyclopaedia Universalis* or, in a neighbouring field, my talks at the Bourbaki Seminar (1952–1953) on Hecke's work (zeta functions of number fields, modular functions), the first of their kind in France in a field that had not yet been modernized, attest to this. One of the members of the group during this period, Charles Pisot, was a transcendental number specialist, a subject not very much in the Bourbaki line for the time.

As for the other sciences, we had far less prejudices against geology than shown nowadays by M. Claude Allègre, a great specialist of the subject and a recent education minister, for our version of mathematics: we merely ignored him for obvious reasons. Generally speaking, we did not consider it our duty to provide experimentalists with mathematics in its traditional form, which they had most often learnt in their youth and were determined to preserve. In fact, their physics has become in many aspects as abstract as our mathematics and it is hard to see why we should have had to grant them the exclusivity of " modernism "

The conclusion that follows from all this seems to me to be that it is above all to Henri Cartan and to the enthusiasm of the members of the Bourbaki group for "modern" mathematics that the French mathematical school owes its post-war recovery of the position it had lost since Poincaré, Picard and Lebesgue.

Another rarely acknowledged factor needs to be mentioned. The weakness and isolation of the French school before 1939 is often explained by referring to the Great War, its massacres and to the hostility towards Germany that continued long after 1918 in particular because of Émile Picard. On the contrary, its renewal after 1945 is perhaps also due to the fact that, during the war, the French – including Leray in his Oflag, including people like Schwartz, Samuel and the young Grothendieck threatened by antisemite policies, including Weil and Chevalley in the USA or in Brazil – had nothing else to do apart from "real" mathematics at a time when their German, English, American, Russian, etc. contemporaries were engaged in doing work for war purposes of a far lower quality than they were capable of, like Polish Jews, were ending their lives in Nazi concentration camps.

After 1945, mainly in the United States, but also in Japan, Germany, the USSR where an excellent school of functional analysis had flourished for a long time, many mathematicians who, at first, *were unaware of even the existence of Bourbaki* but not of "abstract" and "modern" mathematics which it is far from having invented, returned or converted to this type of mathematics where the prospects were enormous; for example in algebra, van der Waerden's books and later those of Birkhoff-MacLane have probably exerted a greater worldwide influence than those of Bourbaki; in fact, it is in van der Waerden that Koszul and I learnt algebra at the École Normale.

As I have explained in the postscript to vol. II, it is also thanks to the support of the military and of industrialists after 1945 that a "militant gang", far more influential than us and much more keen to propagate applied mathematics was formed; the latter is now on the point of dominating our science.

They are also starting to dominated the minds of some eminent pure mathematicians. When in the year 2000, "year of mathematics", during a religious ceremony, a journalist questioned Alain Connes (Fields medallist,Collège de France) he replied as follows according to *Le Monde* of 25 May 2000:

> Euclid's questioning was answered by research on non-Euclidean geometry, which stimulated Riemannian Geometry, which in turn inspired Albert Einstein for his work on space-time and general relativity used to refine the global satellite positioning system (GPS).

It would not have been in vain to add that the GPS, like inertial guidance earlier, was developed by the American military for their own planes, ships and ground vehicles and for locating with extreme precision enemy objects; their intention was not to help walkers in the Fontainebleau forest or in the Great Erg nor to help taxi drivers navigate in Chicago *suburbs*. Without American military funding, no civilian firm would have began such an extravagant project needing hundreds of billions in investments and decades of technical progress regarding missiles, satellites and telecommunications. The fact that the GPS is now available in the civilian sector does not change anything to the fact that we ought to be able to find better illustrations of the usefulness of mathematics from Euclid to Einstein than the guidance system of the most terrifying armaments ever invented by humanity. Because this is what the GPS also continues to be 2001 even if, for the sake of the cause, we prefer, as always, to cover up that bosom, which we can't endure to look on.

2 – Differential Calculus of n Variables

(i) *Differential functions.* Let f be a function defined on an open subset U of a n-dimensional real Cartesian space[10] E with values in a p-dimensional Cartesian space F, for example $\mathbb{C}$ if $p = 2$. Like in the case $n = 2$ recalled in chapter VIII, f is said to be *differentiable* at $x \in U$ if there is a linear map $u : E \longrightarrow F$ such that, for sufficiently small $h \in E$,

$$(2.1) \qquad f(x + h) = f(x) + u(h) + o(h),$$

---

[10] In fact, almost everything applies to functions defined on and with values in Banach spaces.

the symbol $o(h)$ denoting any function such that the ratio[11] $\|o(h)\|/\|h\|$ approaches 0 as h tends to 0. The linear map u is unique[12] because, for a *linear* function, the relation $u(h) = o(h)$ implies that $u = 0$: for any $h \in E$, the ratio $\|u(th)\|/\|th\|$ must approach 0 as $t \in \mathbb{R}$ tends to 0 though it is independent from it. In (1), u is called the *differential* or the *derivative* of f at x, or else the *tangent linear map* to f at x. As it depends on the point x, it is written $df(x)$ or $f'(x)$, its value at a vector h being written, $df(x; h)$ or $f'(x)h$. It is also given by

$$(2.2) \qquad f'(x)h = df(x; h) = \text{value of } \frac{d}{dt} f(x + th) \text{ for } t = 0 .$$

And so, more generally,

$$(2.2') \qquad\qquad df(x + th; h) = \frac{d}{dt} f(x + th)$$

since the left hand side is the derivative of the function $s \mapsto f[(x+th)+sh] = f[x + (t + s)h]$ at $s = 0$. For any x, obviously

$$(2.3) \qquad df(x ; h) = f(h) \quad \text{and} \quad f'(x) = f \quad \text{if } f \text{ is linear}.$$

Definition (2) is related to that of partial derivatives. If a basis (a_i) for the vector space considered is chosen and if, x^i generally denotes the coordinates of a point x, so that $f(x)$ becomes a function of these n real variables, the partial derivatives of f at x are the vectors[13]

$$(2.4) \qquad\qquad D_i f(x) = \text{value of } \frac{d}{ds} f(x + sa_i) \text{ for } s = 0$$
$$= df(x; a_i) = f'(x)a_i$$

of F obtained by differentiating the function

$$(2.5) \qquad\qquad s \longmapsto f\left(x^1, \ldots, x^{i-1}, x^i + s, x^{i+1}, \ldots, x^n\right)$$

with respect to S at $s = 0$. Having said that :

(a) The existence of $df(x)$ implies that of partial derivatives $D_i f(x) = df(x; a_i) \in F$ at x as well as the relation

$$(2.6) \qquad\qquad df(x; h) = D_i f(x)h^i ,$$

---

[11] Write $\|h\|$ for the norm of a vector h defined by any reasonable formula.

[12] and, in the case of Banach spaces, *continuous*, since (1) shows that $\|u(h)\|$ remains bounded when h remains in a sufficiently small ball centered at 0.

[13] The notation D_i indicates differentiation with respect to the ith variable. Its name does not need to be specified. Some authors write ∂_i instead of D_i. There is, of course, also Jacobi's notation $\partial/\partial x^i$, which I will write d/dx^i, a notation that can be easily typed and is self-explanatory.

where the scalars h^i are placed on the right of the vectors $D_i f(x)$ contrary to tradition,[14]

(b) $df(x)$ exists if and only the partial derivatives $D_i f$ exist and are continuous in the neighbourhood[15] of x (Chap. III, §5, n° 20).

(6) reduces to numerical expressions by choosing a basis (b_j) for F and setting $f(x) = f^j(x)b_j$, whence

$$(2.7) \qquad\qquad D_i f(x) = D_i f^j(x)b_j$$

since the b_j are independent of x; as a result,

$$(2.8) \qquad\qquad f'(x)h = df(x; h) = D_i f^j(x)h^i b_j .$$

The coordinates of the vector $df(x; h) \in F$ are, therefore, the numbers

$$(2.9) \qquad\qquad df(x; h)^j = D_i f^j(x)h^i = df^j(x; h) .$$

This is the value at the vector h of the differential in x of the function f^j.

When the $D_i f(x)$ exist and are continuous on U, f is said to be *of class C^1* on U, etc.

$$(2.10) \qquad\qquad D_i D_j f(x) = D_j D_i f(x)$$

if f is of class C^2 and in fact, using weaker assumptions (Chap. III, §5, n° 23).

The $n \times p$ matrix whose entries are the $D_i f^j(x)$ is the *Jacobian matrix* of f at x. It is that of the linear map $f'(x)$ with respect to the two chosen bases of E and F since

$$f'(x)a_i = D_i f^j(x)b_j .$$

The rank (n° 1, (i)) of this linear map is the *rank of f at x*. As the determinants of the square sub-matrices of the Jacobian matrix are continuous functions of x if f is C^1, if f is of rank r at x, then its rank is clearly $\geq r$ in the neighbourhood of x. In the case of map from E to itself, its determinant can be associated to $f'(x)$. This is the *Jacobian*

$$(2.11) \qquad\qquad J_f(x) = \det f'(x) = \det \left(D_i f^j(x) \right)$$

of f at the point x; it was earlier called the *functional determinant* of the f_j at x and was written $D(f^1, \ldots, f^n)/D(x^1, \ldots, x^n)$ or $D(f)/D(x)$ for short.

Using the differential notation $df = f'(x)dx$ as in the case of a real variable is possible and in practice very useful. Clearly, the differential of the

---

[14] It suffices to set once and for all that $th = ht$ if h is a vector and t a scalar: there is no problem since the field $\mathbb{R}$ is commutative.

[15] In a topological space, a relation involving a variable x holds *in the neighbourhood of a* if there is an open subset U containing a such that it holds for all $x \in U$ (Chap. II, §1, n° 3 for the case of $\mathbb{R}$ or $\mathbb{C}$).

coordinate function $u^i : x \mapsto x^i$ is – see (3) – at each point of E a linear functional $u^i : h \mapsto h^i$. For a complex valued function f, the formula $df(x; h) = D_i f(x) h^i$ can, therefore, also be written

$$df(x; h) = D_i f(x) du^i(x; h).$$

This shows that, the relation $df(x) = D_i f(x) du^i(x)$ between $df(x)$ and the $du^i(x)$ – linear functions of the ghost vector h – holds in the complex dual of E. But as $du^i(x)$ is in fact independent of x, it may as well be written du^i; and as u^i denotes the function $x \mapsto x^i$, its differential may as well be written dx^i. So $df = D_i f.dx^i$ for short.

Leibniz would not have had any difficulties explaining that the differential of f at x is, as in the case of $\mathbb{R}$, the increment of f when the variable x undergoes an infinitesimal vector increment dx:

$$df(x; dx) = f(x + dx) - f(x) = f'(x) dx.$$

A priori this formulation is not well-defined, but it is often convenient to use it in order to quickly recover results. This is why physicists like it despite its metaphysical character. For example, the formula

$$f(x + h.dt) = f(x) + f'(x) h.dt,$$

which holds for a given vector h and an "infinitesimally small" dt, immediately shows that $f'(x)h$ or $df(x; h)$ is the derivative of the function $t \mapsto f(x + th)$ for $t = 0$.

This can be justified provided the symbol dx is understood in a different way. The differential at any point of the identity map $id : x \mapsto x$ is just $h \mapsto h$; since it does not depend on the point at which it is calculated, it may as well be written $dx(h)$ rather than $d(id)(x; h)$ as it should theoretically be the case. The expression $df(x; h)$ or $f'(x)h$ can then be written $df[x; dx(h)]$ or $f'(x)dx(h)$. So, the expression $df(x; dx) = f'(x)dx$ is shorthand for the differential of f at the point x. This is all in appearance quite subtle, and in reality tautological, but it is sometimes useful in order to intuitively understand the formulas; the following point will illustrate this.

(ii) *Multivariate chain rule.* This is the formula that allows first order differential calculations to be reduced to linear algebra calculations; its importance cannot be overstated. Let E, F, G be Cartesian spaces, U and V open subsets of E, and F respectively, $f : U \longrightarrow V$ and $g : V \longrightarrow G$ two C^1 maps, and let us consider the composite map

$$p = g \circ f : U \longrightarrow G.$$

It is also a C^1 map and, for any $h \in E$,

(2.12) $$dp(x; h) = dg\left[f(x); df(x; h)\right]$$

or, with a different notation,

$$(2.13) \qquad\qquad p'(x) = g'[f(x)] \circ f'(x),$$

i.e. the composition of the tangent maps $f'(x) : E \longrightarrow F$ and $g'[f(x)] : F \longrightarrow G$ to f and g at $x \in U$ and $f(x) \in V$. This is the theorem for finding the derivative of a composite function. It will constantly be used in this chapter. In Leibniz's notation, (13) is written

$$(2.13') \qquad dp(x; dx) = dg(y; dy) \quad \text{with} \quad y = f(x), \quad dy = f'(x)dx.$$

(13) can easily be remembered by observing that it is the one and only conceivable formula having a meaning given the degree of generality of the situation : we want to find a *linear* map $p'(x) : E \longrightarrow G$, and as only the linear maps available are $f'(x) : E \longrightarrow F$ and $g'(y) : F \longrightarrow G$, their composite $g'(y) \circ f'(x) : E \longrightarrow G$ is the possible candidate. Moreover, as the result sought should not depend on x, y must depend on it. The only possibility proposed by the data being the substitution of y by $f(x)$, (13) follows. This is the beauty of "intrinsic" or "absolute" arguments : the formulas to be proved are imposed by the very nature of the objects considered (and they are correct). Leibniz would have explained that

$$p(x) + p'(x)dx = p(x + dx) = g\,[f(x + dx)] = g\,[f(x) + f'(x)dx] =$$
$$= g(y + dy) = g(y) + g'(y)dy,$$

where $y = f(x)$ and $dy = f'(x)dx$; hence (13'). Though this argument makes no sense if dx is interpreted in the same manner as the author of *Théodicée*, it leads as easily to the result in the general case as for functions of only one real variable. The important thing is not to become puzzled to the point of thinking, like Leibniz and the physicists of past times, that this calculation is a genuine proof.[16]

When $E = F = G$ in the above, the Jacobians (11) of f, g and p can be considered. The theorem on products of determinants and (13) then show that

$$(2.14) \qquad\qquad J_{g \circ f}(x) = J_g\,[f(x)]\,J_f(x).$$

(12) can be written is an explicit form in terms of numerical functions. Choosing a basis $(b_j)_{1 \leq j \leq p}$ for F, a basis $(c_k)_{1 \leq k \leq q}$ for G and setting

$$(2.15) \qquad f(x) = f^j(x)b_j, \quad g(y) = g^k(y)c_k, \quad p(x) = p^k(x)c_k$$

---

[16] Naturally, physicists have a different conception of proofs from mathematicians: if sloppy mathematical arguments provide them with the formulas confirmed by their experiments, for them, the formulas have been proved.

with numerical valued functions $f^j(x)$, $g^k(y)$ and $p^k(x)$, we get

(2.16) $$p^k(x) = g^k [f(x)] .$$

On the other hand, relations (6) and (8) show that, for $h \in E$,

(2.17) $$dp(x; h) = dg [y; df(x; h)^j b_j] = dg (y; b_j) df(x; h)^j =$$
$$= D_j g^k(y) c_k . D_i f^j(x) h^i .$$

Since by (8) applied to p, $dp(x; h) = D_i p^k(x) h^i c_k$,

(2.18) $D_i \{g^k [f(x)]\} = D_i p^k(x) = D_i f^j(x) . D_j g^k(y)$ where $y = f(x)$

finally follows. This expresses (13) in terms of matrices: the Jacobian matrix of p at x is the product of the Jacobian matrix of f at x and that of g at $y = f(x)$. The left hand side of (18) denotes the effect of the operator $D_i = d/dx^i$ on the function $x \mapsto g^k[f(x)]$, not to be confused with $D_i g^k[f(x)]$, which is the value of the function $D_i g^k$ at $f(x)$; this value is not usually well-defined since the function g^k depends on y and not on x. For example, the second expression, $D_i p^k(x)$, is the value of the function $D_i p^k$ at x, and $D_j g^k(y)$ the value of the la fonction[17] $D_j g^k$ at y, where $D_j = d/dy^j$. The presence of a punctuation point in the expression $D_i f^j(x) . D_j g^k(y)$ indicates that the operator D_i applies only to $f^j(x)$ and not to $f^j(x) D_j g^k(y)$. These conventions will be systematically used in order to avoid confusion.

These formulas can be simplified if the function g is real valued; this is then also the case of p and we get

(2.19) $$D_i \{g [f(x)]\} = D_j g [f(x)] . D_i f^j(x) .$$

In particular, if (case $E = \mathbb{R}$) there is map, written $t \mapsto \mu(t)$ rather than f, from an interval of $\mathbb{R}$ to the domain of definition V of g. So setting $D = d/dt$,

(2.20) $$D \{g [\mu(t)]\} = dg [\mu(t); \mu'(t)] = D_j g [\mu(t)] . D \mu^j(t)$$

at each point t where $D\mu(t) = \mu'(t) \in F$ exists.

(iii) *Partial differentials.* Seemingly more complicated composite functions often need to be considered. For example,

(2.21) $$p(u) = g [x(u), y(u), z(u)] ,$$

where u varies in an open subset Ω of a Cartesian space, where the functions x, y, z map Ω to open subsets U, V, W of three Cartesian spaces E, F, G and where g is defined on the open subset $U \times V \times W$ of the Cartesian space

---

[17] Generally, the symbol $D_i f$ always denotes the partial derivative of the function f with respect to the ith variable on which it depends, irrespective of the letters used to denote them.

$E \times F \times G$. To compute dp, introduce the *partial differentials* of the function $g(x, y, z)$ with respect to x, y and z. The partial differential $d_1g[(x; h), y, z]$, which depends linearly on an additional variable $h \in E$, is obtained by fixing $y \in V$ and $z \in W$ and by differentiating the map $x \mapsto g(x, y, z)$; hence, by definition,

$$(2.22) \qquad d_1g\,[(x; h), y, z] = \frac{d}{dt}g(x + th, y, z) \quad \text{for } t = 0$$

or, in Leibniz style,

$$(2.22') \qquad d_1g\,[(x; dx), y, z] = g(x + dx, y, z) - g(x, y, z)\,.$$

The differentials $d_2g[x, (y; k), z]$ and $d_3g[x, y, (z; l)]$ are similarly defined, the letters k and l denoting vector variables in F and G. As g is defined on an open subset of $E \times F \times G$, its (total) differential depends on a vector varying in this space, i.e. on three vectors $h \in E$, $k \in F$, $l \in G$. By definition, it is obtained by considering the value of g at the point $(x, y, z) + t(h, k, l) = (x+th, y+tk, z+tl)$ and by differentiating the result at $t = 0$:

$$(2.23) \quad dg\,[(x, y, z); (h, k, l)] = \frac{d}{dt}g(x + th, y + tk, z + tl) \quad \text{for } t = 0$$

$$(2.24) \quad dg\,[(x, y, z); (h, k, l)] = d_1g\,[(x; h), y, z] + d_2g\,[x, (y; k), z] + \\ + d_3g\,[x, y, (z; l)]$$

is easily seen to follow. The left hand side is indeed a linear function of the vector $(h, k, l) \in E \times F \times G$; as

$$(h, k, l) = (h, 0, 0) + (0, k, 0) + (0, 0, l)\,,$$

it is, therefore, equal to $dg[(x, y, z); (h, 0, 0)]+$ etc. But, by definition,

$$dg\,[(x, y, z); (h, 0, 0)] = \frac{d}{dt}g\,[(x, y, z) + t(h, 0, 0)] =$$
$$= \frac{d}{dt}g(x + th, y, z) \quad \text{for } t = 0\,,$$

an expression equal to $d_1g[(x; h), y, z)]$ by definition. This gives (24). As an aside, the following mistake should be avoided: the left hand side of (24) is a *linear* function of the vector (h, k, l) in the vector space $E \times F \times G$, and not a *trilinear* function of the vectors $h \in E$, $k \in F$, $l \in G$.

For example, suppose that, as in the case of a tensor, $g(x, y, z)$ is a trilinear function of x, y, z. Since $x \mapsto g(x, y, z)$ is linear, by (3), its differential d_1g is just $dx \mapsto g(dx, y, z)$. Hence

$$(2.25) \qquad dg = g(dx, y, z) + g(x, dy, z) + g(x, y, dz)$$

or, more explicitly,

$$(2.25') \qquad dg\left[(x, y, z); (h, k, l)\right] = g(h, y, z) + g(x, k, z) + g(x, y, l).$$

More specifically, suppose that the variables x, y, z are $n \times n$ matrices or linear operators on a Cartesian space and that $g(x, y, z) = xyz$; we then get the formula

$$dg = dx.yz + xdy.z + xydz,$$

where the order of the terms need to be carefully respected; more explicitly,

$$dg\left[(x, y, z); (h, k, l)\right] = hyz + xkz + xyl$$

where, like x, y, z, h, k, l are matrices or linear operators; the spaces E, F, G in (24) are here identical to the vector space $\mathcal{L}(M)$ of linear maps from M to itself.

Having done this, we can return to the computation of the differential of the composite function $p(u) = g[x(u), y(u), z(u)]$ we started with. By the multivariate chain rule, it can be obtained by replacing in dg, the variable (x, y, z) and its differential (dx, dy, dz) by their expressions in terms of u; and so, without assuming g to be multilinear,

$$\begin{aligned} dp(u; du) = \ & d_1 g\left\{[x(u); x'(u)du], y(u), z(u)\right\} + \\ & + d_2 g\left\{x(u), [y(u); y'(u)du], z(u)\right\} + \\ & + d_3 g\left\{x(u), y(u), [z(u); z'(u)du]\right\}. \end{aligned}$$

Having done that, replace du by h. If g is *multilinear*, the result simplifies by (25):

$$\begin{aligned} (2.26) \qquad dp(u; h) = \ & g\left[x'(u)h, y(u), z(u)\right] + g\left[x(u), y'(u)h, z(u)\right] + \\ & + g\left[x(u), y(u), z'(u)h\right]. \end{aligned}$$

If, moreover, the variable u is real, in which case so is h as well, then, because of the multilinearity of g, h becomes a common factor, and since $dp(u; h) = p'(u)h$ we get

$$\begin{aligned} (2.27) \qquad p'(u) = \ & g\left[x'(u), y(u), z(u)\right] + g\left[x(u), y'(u), z(u)\right] + \\ & + g\left[x(u), y(u), z'(u)\right] \end{aligned}$$

as if it was a matter of differentiating a product $x(u)y(u)z(u)$; this is the key point. Besides, if it is a genuine product of functions whose values are, for example, linear operators or $n \times n$ matrices, we retrieve the classical formula

$$\frac{d}{du} x(u)y(u)z(u) = x'(u)y(u)z(u) + x(u)y'(u)z(u), +x(u)y(u)z'(u)$$

where, once again, the order of the factors is essential.

(iv) *Diffeomorphisms.* Let U be an open subset of a Cartesian space E and f a map from U to a Cartesian space F of the same dimension as E. Suppose that f maps U on an open subset of F. f is said to be a *diffeomorphism* of class C^p from U to V if f as well as the inverse map $g = f^{-1} : V \longrightarrow U$ are bijective and of class C^p. Clearly, for $y = f(x)$, the linear maps $f'(x)$ and $g'(y)$ are mutually inverse. If $E = F$, then

$$(2.28) \qquad J_g(y)J_f(x) = 1 \quad \text{for} \quad y = f(x), \quad g = f^{-1}$$

follows and in particular, $J_f(x) \neq 0$ for $x \in U$.

Conversely, let us start with a C^p map f from an open subset U of E to F, with $\dim(E) = \dim(F)$, and suppose that $f'(x)$ is invertible for all $x \in U$. The local inversion theorem (Chap. III, § 5, n° 24, Theorem 24), whose proof in dimension n is similar to the one in dimension 2, tells us that for all $x \in U$, there is an open neighbourhood $U(x)$ of x homeomorphically mapped by f onto an open neighbourhood V of $y = f(x)$, the inverse map from $V(y)$ to $U(x)$ also being C^p. If that is the case for all $x \in U$, then the image $V = f(U)$ is open, and more generally so is the image of any open subset of U. If, moreover, f is injective not only in the neighbourhood of each point, but globally, and so is a bijection from U onto V, then the inverse map $f^{-1} : V \longrightarrow U$ can be considered; it is C^p, so that f is a diffeomorphism.

When we will prove the change of variable formula for a multiple integral, we will need to consider a bounded open subset U of a Cartesian space E, its compact closure A and a map f from A to E or, more generally, to a Cartesian space F. f will be said to be *of class C^p on A* if f is of class C^p on U and if the partial derivatives of order $\leq p$ of the f^j extend to continuous functions on A. Since A is compact, this is the case if and only if these derivatives are uniformly continuous on U (Chap. V, § 2, n° 8, corollary 2 of Theorem 8 generalized to n variables). So the traditional notation for extensions to A of partial derivatives will continue to be used for the vector $D_i f(a)$ with coordinates $D_i f^j(a)$, for the linear maps $f'(a) : h \mapsto D_i f(a) h^i$ and the Jacobian $J_f(a)$ will be defined in the obvious manner for all $a \in A$.

When $\dim E = \dim F$, f will be said to be a *diffeomorphism of class C^p from A onto $B = f(A)$* if, moreover, (i) f is bijective, in which case f is a homeomorphism from A onto the compact set $B = f(A)$, (Chap. III, § 3, n° 1, Theorems 11 and 12), (ii) f is a C^p diffeomorphism from U onto an open set $V \subset F$, and so $B = \bar{V}$, (iii) the inverse map $f^{-1} : B \longrightarrow A$ is C^p in the sense defined above.

Conditions (ii) and (iii) require

$$(2.29) \qquad\qquad J_f(a) \neq 0 \quad \text{for all} \quad a \in A$$

since the two sides of (28) are continuous functions on A. Conversely, if (29) holds, (ii) follows by (i) and the local inversion theorem, and (iii) is a consequence of the fact that, if $M_f(x)$ is the Jacobian matrix of f at $x \in U$, the entries of its inverse, i.e. the derivatives of f^{-1}, are the quotients of minors of $M_f(x)$ by $J_f(x)$ (Cramer formulas), and so extend by continuity to B when (29) holds.

The situation described above is encountered when f is the restriction to A of a diffeomorphism defined on an open set containing A, but the converse is most dubious: for a function defined on a compact set A to have a C^1 extension on an open set containing A, it must satisfy more restrictive conditions than those imposed above.[18]

(v) *Immersions, submersions, subimmersions.* In section (iv), f was assumed to map an open subset U of E to a Cartesian space of the same dimension as E and the tangent maps $f'(x)$ were assumed to be invertible. This assumption has no meaning anymore if $\dim(F) \neq \dim(E)$ and even if $\dim(E) = \dim(F)$, the important notion is that of the rank of f at x, defined in section (i), i.e. the dimension of the vector subspace $\operatorname{Im} f'(x) = f'(x)E$ of F. It can be written $rg_x(f)$; as it is computed by using minors of the Jacobian matrix, it is a lower semi continuous function of x: for any M, the relation $rg_x(f) > M$ defines an open subset of U. If $rg_x(f) = \dim(E)$, i.e. if $f'(x)$ is injective, f is said to be an *immersion* at x; if, conversely, $rg_x(f) = \dim(F)$, i.e. if $f'(x)$ is surjective, f is said to be a *submersion* at x. If, more generally, the rank of f is constant in the neighbourhood of x, f is said to be a *subimmersion* at x. These notions will arise later in relation to submanifolds of a Cartesian space.

3 – Calculations in Local Coordinates

(i) *Diffeomorphisms and local charts.* The notion of a diffeomorphism is similar to that of a *curvilinear coordinate system* (meaning: global) in an open subset U of an n-dimensional Cartesian space E. Such a system is a family of n functions $\varphi^i : U \longrightarrow \mathbb{R}$ of class C^1, at least such that the differentiable functions on any open set $V \subset U$ are precisely those that can be expressed in a differentiable manner by using "coordinates" $\varphi^i(x) = \xi^i$ de x; more precisely, the map

$$\varphi : x \longmapsto \left(\varphi^1(x), \ldots, \varphi^n(x)\right)$$

from U to $\mathbb{R}^n$ is required to be a diffeomorphism from U to an open subset of $\mathbb{R}^n$. In what follows, we will write $x = f(\xi)$, or sometimes $x(\xi)$ if no confusion follows, for the point of U corresponding to point $\xi \in U' = \varphi(U)$, so that the map $f : U' \longrightarrow U$ is the inverse of φ; then

$$(3.1) \qquad\qquad f'(\xi) = \varphi'(x)^{-1},$$

the two sides being calculated at corresponding points $x \in U$ and $\xi \in U'$.

---

[18] See for example Dieudonné, *Eléments d'analyse*, vol. 3, XVI.4, problems 4, 5, 6 (Whitney's theorems), which deals with the case of C^r maps.

At least among mathematicians, the expression "curvilinear coordinates" has long fallen into disuse. In a Cartesian space E, the term *chart* is preferable for any diffeomorphism φ from an open set $U \subset E$ onto an open subset of a Cartesian space; (U, φ) will denote such a chart. If $a \in U$, (U, φ) is also said to be a *local chart* of E at the point a; it is often convenient to assume $\varphi(a) = 0$.

The following result immediately shows the usefulness of local charts:

Theorem 1. *Let f be a $\mathbb{R}^q$-valued C^s map defined in the neighbourhood of 0 in $\mathbb{R}^p$ and such that $f(0) = 0$. Suppose that the rank r of f is constant in the neighbourhood of 0. Then there are C^s local charts (U, φ) of $\mathbb{R}^p$ at 0 and (V, ψ) of $\mathbb{R}^q$ at 0 such that*

$$\psi \circ f \circ \varphi^{-1}(\xi) = (\xi^1, \dots, \xi^r, 0, \dots, 0)$$

for any $\xi \in U$.

An equivalent formulation: setting $y = f(x)$ and denoting by $\xi^i (1 \leq i \leq p)$ the coordinates of ξ in the chart (U, φ) and by $\eta^j (1 \leq j \leq q)$ those of y in the chart (V, ψ),

$$\eta^j = \xi^j \ (1 \leq j \leq r), \quad \eta^j = 0 \ (r + 1 \leq j \leq q) .$$

To prove the theorem, first note that, up to a permutation of the canonical coordinates in $\mathbb{R}^p$ and $\mathbb{R}^q$, $D(f^1, \dots, f^r)/D(x^1, \dots, x^r) \neq 0$ may be assumed to hold at 0, hence also in a neighbourhood of 0. Then consider the map

$$\varphi\left(x^1, \dots, x^p\right) = \left(f^1(x), \dots, f^r(x), x^{r+1}, \dots, x^p\right)$$

from the latter to $\mathbb{R}^p$. Its Jacobian matrix is of the form

$$\begin{pmatrix} A & 0 \\ ? & 1_{p-r} \end{pmatrix}$$

where A is that of $f^1, \dots, f^r$ with respect to $x^1, \dots, x^r$. It is, therefore, invertible like A in the neighbourhood of 0. As a result, φ is a diffeomorphism from an open neighbourhood U of 0 to an open neighbourhood U' of 0. So, we get a chart (U, φ) of $\mathbb{R}^p$ at 0 for which

$$(3.2') \qquad f \circ \varphi^{-1}(\xi) = (\xi^1, \dots, \xi^r, g^{r+1}(\xi), \dots, g^q(\xi))$$

with new functions $g^i (r + 1 \leq i \leq q)$ defined on U'. Denoting the Jacobian matrix of $g^{r+1}, \dots, g^q$ with respect to $\xi^{r+1}, \dots, \xi^q$ by D, that of (2') is of the form

$$\begin{pmatrix} 1_r & ? \\ 0 & D \end{pmatrix}$$

As its rank is by assumption r, all the entries of D are zero.[19] This mean that, in the neighbourhood of 0, the g^i only depend on the first r variables ξ^i. The g^i being defined in the neighbourhood of 0, the expression

(3.2")
$$\psi(y) = \left(y^1, \ldots, y^r, y^{r+1} - g^{r+1}\left(y^1, \ldots, y^r\right), \ldots, y^q - g^q\left(y^1, \ldots, y^r\right)\right)$$

is well-defined for all $y \in \mathbb{R}^q$ in the neighbourhood of 0. The Jacobian matrix of ψ is of the form

$$\begin{pmatrix} 1_r & ? \\ 0 & 1_{n-r} \end{pmatrix}$$

and so is invertible. As a result, ψ is a diffeomorphism from a neighbourhood of 0 onto an open subset V' of $\mathbb{R}^q$, giving a chart (V, ψ) of $\mathbb{R}^q$ at the origin. We can obviously assume $f(U) \subset V$ and then compute the map

$$\psi \circ f \circ \varphi^{-1} : U' \longrightarrow V'$$

which gives f in the charts that have been obtained. This amounts to replacing $y^1, \ldots, y^q$ in (2") by the expressions $\xi^1, \ldots, g^q(\xi)$ occurring on the right hand side of (2'), which replaces $y^{r+i} - f^{r+i}(y^1, \ldots, y^r)$ by $f^{r+i}(\xi^1, \ldots, \xi^r) - f^{r+i}(\xi^1, \ldots, \xi^r) = 0$; hence

$$\psi \circ f \circ \varphi^{-1}(\xi) = (\xi^1, \ldots, \xi^r, 0, \ldots, 0) ,$$

proving the theorem.

(ii) *Moving frames and tensor fields.* Consider a chart (U, φ) and the inverse $f : U' \longrightarrow U$ of the map φ. For any $\xi \in U' = \varphi(U)$, the tangent map $f'(\xi)$ transforms the canonical basis (e_i) of $\mathbb{R}^n$ into a basis $(a_i(\xi))$ of E which depends both on the point $x = f(\xi)$ and on the chart φ, namely

(3.3)
$$a_i(\xi) = f'(\xi)e_i = df\,(\xi; e_i) = D_i f(\xi) .$$

This is the partial derivative of the map $f : U' \longrightarrow E$ at the point $\xi = \varphi(x) \in \mathbb{R}^n$. So, by (1),

(3.3')
$$\varphi'(x)a_i(\xi) = e_i .$$

For example, denoting by (x, y) the usual Cartesian coordinates and setting U to be the open subset $\mathbb{R}^2 - \mathbb{R}_-$ of $\mathbb{R}^2$ already encountered in Cauchy theory, set

$$x = r.\cos\theta , \quad y = r.\sin\theta \quad \text{with } r > 0 \text{ and } |\theta| < \pi .$$

---

[19] Calculate the determinant of order $r + 1$ obtained by adding a row and a column to the matrix 1_r.

Then the map φ is $(x, y) \mapsto (r, \theta)$ and it is a diffeomorphism from U onto the open subset of $\mathbb{R}^2$ defined by the inequalities imposed on r and θ; its inverse map is $f(r, \theta) = (r. \cos\theta, r. \sin\theta)$. Then, to say that a function p defined on an open set $G \subset U$ is of class C^1 means that

$$p(x, y) = P(r, \theta)$$

with a C^1 function P defined on the open subset $\varphi(G)$. The differential of f is $(\cos\theta.dr - r\sin\theta.d\theta, \sin\theta.dr + r\cos\theta.d\theta)$ and, for $\xi = (r, \theta)$, the vectors $a_i(\xi)$ are obtained by replacing $(dr, d\theta)$ either by $(1, 0)$ or by $(0, 1)$; hence

$$a_1(\xi) = (\cos\theta, \sin\theta),$$
$$a_2(\xi) = (-r\sin\theta, r\cos\theta).$$

Exercise. In $\mathbb{R}^3 - \{0\}$, use *spherical coordinates* defined by

$$x = r\cos\varphi.\cos\theta, \quad y = r\sin\varphi.\cos\theta, \quad z = r\sin\theta,$$

with $r > 0$, $0 \le \varphi < 2\pi$, $|\theta| \le \pi/2$. (It is not exactly a diffeomorphism onto an open set, but this does not matter). Calculate the $a_i(\xi)$.

The basic idea of classical tensor analysis is to use for all calculations at the point $x = f(\xi)$ what Élie Cartan called a *moving frame* $(a_i(\xi))$, instead of a basis for E chosen once for all. This amounts to regarding a function of $x \in U$ as a function of the corresponding point $\xi = \varphi(x)$ of U' and to calculating in the canonical basis of $\mathbb{R}^n$; we will see (§4) that this is also what we are forced to do in "curved spaces", i.e. the differential manifolds of the modern theory, since they are not contained in a Cartesian space whose basis we could choose.

For example, let us consider a vector $h \in E$ and calculate its components $h^i(\xi)$ with respect to the basis $a_i(\xi) = f'(\xi)e_i$ attached to the point $x = f(\xi)$. Set $\varphi(x) = \varphi^i(x)e_i$, so the coordinates of x in the chart (U, f) are $\xi^i = \varphi^i(x)$; taking (3) and $f'(\xi) = \varphi'(x)^{-1}$ into account,

$$h = f'(\xi)\varphi'(x)h = f'(\xi)d\varphi(x; h) = f'(\xi)d\varphi^i(x; h)e_i = d\varphi^i(x; h)a_i(\xi).$$

This gives the coordinates $h^i(\xi) = d\varphi^i(x; h)$ sought and the relation

(3.4) $$h = d\varphi^i(x; h)a_i(\xi)$$

that Élie Cartan, following Leibniz, used to write as

(3.4') $$dx = a_i(\xi)d\xi^i,$$

the expression $d\xi^i$ in these formulas being the differential of the ith "curvilinear" coordinate of $x \mapsto \xi^i = \varphi^i(x)$ considered a function of x This relation merely expresses the fact that the $a_i(\xi)$ are the partial derivatives $D_i f(\xi)$ with respect to the coordinates ξ^i of the inverse map $x = f(\xi)$ of φ.

In fact, Élie Cartan used more general moving frames than those defined above from local charts; for him, it was merely a basis $(a_i(x))$ of E depending on a point $x \in E$ and whose vectors were functions of x as much differentiable as necessary. We will return to this a bit later in our discussion of differential forms, where it is essential to understand their calculations.

The notion of a local chart allows us to understand Rene Lagrange's (i.e. Ricci's and Levi-Civita's) "tensors", mentioned in n° 1, (ii), and foremost in the case of a Cartesian space E since that is all we know until further notice.

Suppose that there is a tensor field on an open subset X or E, for example a function $T(x; h, k, u)$ that is multilinear in $h, k \in E$ and $u \in E^*$, for all $x \in X$. Consider a chart (U, φ) with $U \subset X$. Write f for the inverse map of φ and for $x = f(\xi)$, let $(a_i(\xi))$ be, as above, the image basis under $f'(\xi)$ of the canonical basis of $\mathbb{R}^n$; write $(a^i(\xi))$ for the dual basis of E^*. For $x = f(\xi) \in U$, let

$$(3.5) \qquad T_{ij}^k(\xi) = T\left[x; a_i(\xi), a_j(\xi), a^k(\xi)\right]$$

be the components of the tensor $\mathsf{l}T(x) : (h, k, u) \mapsto T(x; h, k, u)$ with respect to the basis $(a_i(\xi))$ of E. For the founders of tensor calculus, a tensor was simply a system of components $(T_{ij}^k(\xi))$ attached to each point x and to each local chart. To understand the change of coordinate formulas that these components (5) are subject to, the big question is how to calculate the vectors $a_i(\xi)$ occurring in (5) in terms of similar vectors with respect to another chart. The problem being local, the latter can be assumed to be defined in the open subset U, hence of the form (U, ψ), where ψ is another diffeomorphism from U onto an open subset of $\mathbb{R}^n$. Hence every $x \in U$ has two types of coordinates, namely the points $\xi = \varphi(x)$ and $\eta = \psi(x)$ of $\mathbb{R}^n$, and at each point $x \in U$ there are two frames; write

$$(3.3") \qquad a_i(\xi) = f'(\xi)e_i \quad \text{and} \quad b_\alpha(\eta) = g'(\eta)e_\alpha$$

for these moving frames, using Roman (resp. Greek) indices in the first (resp. second) chart, following the author of *Absolute Differential Calculus*. As φ and ψ are diffeomorphisms from U onto open subsets V and W of $\mathbb{R}^n$, there are pairwise inverse diffeomorphisms

$$\theta : V \longrightarrow W, \quad \rho : W \longrightarrow V$$

such that

$$\psi = \theta \circ \varphi, \quad \varphi = \rho \circ \psi,$$
$$g = f \circ \rho, \quad f = g \circ \theta;$$

this means that the coordinates $\xi^i = \varphi^i(x)$ of a point x in the char (U, φ) and its coordinates $\eta^\alpha = \psi^\alpha(x)$ in the chart (U, ψ) are connected by the formulas

$$\eta^\alpha = \theta^\alpha\left(\xi^1, \ldots, \xi^n\right), \quad \xi^i = \rho^i\left(\eta^1, \ldots, \eta^n\right).$$

The relations

$$\theta'(\xi)e_i = \theta_i^\alpha(\xi)e_\alpha\,, \quad e_\alpha = \rho_\alpha^i(\eta)e_i$$

also hold. Their coefficients

(3.6) $$\theta_i^\alpha(\xi) = D_i\theta^\alpha(\xi)\,, \quad \rho_\alpha^i(\eta) = D_\alpha\rho^i(\eta)$$

are the partial derivatives of the changes of coordinates.
    Having said that,

(3.7) $$\psi'(x) = \theta'(\xi) \circ \varphi'(x)\,, \quad \varphi'(x) = \rho'(\eta) \circ \psi'(x)\,,$$
(3.7') $$g'(\eta) = f'(\xi) \circ \rho'(\eta)\,, \quad f'(\xi) = g'(\eta) \circ \theta'(\xi)\,.$$

Using (3"),

(3.8') $$a_i(\xi) = f'(\xi)e_i = g'(\eta)\theta'(\xi)e_i = g'(\eta)\theta_i^\alpha(\xi)e_\alpha = \theta_i^\alpha(\xi)b_\alpha(\eta)\,,$$

and similarly

(3.8") $$b_\alpha(\eta) = \rho_\alpha^i(\eta)a_i(\xi)\,.$$

It remains to show how the covectors $a^k(\xi)$ occurring in (5) transform. Now, we know that in a vector space, if the basis (a_i) is transformed into to the basis (b_α) by $b_\alpha = c_\alpha^i a_i$, then the dual basis (b^α) is transformed into the dual basis (a^i) by $a^i = c_\alpha^i b^\alpha$. Hence here,

(3.9') $$a^k(\xi) = \rho_\alpha^k(\eta)b^\alpha(\eta)\,,$$
(3.9") $$b^\alpha(\eta) = \theta_k^\alpha(\xi)a^k(\xi)\,.$$

Having done that, transformation formulas for the components (5) of the tensor field T follow immediately by applying relation (1.9) to the trilinear form $T(x)$; clearly,

(3.10) $$T_{\alpha\beta}^\gamma(\eta) = \rho_\alpha^i(\eta)\rho_\beta^j(\eta)\theta_k^\gamma(\xi)T_{ij}^k(\xi)\,.$$

These are the mysterious transformation formulas of tensors into curvilinear coordinates that the founders of the theory used to state without proof. Note that these calculations respect the tensor calculus rules formulated in (ii) of n° 1.

    Exercise. Let F be a C^1 function on E; show that the functions $p_i(\xi) = D_i\{F[f(\xi)]\}$ are the components of a tensor field of type $(0,1)$ by verifying that they satisfy the transformation formula (10) for type $(1,0)$. [The tensor field in question is obviously the function $(x, h) \mapsto dF(x; h)$].

    Note that in these calculations, the fact that the moving frames $(a_i(\xi))$ and $(b_\alpha(\eta))$ are associated to charts is of little importance; in all cases, formulas similar to (8') and (9') and hence to (10) exist, using local charts has no other purpose than to express the coefficients ρ and θ as partial derivatives of the change of curvilinear coordinate formulas.

(iii) *Covariant derivatives on a Cartesian space.* A "differential" or a "covariant derivative" can be assigned to any C^1 tensor field T. It is also written T or, in classical Riemannian geometry, ∇T, where the sign ∇, "nabla", supposed to come from Pharaonic Egyptian, must have been chosen by the inventors of tensor calculus to highlight its esoteric character; this operation takes a vector field of (p, q) to one of type $(p + 1, q)$. When calculating in a chart, a first idea that comes to mind is to go, for example, from a tensor field T of type $(2, 1)$ whose coefficients (5) depend on the coordinates ξ^i of the point x, to a tensor field of type $(3, 1)$ whose coefficients are reportedly the functions $D_i T_{jk}^h(\xi)$, where $D_i = d/d\xi^i$. This is a bad idea because differentiating formulas (10) would give rise to second derivatives of the ξ with respect to the η, which would not lead us to the components of a tensor. It is better, as always, to argue geometrically: since T associates the trilinear function $(k, l, u) \mapsto T(x; k, l, u)$ to every $x \in U$, a quadrilinear function can be deduced by differentiating with respect to x for given k, l, u; this is the *covariant derivative* T' of the tensor field T defined by the formula

$$(3.11) \qquad T'(x; h, k, l, u) = \frac{d}{dt} T(x + th; k, l, u) \quad \text{for } t = 0,$$

$$= d_1 T\left[(x; h), k, l, u\right]$$

in accordance to (2.22), with $h, k, l \in E$ and $u \in E^*$. The result is easily calculated by using a basis (a_i) for E *independent of x* . Indeed, then

$$T(x; k, l, u) = T_{pq}^r(x) k^p l^q u_r$$

with coefficients $T_{pq}^r(x) = T(x; a_p, a_q, a^r)$, and so obviously

$$T'(x; h, k, l, u) = d T_{pq}^r(x; h) k^p l^q u_r = D_i T_{pq}^r(x) h^i k^p l^q u_r \,,$$

where $D_i = d/dx^i$. Therefore, in this case, the coefficients of T' are indeed the partial derivatives of the coefficients of T with respect to the Cartesian coordinates of x. But as we want to use the *variable* basis $(a_i(\xi))$ for all calculations at the point $x = f(\xi)$, we have to argue differently.

Applying the definition of T' for $x = f(\xi)$, $h = a_i(\xi)$, $k = a_p(\xi)$, $l = a_q(\xi)$ and $u = a^r(\xi)$, we need to calculate[20]

$$(3.12) \qquad \nabla_i T_{pq}^r(\xi) = T'\left[x; a_i(\xi), a_p(\xi), a_q(\xi), a^r(\xi)\right] =$$

$$= \frac{d}{dt} T\left[x + t a_i(\xi); a_p(\xi), a_q(\xi), a^r(\xi)\right]$$

---

[20] The traditional notation $\nabla_i T_{pq}^r$ indicates that we can go from the components of T to those of T' by "partial covariant differentiations" similar but not identical to classical partial differentiations with respect to the variables ξ^i.

for $t = 0$. For this, let us differentiate the function

$$T_{pq}^r(\xi) = T\left[f(\xi); a_p(\xi), a_q(\xi), a^r(\xi)\right]$$

with respect to ξ by applying the general formulas (2.25) and (2.27) stated at the end of n° 2. The result is the sum of fours partial differentials with respect to the four variables on which $T(x, h, k, u)$ depends, it being understood that, in these differentials, the functions $f(\xi), a_p(\xi), \ldots$ and their differentials with respect ξ to will have to be substituted to x, h, k, u and to their differentials, as if we had to differentiate the product $f(\xi)a_p(\xi)a_q(\xi)a^r(\xi)$. Differentiating $T(x; k, h, u)$ with respect to x, by definition, we get $d_1T[(x; dx), h, k, u] = T'(x; dx, h, k, u)$ by (11); the first term of the result sought is, therefore,

$$T'\left[f(\xi); df(\xi), a_p(\xi), a_q(\xi), a^r(\xi)\right] = T'\left[f(\xi); f'(\xi)d\xi, a_p(\xi), a_q(\xi), a^r(\xi)\right].$$

In the three other terms, we differentiate a *linear* function; hence it is sufficient to replace the variable in T, for example, $a_p(\xi)$, which depends on ξ, by its differential $da_p(\xi; d\xi)$. Replacing $d\xi$ by the variable $h \in \mathbb{R}^n$ on which the differential of a function of $\xi \in \mathbb{R}^n$ depends, we, therefore, finally get

$$\begin{aligned}
dT_{pq}^r(\xi; h) = {} & T'\left[x; df(\xi; h), a_p(\xi), a_q(\xi), a^r(\xi)\right] + \\
& + T\left[x; da_p(\xi; h), a_q(\xi), a^r(\xi)\right] + \\
& + T\left[x; a_p(\xi), da_q(\xi; h), a^r(\xi)\right] + \\
& + T\left[x; a_p(\xi), a_q(\xi), da^r(\xi; h)\right].
\end{aligned}$$

As we are interested in coefficient (12) of T', in the above, we need to take $h = e_i$ since $df(\xi; h) = a_i(\xi)$ on the left hand side of the previous formula. $da_p(\xi; e_i) = D_i a_p(\xi)$, etc. remain to be expressed in terms of the $a_j(\xi)$ and $a^j(\xi)$ themselves; for this, set

(3.13) $$D_i a_p(\xi) = \Gamma_{ip}^j(\xi)a_j(\xi), \quad D_i a^p(\xi) = \Delta_{ij}^p(\xi)a^j(\xi)$$

with numerical coefficients to be determined, i.e. the *Christoffel symbols*. In view of (12) and the multilinearity of T, we get

$$D_i T_{pq}^r(\xi) = \nabla_i T_{pq}^r(\xi) + \Gamma_{ip}^j(\xi)T_{jq}^r(\xi) + \Gamma_{ir}^j(\xi)T_{pj}^r(\xi) + \Delta_{ij}^r(\xi)T_{pr}^j(\xi),$$

and so, omitting the ξ, the formula

(3.14) $$\nabla_i T_{pq}^r = D_i T_{pq}^r - \Gamma_{ip}^j T_{jq}^r - \Gamma_{iq}^j T_{pj}^r - \Delta_{ij}^r T_{pq}^j.$$

Note that, since $a_j(\xi) = f'(\xi)e_j = D_j f(\xi),$[21]

$$D_i a_j(\xi) = D_i D_j f(\xi) = D_j D_i f(\xi) = D_j a_i(\xi),$$

---

[21] If f is C^2, a harmless assumption in a context limited to quasi-formal calculations.

and so

(3.15) $$\Gamma_{ij}^k = \Gamma_{ji}^k.$$

On the other hand,

$$\Delta_{ij}^p = -\Gamma_{ij}^p$$

for the following reason. Set

$$u(h) = B(h, u) \quad \text{for } h \in E \text{ and } u \in E^*.$$

This gives a bilinear function of h and u, so that

$$D_i\{B\,[h(\xi), u(\xi)]\} = B\,[D_ih(\xi), u(\xi)] + B\,[h(\xi), D_iu(\xi)]$$

for any functions $h(\xi)$ and $u(\xi)$. Since $B(a_j, a^p) = \delta_j^p$ by definition of the dual basis,

$$0 = B\,(D_ia_j, a^p) + B\,(a_j, D_ia^p) = \Gamma_{ij}^q B\,(a_q, a^p) + \Delta_{ik}^p B\,(a_j, a^k)$$
$$= \Gamma_{ij}^q \delta_q^p + \Delta_{ik}^p \delta_j^k = \Gamma_{ij}^p + \Delta_{ij}^p$$

as announced. Hence the final formula

(3.16) $$\nabla_i T_{pq}^r = D_i T_{pq}^r - \Gamma_{ip}^j T_{jq}^r - \Gamma_{iq}^j T_{pj}^r + \Gamma_{ij}^r T_{pq}^j,$$

where $D_i = d/d\xi^i$. These are the famous Christoffel formulas (1869), which generalize in an obvious way to tensor fields of arbitrary type.

Exercise. Are the Γ_{ij}^k components of a tensor field?

Index calculators did not leave it at that for their aim was differential geometry on "non-Euclidean curved" spaces, and in particular on vector spaces where distances are computed by non-Euclidean formulas. Using their terminology, and as it has already been alluded to in n° 1, (ii), suppose that at every point of E there is simple formula for calculating the length ds of the vector connecting some $x \in E$ to an infinitesimally near point $x + dx$. Since in an infinitesimal neighbourhood of x we want the geometry of the space to be, as a first approximation, like that of an Euclidean space, we take the square ds^2 to be a quadratic form

(3.17) $$ds^2 = g_{ij}(\xi)d\xi^i d\xi^j$$

in the coordinates $d\xi^i$ of dx, with a function $g_{ij} = g_{ji}$ depending on the point considered; the right hand side must obviously always be > 0 for $dx \neq 0$, in other words

(3.18) $$g_{ij}(\xi)h^i h^j > 0$$

for any non-zero scalars h^i. As all this being irrelevant if ds^2 depends on the chosen coordinate system, the $g_{ij}(\xi)$ must be the components in the latter of a tensor field of type $(2,0)$ on E and given by

$$(3.19) \qquad g(x; h, k) = g_{ij}(\xi) h^i k^j \text{ if } h = h^i a_i(\xi), \quad k = k^j a_j(\xi),$$

which implies transformation formulas for the g_{ij} similar to (10) when changing curvilinear coordinates; (19) can be interpreted as a Hilbert scalar product depending on x and applying to vectors with starting point x.

In the simplest case, namely that of traditional Euclidean geometry, let us once for all fix a scalar product denoted by $(\,|\,)$; then $g(x; h, k) = (h|k)$ for $x, h, k \in E$; in the local chart (U, φ), the components of this tensor field at the point $x = f(\xi)$ are the functions

$$(3.20) \qquad g_{ij}(\xi) = \Big(a_i(\xi) | a_j(\xi) \Big)$$

and ds^2 is given by the formula

$$ds^2 = \big(a_i(\xi) d\xi^i | a_j(\xi) d\xi^j \big) = g_{ij}(\xi) d\xi^i d\xi^j .$$

The Christoffel symbols Γ_{ij}^k can then be calculated by using the g_{ij}. To do this, differentiate g_{jk} with respect to ξ^i. Writing $a_i(\xi)$ as a_i for short,

$$D_i g_{jk} = (D_i a_j | a_k) + (a_j | D_i a_k) = \Gamma_{ij}^p (a_p | a_k) + \Gamma_{ik}^p (a_j | a_p) =$$
$$= \Gamma_{ij}^p g_{pk} + \Gamma_{ik}^p g_{pj} = \Gamma_{ijk} + \Gamma_{ikj}$$

where we set

$$(3.21) \qquad \Gamma_{ijk} = g_{pk} \Gamma_{ij}^p = \Gamma_{jik} .$$

An an aside, note that this calculation says that the covariant derivative

$$\nabla_i g_{jk} = D_i g_{jk} - \Gamma_{ik}^p g_{jp} - \Gamma_{ij}^p g_{pk}$$

of the tensor field g is zero; this comes as little surprise since

$$g'(x; h, k, l) = \frac{d}{dt} g(x + th; k, l) = \frac{d}{dt} (k|l) \quad \text{for } t = 0$$

is the derivative of a function independent of t.

Let us then write the relations

$$\Gamma_{ijk} + \Gamma_{ikj} = D_i g_{jk}, \ \Gamma_{jki} + \Gamma_{jik} = D_j g_{ik}, \ \Gamma_{kij} + \Gamma_{kji} = D_k g_{ij};$$

by adding the first two and subtracting the third, by (21),

$$(3.22) \qquad \Gamma_{ijk} = \frac{1}{2} \big(D_i g_{jk} + D_j g_{ik} - D_k g_{ij} \big)$$

immediately follows. This is a well-known formula allowing the Γ_{ijk} to be calculated in terms of the derivatives of the g_{ij}; by (21), we go from this to the Γ_{ij}^k by inverting the matrix with entries g_{ij}.

Ricci and his assistant Levi-Civita (who was later to be one of the big specialists of fluid mechanics) had explained this and many other things in the more subtle case of arbitrary ds^2, in particular in a long memoir in French published in the famous German journal (*Méthodes du calcul différentiel absolu*, Math. Annalen, 1901) – an excellent example of scientific internationalism at a time when saber rattlers, as they were called, held sway everywhere. Albert Einstein had to assimilate everything to build his general relativity about a decade later; not being as great a virtuoso of tensor calculus as the Italians, he encountered some problems with them.[22] The subsequent development of Riemannian geometry has made it possible to get rid of a large part of these coordinate calculations from the theory – though this has not made it easier to understand... –, but when he wanted to deduce that light rays were deviated in the neighbourhood of an intense gravitational field, Einstein was obliged to make everything explicit to obtain numerical results, verifiable by experiments, a task that mathematicians are generally exempted from. Finally, as in many other fields, the modern theory arrived after the "profusion of indices" which, for want of better, can still serve as exercises for the reader and from which Dieudonné himself did not fully escape at the end of vol. 3 of his *Eléments d'analyse*.

---

[22] For a history of the subject, see Karin Reich, *Die Entwicklung des Tensorkalküls. Vom absoluten Differentialkalkül zur Relativitätstheorie* (Birkhauser, 1989); a chapter on the subject can also be found in T. Hawkins, *Emergence of the Theory of Lie Groups. An Essay in the History of Mathematics 1869–1926* (Springer-Verlag, 2000), which covers a far wider field. I forego citing modern presentations of Riemmanian geometry; there are too many and they do not throw any light on the history of the subject.

§ 2. Differential Forms of Degree 1

4 – Differential Forms of Degree 1

As seen in Chapter VIII, finding a primitive F of a holomorphic function f on an open set $U \subset \mathbb{C} = \mathbb{R}^2$ amounts to constructing a C^1 function in U such that $dF = f(z)dz = f(z)dx + if(z)dy$, in other words such that

$$D_1 F = f, \quad D_2 F = if.$$

A somewhat more general problem is to write as dF an expression like

$$(4.1) \qquad\qquad \omega = p(x,y)dx + q(x,y)dy,$$

i.e. a *differential form* of degree 1 on U, where p and q, its *coefficients*, are given functions; the coefficients are assumed to be always continuous and in order to avoid serious complications, it is better to assume they are of class C^1 at least on U.

The aim is, therefore, to find C^1 functions F on U satisfying

$$(4.2) \qquad\qquad D_1 F = p, \quad D_2 F = q.$$

If such a *primitive* F of ω exists in U, ω is said to be an *exact* diffferential; if U is connected, F is unique up to an additive constant (Chap. III, n° 21, consequence of the mean value theorem for several variables).

If p and q are C^1, in which case F is C^2, the relation $D_2 D_1 F = D_1 D_2 F$ requires

$$(4.3) \qquad\qquad D_1 q = D_2 p;$$

if this necessary but not always sufficient condition is satisfied, $pdx + qdy$ is said to be a *closed* differential. This terminology is inherited from algebraic topology and "Stokes" type integration formulas. In the holomorphic case $(p = f, q = if)$, (3) is just Cauchy's holomorphic condition.

Another case: it was observed in Chap. VII, n° 24 that if H is a real harmonic function on U, finding a holomorphic function with real part H amounts to finding a primitive of the holomorphic function $D_1 H - iD_2 H$, hence of the differential form

$$(4.4) \qquad (D_1 H - iD_2 H)(dx + idy) = dH - i(D_2 H dx - D_1 H dy),$$

hence of the differential form $D_2 H dx - D_1 H dy$; the latter is closed since, by assumption, $\Delta H = 0$, where $\Delta = D_1 D_1 + D_2 D_2$ is the de Laplace operator $d^2/dx^2 + d^2/dy^2$.

There are similar problems in arbitrary dimension n. A differential form (of degree 1) on an open subset U of an n-dimensional space E with chosen basis (a_i) is for the moment a purely symbolic expression

$$(4.5) \qquad \omega = p_1(x)dx^1 \ldots + p_n(x)dx^n = p_i dx^i \,,$$

where the p_i are given possibly complex or even vector-valued functions of class C^1 at least on U, and where the x^i are the coordinates of $x \in U$ with respect to the given basis of E; this definition has no "absolute" meaning if the changes undergone by the p_i under a change of basis is not made clear, but a direct definition of differential forms can be given, which gets rid of this problem.

Indeed, if there a function F of class at least C^1 on U, its differential is a function $dF(x; h)$ of a point $x \in U$ and of a vector $h \in E$. For fixed x, this function is linear in h. The natural generalization is therefore to consider functions $w(x; h)$ subject to the same conditions, in other words tensor fields of type $(0, 1)$ or – what amounts to the same since a linear functional

$$\omega(x) : h \longmapsto \omega(x; h)$$

on E, i.e. an element of E^* is associated to each $x \in U$ – maps from U to E^* (or to $E_{\mathbb{C}}^*$ since complex valued forms are also considered). Conversely, once a basis for E is chosen, thanks to this definition, any map $w : U \longrightarrow E^*$ can be written in the form (5). To see this, first write that

$$(4.6) \ \omega(x; h) = \omega\left(x; h^i a_i\right) = h^i \omega\left(x; a_i\right) = p_i(x) h^i \ \text{where} \ p_i(x) = \omega(x; a_i) \,.$$

The expression $p_i(x) dx^i$ is then justified by precisely the same reasons as those given at the end of section (i) of n° 2: write that $h^i = dx^i(x; h)$. Conversely, it suffices to associate the function

$$(4.7) \qquad \omega(x; h) = p_i(x) h^i \quad (x \in U \,, \ h \in E)$$

to (5) to reduce to the new definition.

As in n° 3, (ii), w could be written in an arbitrary local chart (U, φ). The latter defines a basis $(a_i(\xi))$ of E at each $x \in U$, where $\xi = \varphi(x)$; as seen in (3.4), any vector $h \in E$ can then be written

$$h = h^i(\xi) a_i(\xi) = a_i(\xi) d\xi^i(x; h)$$

with coordinates depending both on the chart and the point x considered. Hence

$$\omega(x; h) = p_i(\xi) h^i(\xi) = p_i(\xi) d\xi^i(x; h) \,,$$

where the $p_i(x) = \omega[x; a_i(\xi)]$ are the components of the tensor field w in the chart considered; so, following Leibniz, the expression for w in the chart (U, φ) is

$$(4.8) \qquad \omega(x; dx) = p_i d\xi^i \,.$$

Under chart change, the coefficients $p_i(\xi)$ are transformed like those of a tensor of type $(0, 1)$:

$$(4.9) \qquad\qquad p_\alpha(\eta) = \rho^i_\alpha(\eta) p_i(\xi) \,,$$

where the derivative $\rho^i_\alpha(\eta) = d\xi^i/d\eta^\alpha$ is calculated at the point $\eta(x)$. We only mostly argue in terms of standard Cartesian coordinate, but using arbitrary local chart cannot be avoided when extending the theory to differentiable manifolds.

Finally note the similarity – and not the identity – between vector fields and differential forms: a vector field on an open set $U \subset E$ associates to each $x \in U$ a vector of E (i.e. is a map from U to E), whereas a differential form associates to each $x \in U$ a covector of E (i.e. is a map from U to the possibly complex dual E^* of E). This is the difference between tensor fields of type $(1,0)$ and $(0,1)$.

5 – Local Primitives

(i) *Existence: calculations in terms of coordinates.* Having said that, and considering a Cartesian space E equipped with a basis (a_i), let us return to the search for a primitive F of a differential form $\omega = p_i(x)dx^i$ of class C^1. We will consider an open connected subset G, i.e. a domain, as the general case can be reduced to it in an obvious way. In all that follows, x^i will denoted the coordinates of a point or a vector with respect to the basis taken and we set $D_i = d/dx^i$ as usual.

Relations

$$(5.1) \qquad\qquad D_i F = p_i \quad \text{for} \quad 1 \le i \le n$$

obviously require

$$(5.2) \qquad\qquad D_j p_i - D_i p_j = 0$$

for all i and j, which in reality gives $n(n-1)/2$ independent relations, for example those for which $i > j$; if these necessary conditions hold, ω is said to be *closed*.

Exercise. Show that this definition is independent of the choice of the basis (a_i). Does relation (2) hold in an arbitrary local chart?

In the case of holomorphic functions, we know there are always *local* primitives; this remains true in the general case, a result already known to Euler for two variable forms. It can be proved as in Chapter VIII, n° 2, (i), formula (2), though not quite so easily.

First, let F be a C^2 function defined at least on the ball $B : \|x\| < R$ centered at O in E [to argue in the neighbourhood of an arbitrary point a, consider $F(a+x)$]. For given $x \in B$, the function $t \mapsto F(tx)$ is defined on an open subset of $\mathbb{R}$ containing $I = [0,1]$. Setting $D = d/dt$, the multivariate chain rule shows that

$$(5.3) \qquad\qquad D\{F(tx)\} = D_i F(tx).x^i \,.$$

The FT then shows that

$$(5.4) \qquad F(x) = F(0) + \int_0^1 D_i F(tx).x^i dt \, .$$

Hence, if a closed differential $\omega = p_i(x)dx^i$ with coefficients of class C^1 – or even C^0, but the following arguments fall apart in this case – admits a primitive F on the ball B where it is defined, then for all $x \in B$, up to an additive constant

$$(5.5) \qquad F(x) = \int_0^1 p_i(tx)x^i dt = \int_0^1 \omega(tx; x)dt$$

necessarily. $D_j F = p_j$ for all j remains to be check that. For this, the right hand side of (5) needs to be differentiated with respect to x^j. Now, the p_i being C^1, the function de $(t, x) \in I \times G$ being integrated has continuous first order partial derivatives with respect to the x^i, and so

$$(5.6) \qquad D_j F(x) = \int_0^1 D_j \left\{ p_i(tx)x^i \right\} dt =$$

$$= \int_0^1 D_j \left\{ p_i(tx) \right\} x^i dt + \int_0^1 p_i(tx) D_j x^i dt =$$

$$= \int_0^1 D_j p_i(tx).tx^i dt + \int_0^1 p_i(tx) \delta_j^i dt =$$

$$= \int_0^1 D_j p_i(tx).tx^i dt + \int_0^1 p_j(tx)dt \, ,$$

where $\delta_j^i = D_j x^i$. The Kronecker delta already mentioned in (1.8), is equal to 1 or 0 according to whether i and j are equal or not. But, by (2) and by (2.20) applied to p_j

$$(5.7) \qquad D_j p_i(tx).tx^i = D_i p_j(tx).tx^i = tD \left\{ p_j(tx) \right\} \, .$$

As $D = d/dt$, integration by parts is possible, and so

$$D_j F(x) = tp_j(tx) \Big|_0^1 - \int_0^1 p_j(tx)dt + \int_0^1 p_j(tx)dt = p_j(x) \, ,$$

which solves the problem.

This calculation applies to any star-shaped open set G, i.e. in which there is some $a \in G$ such that, for all $x \in G$, the line segment $[a, x]$ is contained in G. In a star-shaped open subset[23] of a Cartesian space, all closed differential forms of class C^1 have a primitive, i.e. are exact. In particular, in a star domain $G \subset \mathbb{C}$, any real harmonic function is the real part of a holomorphic function on G – a result already shown in Chapter VIII –, namely

---

[23] or simply connected one, as will be shown later.

$$(5.8) \qquad f(z) = H(x,y) - i \int_0^1 [D_2 H(tx, ty)x - D_1 H(tx, ty)y]\, dt$$

assuming G is star-shaped with respect to the point $a = 0$: it suffices to apply (4) to the form (4.4).

Exercise. Check directly that function (8) is holomorphic by giving an explicit proof of the general theorem.

(ii) *Existence of local primitives: intrinsic formulas.* We show how to obtain the previous result without using coordinates. This method is far less stupid than the first one, though it is only a camouflage, but it has the advantage of holding for differential forms on Banach spaces.[24] Whether it is easier to understand than the former one is left to the reader to judge.

Let us start again with a closed differential form ω of class C^1 on an open set G star-shaped with respect to the point 0 and suppose that it admits a primitive F on G with $F(0) = 0$. For all $x \in G$, the derivative of $t \mapsto F(tx)$ at the point t is the derivative of $s \mapsto F[(t+s)x] = F(tx + sx)$ at $s = 0$; it is, therefore, $dF(tx; x) = \omega(tx; x)$. Then, as above, the FT shows that

$$(5.9) \qquad F(x) = \int_0^1 \omega(tx; x)\, dt\,.$$

So it all amounts to verifying that, if ω is closed, the formula does indeed define a function such that $dF = \omega$.

Since $dF(x; h)$ is the derivative of $s \mapsto F(x + sh)$ at $s = 0$, $dF(x; h)$ is obtained by taking $s = 0$ in the following calculation:

$$(5.10) \quad dF(x; h) = \frac{d}{ds} \int_0^1 \omega(tx + tsh; x + sh)\, dt =$$

$$= \frac{d}{ds} \int_0^1 \omega(tx + tsh; x)\, dt + \frac{d}{ds} s \int_0^1 \omega(tx + tsh; h)\, dt\,,$$

where linearity of $h \mapsto \omega(y; h)$ for given y has been used. Differentiating under the $\int$ sign raises no difficulty. To conveniently formulate the result, the covariant derivative [see (3.11)]

$$(5.11) \qquad \omega'(x; h, k) = \frac{d}{ds}\omega(x + sh; k)\Big|_{s=0} = D_i p_j(x) h^i k^j$$

proves useful if $\omega = p_i(x)dx^i$. Having done that, let us return to the last two integrals of (10) and differentiate under the $\int$ sign; taking (11) into account and setting $s = 0$ in the result,[25]

$$(5.12) \qquad \int_0^1 \omega'(tx; h, x)t\, dt + \int_0^1 \omega(tx; h)\, dt\,.$$

---

[24] See Henri Cartan, *Calcul différentiel* (Hermann).
[25] The derivative of a function of the form $sf(s)$ at $s = 0$ is equal to $f(0)$.

Integrating by parts the second term, it becomes

$$\omega(tx; h)t\Big|_0^1 - \int_0^1 \frac{d}{dt}[\omega(tx; h)].t\,dt = \omega(x; h) - \int_0^1 \dots .$$

But

$$\frac{d}{dt}\omega(tx; h) = \frac{d}{ds}\omega(tx + sx; h)\Big|_{s=0} = \omega'(tx; x, h),$$

so that finally,

(5.13) $dF(x; h) = \omega(x; h) + \displaystyle\int_0^1 [\omega'(tx; h, x) - \omega'(tx; x, h)]\,dt .$

This involves what is called the *exterior derivative*

(5.14) $d\omega(x; h, k) = \omega'(x; h, k) - \omega'(x; k, h)$

of the form ω; for given x, it is an alternating or antisymmetric bilinear form of the vectors h, k. Without going further into this topic which we will return to later, we find that the formula

(5.13') $dF(x; h) = \omega(x; h) - \displaystyle\int_0^1 d\omega(tx; x, h)\,dt$

holds without any assumptions on ω. On the other hand, setting

$$\omega(x; h) = p_i(x)h^i ,$$

it is easy to first calculate

$$\omega'(x; h, k) = dp_j(x; h)k^j = D_i p_j(x)h^i k^j ,$$
$$\omega'(x; k, h) = dp_i(x; k)h^i = D_j p_i(x)h^i k^j ,$$

and then

(5.15) $d\omega(x; h, k) = (D_i p_j - D_j p_i)\, h^i k^j .$

This formula shows that

(5.16) ω is closed $\iff d\omega = 0 .$

Relation (13') then reduces to

$$dF(x; h) = \omega(x; h) ,$$

ending the proof.
    Relation

$$\omega'(x; h, k) = \omega'(x; k, h)$$

is still well-defined in Banach spaces where calculations in terms of coordinates is no longer possible; it then serves as a definition for closed forms.
    Exercise. Suppose $\omega = dF$; calculate $\omega'(x; h, k)$ directly, without coordinates, and show that (16) reduces to the formula $d^2/dsdt = d^2/dtds$.

6 – Integration Along a Path. Inverse Images

(i) *Integrals of a differential form.* Everything that has been said in Chapter VIII about integrals of holomorphic functions generalizes, with some small adjustments, to differential forms. So we will remain brief.

Consider a differential form $\omega = p_i(x)dx^i$ of degree 1 on a domain G in a finite-dimensional vector space E and assume it has a primitive F on G. Connect $a \in G$ to $b \in G$ by a *path* $\mu : [0,1] = I \longrightarrow G$ of class C^1 or, more generally, an *admissible* path or *of class* $C^{1/2}$, i.e. such that the coordinates $\mu^i(t)$ of $\mu(t)$ are primitives of regulated functions. Writing D for the differentiation operator with respect to t, the multivariate chain rule (2.20) shows that outside a countable set of values of t where $\mu(t)$ is not differentiable (i.e. has different right and left derivatives),

$$(6.1) \ D\left\{F\left[\mu(t)\right]\right\} = dF\left[\mu(t); \mu'(t)\right] = D_iF\left[\mu(t)\right].D\mu^i(t) = p_i\left[\mu(t)\right]D\mu^i(t),$$

where $p_i = D_iF$. The result is a regulated function of t since the $p_i[\mu(t)]$ are continuous and the $D\mu^i(t)$ are regulated; hence (FT)

$$(6.2) \qquad F(b) - F(a) = \int_I p_i\left[\mu(t)\right]D\mu^i(t)dt = \int_I \omega\left[\mu(t); \mu'(t)\right]dt.$$

The right hand side of (2), which is well-defined for any form ω, is by definition the *integral of ω along μ*, denoted by any one of the following

$$(6.3) \qquad \int_\mu \omega = \int_\mu p_i(x)dx^i = \int_I \omega\left[\mu(t), \mu'(t)\right]dt = \int_I p_i\left[\mu(t)\right]d\mu^i(t)$$

with, finally, Stieltjes integrals as in Chapter VIII, n° 2, (iii).

As before, these calculations show that a closed form of class C^1 has a primitive on G if and only if its integral along a path does not depend on its endpoints or, equivalently, that its integral along any closed path is zero. In fact, the result can be generalized to all forms of class C^0 in the following manner.

Assume that the stated condition holds; then for any $a, b \in G$, we can write

$$\int_a^b \omega$$

for the integral of ω along any path connecting a to b in G; there is no ambiguity. The traditional formula (Chap. V, §3, n° 12)

$$\int_a^b = \int_a^c + \int_c^b$$

continues to hold since it is possible to go from a to b through c... Having said this, set

$$F(x) = \int_a^x \omega\,,$$

where $a \in G$ is chosen once for all. Then, for given x and sufficiently small h,

$$F(x+h) - F(x) = \int_x^{x+h} \omega = \int_0^1 \omega(x+th;h)dt = h^i \int_0^1 p_i(x+th)dt$$

follows. This can be seen by integrating along the line segment $[x, x+h]$ in a disc centered at x contained in G. But as ω is of class C^0, for given x and for any $r > 0$, there is $r' > 0$ such that $\|h\| < r'$ implies $|\omega(x+th;h) - \omega(x;h)| < r\|h\|$ for all $t \in [0,1]$, and so

$$F(x+h) - F(x) = \omega(x;h) + o(h)$$

and $dF(x;h) = \omega(x;h)$, qed. In conclusion:

Theorem 2. *For any differential form ω of degree 1 and class C^0 on a domain G, the following conditions are equivalent:*

(i)   *ω is an exact differential;*
(ii) *the integral of ω along any admissible path in G does not depend on its endpoints;*
(iii) *the integral of ω along any closed admissible path in G is zero.*

The comments of Chapter VIII, n° 2, (ii) on changes of parametrization for a path, "opposite" paths, additivity of the integral when two paths are adjoined, etc. can be made to apply without any changes to the general case.

(ii) *Inverse image of a differential form.* It is more useful to observe that formula (3) defining a curvilinear integral suggests an operation on differential forms generalizing the composition of maps and which will be later generalized to forms of arbitrary degree. For this, consider the open subsets U and V in the Cartesian spaces E and F and a map $f : U \longrightarrow V$. It associates to each function q on V, a composite function $p = q \circ f$ on U, given by

$$p(x) = q\,[f(x)]\,.$$

If q and f are of class C^1, in which case the same holds for p, the differentials $\omega = dp$ and $\varpi = dq$ of p andt q are connected by the multivariate chain rule, which becomes here

(6.4) $\omega(x;h) = \varpi\,[f(x); f'(x)h]\,.$

Now, the right hand side of (4) is well-defined for *any* form ϖ on V and, for given x, is a linear map of $f'(x)h$ and hence of h. So by relation (4), a differential form ω on U can be deduced from ϖ and f, the *inverse image*[26] *of ϖ under f*, which could also be called the composition of ω and f; the nature of this operation is imposed by the necessity of making it well-defined. Some authors denote it by $f^*(\varpi)$ or $^tf(\varpi)$; this barbarian notation presents occasional advantages, but no one has ever denoted the composite $q \circ f$ of two functions by f^*q or $^tf(q)$. So we will denote the inverse image by the notation

$$(6.5) \qquad \varpi \circ f : (x; h) \longmapsto \varpi\,[f(x); f'(x)h]$$

which, in Leibniz style, can be written

$$(6.6) \qquad \varpi \circ f(x; dx) = \varpi\,[f(x); df(x)] = \varpi\,[f(x); f'(x)dx]$$

in a similar way as the differentiation theorem of composite functions; conversely, it can now be written as

$$(6.7) \qquad dg \circ f = d(g \circ f)\,.$$

A curvilinear integral can then be interpreted in the following manner. Thanks to the path $\mu : I \longrightarrow E$, ω is transformed into a differential form $\omega \circ \mu$ on I, given more explicitly by

$$\omega \circ \mu(t; dt) = \omega\,[\mu(t); \mu'(t)]\,dt\,.$$

This is precisely the expression integrated over I in order to integrate ω along μ. In dimension one, any differential form can be written $p(t)dt$ with a function $p(t)$, and the extended integral of p over I is just the integral of the form $p(t)dt$ along the rather commonplace path $t \mapsto t$. Hence, in Leibniz style, we get the formula

$$(6.8) \qquad \int_\mu \omega(x; dx) = \int_I \omega \circ \mu(t; dt)\,.$$

Now consider three open subsets U, V and W in the Cartesian spaces E, F and G and two maps $f : U \longrightarrow V$ and $g : V \longrightarrow W$. So there is a composite map $g \circ f : U \longrightarrow W$. Given a form ω on W, a form $\omega \circ g$ on V, then a form $(\omega \circ g) \circ f$ on U can be successively defined or, a form $\omega \circ (g \circ f)$ on U can be directly defined. Like in set theory, in this case,

$$(h \circ g) \circ f = h \circ (g \circ f)$$

trivially holds. We have here an associativity formula

$$(6.9) \qquad (\omega \circ g) \circ f = \omega \circ (g \circ f)\,.$$

---

[26] The "direct" image would consist in deducing from f and from a form ω on U a form ϖ on V. This is impossible to do if f is not a diffeomorphism.

Indeed, start with some $x \in U$. The right hand side can be obtained at this point from $\omega(z; dz)$ by replacing z by $g \circ f(x) = g[f(x)]$ and dz by

$$d(g \circ f)(x; dx) = dg\,[f(x); f'(x)dx]$$

in it. Therefore, the right hand side can be calculated by the following operations: first replace z by $g(y)$ and dz by $dg(y; dy)$, which replaces ω by $\omega \circ g(y; dy)$, then replace y by $f(x)$ and dy by $f'(x)dx$, which replaces $\omega \circ g(y; dy)$ by $(\omega \circ g) \circ f(x; dx)$, giving (9).

In particular, we may suppose that[27] $U = I$, so that f is a path μ in V and $g \circ \mu$ a path in W. Integrating, (8) and (9) immediately show that

$$(6.10) \qquad \int_{g \circ \mu} \omega = \int_{\mu} \omega \circ g.$$

If V is also an interval in $\mathbb{R}$, we recover the fact that, within reasonable conditions, the value of a curvilinear integral is independent of the chosen parametrization.

7 – Effect of a Homotopy on an Integral

(i) *Differentiation with respect to a path.* Like in the very particular case of holomorphic functions discussed in the previous chapter, the fundamental property of the integral of a *closed* (but not necessarily exact) differential form consists in not changing when deforming the path of integration without changing its endpoints, i.e. by a fixed-endpoint homotopy. Here too, the arguments of Chapter VIII provide the method and the results.

Before extending what has been proved in Chapter VIII to differential forms, the differentiation formula with respect to the integration path of Chapter VIII, n° 3, (ii) must first be generalized. In other words, consider an admissible path $\mu : I = [0, 1] \longrightarrow G$ in the open subset G of the Cartesian space E, and an admissible path $\nu : I \longrightarrow E$ in E and differentiate the expression

$$(7.1) \qquad F(\mu + s\nu) = \int_{\mu + s\nu} \omega = \int \omega\,[\mu(t) + s\nu(t); \mu'(t) + s\nu'(t)]\,dt$$

with respect to s, where s is a parameter varying in a sufficiently small interval around 0. But as shorthand for (1), write

$$(7.1') \qquad F(\mu + s\nu) = \int \omega\,(\mu + s\nu; \mu')\,dt + s \int \omega\,(\mu + s\nu; \nu')\,dt$$

$$= \int \omega\,(\mu + s\nu; d\mu) + s \int \omega\,(\mu + s\nu; d\nu)$$

or in coordinates,

---

[27] The fact that I is not open is not important.

$$(7.1") \qquad F(\mu + s\nu) = \int p_i(\mu + s\nu)d\mu^i + s \int p_i(\mu + s\nu)d\nu^i,$$

where the components of the "vector" measures $d\mu$ and $d\nu$, $d\mu^i(t) = D\mu^i(t)dt$ and $d\nu^i(t) = D\nu^i(t)dt$ are the Radon measures. As in Chapter VIII, n° 3, (ii), to justify differentiation under the $\int$ sign, it suffices to check that

(a)   the functions

$$(7.2') \qquad\qquad (s,t) \longmapsto p_i\left[\mu(t) + s\nu(t)\right]$$

   or, equivalently,

$$(7.2") \qquad\qquad (s,t) \longmapsto \omega\left[\mu(t) + s\nu(t); h\right]$$

   are continuous for given h,
(b)   their derivatives with respect to s exist and are continuous functions of (s, t).

The first statement is obvious. Differentiability with respect to s is equally obvious since C^1 functions are being composed with a linear function of s. As for the derivative of $(2")$ with respect to s, by definition (5.11) of a covariant derivative, it is $\omega'[\mu(t) + s\nu(t); \nu(t), h]$.

Continuing to write $\mu, \nu, \ldots$ instead of $\mu(t), \nu(t), \ldots$,

$$(7.3) \quad \frac{d}{ds}F(\mu + s\nu) = \int \omega'\left(\mu + s\nu; \nu, \mu'\right) dt + \int \omega\left(\mu + s\nu; \nu'\right) dt +$$

$$+s \int \omega'\left(\mu + s\nu; \nu, \nu'\right) dt =$$

$$= \int \omega'\left(\mu + s\nu; \nu, \mu' + s\nu'\right) dt + \int \omega\left(\mu + s\nu; \nu'\right) dt$$

thus follow. To imitate the calculations of n° 5 or better those of Chapter VIII, n° 3, we now need to apply the integration by parts formula to the last integral. For given x, the expression $\omega(x; h)$ being a linear function of h, it is identical to its differential with respect to h, so that

$$\frac{d}{dt}\omega\left[x; f(t)\right] = \omega\left[x; f'(t)\right]$$

for any vector-valued function $f(t)$. Given this result, the multivariate chain rule and (5.11) show that

$$\frac{d}{dt}\omega\left(\mu + s\nu; \nu\right) = \omega'\left(\mu + s\nu; \mu' + s\nu', \nu\right) + \omega\left(\mu + s\nu; \nu'\right).$$

As a result, the last integral obtained in (3) can also be written

$$\int \omega\left(\mu + s\nu; \nu'\right) dt = \omega\left(\mu + s\nu; \nu\right)\Big|_{t=0}^{t=1} - \int \omega'\left(\mu + s\nu; \mu' + s\nu', \nu\right) dt.$$

Substituting this result in (3), it follows that

$$\frac{d}{ds}F(\mu + s\nu) = \int_0^1 \{\omega'(\mu + s\nu; \nu, \mu' + s\nu') - \omega'(\mu + s\nu; \mu' + s\nu', \nu)\}\, dt +$$

$$+ \omega(\mu + s\nu; \nu)\Big|_{t=0}^{t=1}.$$

Hence, using the exterior derivative

$$d\omega(x; h, k) = \omega'(x; h, k) - \omega'(x; k, h)$$

introduced in (5.14), we get

$$(7.4)\quad \frac{d}{ds}F(\mu + s\nu) = \omega(\mu + s\nu; \nu)\Big|_{t=0}^{t=1} + \int_0^1 d\omega\,(\mu + s\nu; \nu, \mu' + s\nu')\, dt\,.$$

If the form ω is *closed*, then $d\omega = 0$ as already seen in n° 5 and (4) follows from the relation

$$(7.4')\qquad\qquad \frac{d}{ds}F(\mu + s\nu) = \omega\left[\mu(t) + s\nu(t); \nu(t)\right]\Big|_{t=0}^{t=1}$$

which generalizes formula (3.5) of Chapter VIII.

(ii) *Effect of a homotopy on an integral.* The same consequence as in the previous chapter follow from (4'). If in the open set G where ω is defined, there are two sufficiently near admissible paths μ_0 and μ_1 for the path

$$t \longmapsto (1-s)\mu_0(t) + s\mu_1(t) = \mu_0(t) + s\left[\mu_1(t) - \mu_0(t)\right]$$

to be in G for all $s \in [0,1]$, then formula (4') applies for $\mu = \mu_0$ and $\nu = \mu_1 - \mu_0$. The function $F(\mu + s\nu)$ is, therefore, constant, in both usual cases: μ_0 and μ_1 have the same endpoints or else are both closed. In the general case of two homotopic paths, as in Chapter VIII, replace the given homotopy by a succession of linear homotopies. This gives the result:

Theorem 3. *Let G be a domain in a Cartesian space E and ω a closed differential form of class C^1 on G. The integrals of ω along the two admissible paths μ_0 and μ_1 in G are equal if one of the following conditions holds:*

(a)   *There is a fixed-endpoint homotopy from μ_0 to μ_1 on G;*
(b)   *μ_0 and μ_1 are closed homotopic as closed paths in G.*

From this, we can deduce that if all closed paths in G are homotopic to a constant path, then any closed form on G has a primitive; a domain with this property is said to be *simply connected*.

Homotopy being an equivalence relation, it is then obvious that condition (b) of the theorem always holds. The same is true for condition (a); to see this, start with the closed path $\nu_0 = \mu_0 - \mu_1$. Assuming μ_0 and μ_1 to be parameterized by the interval $[0, 1]$, it can be defined by the formulas

$$t \longmapsto \mu_0(t) \text{ if } 0 \leq t \leq 1, \quad t \longmapsto \mu_1(2 - t) \text{ if } 1 \leq t \leq 2,$$

where the parameter t now varies in $[0, 2]$. Let

$$\sigma : [0, 1] \times [0, 2] \longrightarrow G$$

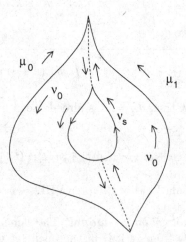

Fig. 7.1.

be a deformation of ν_0 at all times during which the path remains closed, and, for $s = 1$, reduces to a point c. Adjoining to each intermediary path ν_s the paths followed by the endpoints of μ_0 (or μ_1), ν_s can be replaced by the difference between two paths with the same endpoints as μ_0 and μ_1; if we insist on formulas, the first one (which must reduce to μ_0 for $s = 0$) may be defined by

$$
\begin{array}{lll}
& \sigma(3su, 0) & \text{for } 0 \leq u \leq 1/3, \\
(7.5') \qquad u \longmapsto & \sigma(s, 3u - 1) & \text{for } 1/3 \leq u \leq 2/3, \\
& \sigma\,[3s(1 - u), 1] & \text{for } 1/3 \leq u \leq 1;
\end{array}
$$

when t varies from 0 to $1/3$, the corresponding point describes the trajectory followed by the starting point common to both initial paths between "time" $s = 0$ and time s. As u varies over the interval $[1/3, 2/3]$, the point with parameter u traces out the arc $0 \leq t \leq 1$ of $\nu_s(t)$ arising from the deformation of μ_0. Finally, in the interval $[2/3, 1]$, we follow (in the reverse direction) the trajectory of the endpoint common to μ_0 and μ_1. Clearly, the paths thus defined are fixed-endpoint homotopic to μ_0 for all s. Slightly modifying the previous formulas in order to obtain for $1/3 \leq u \leq 2/3$, the arc $1 \leq t \leq 2$ of $\nu_s(t)$ arising from the deformation of μ_1:

(7.5")
$$u \longmapsto \begin{array}{ll} \sigma(3su, 2) & \text{for } 0 \leq u \leq 1/3\,, \\ \sigma[s, 3(1 - u)] & \text{for } 1/3 \leq u \leq 2/3\,, \\ \sigma[3s(1 - u), 1] & \text{for } 2/3 \leq u \leq 1\,, \end{array}$$

define a fixed endpoint-homotopic path to μ_1. When $s = 1$, by assumption, the path ν_s reduces to a point c. Up to parametrization, the paths (5') and (5") are then clearly identical. As they are (fixed-endpoint) homotopic to respectively μ_0 and μ_1, the same also holds for these.

Homotopy theory includes many small calculations of this type. After having done them once or twice, diagrams are substituted for them, to which the reader is in turn free to substitute his small calculations.

Any star domain G with respect to a point a is clearly simply connected: homotheties with centre a and ratio $s \in [0, 1]$ deform closed paths into a unique point. More generally, the same holds for any contractible domain G in the sense of the previous chapter. The main difference with the case of domains in $\mathbb{C}$ is that in $\mathbb{R}^n$, a simply connected domain is not necessarily contractible (counterexample: the open subset bounded by two concentric spheres).

(iii) *The Banach space* $C^{1/2}(I; E)$. As in n° 3, (ii) of Chapter VIII, the computation (3) of the derivative of $F(\mu + s\nu)$ can be interpreted in terms of differential calculus in a Banach space.[28] The vector space $C^{1/2}(I, E)$ of $C^{1/2}$ maps $\mu : I \longrightarrow E$, where E is the ambient Cartesian space, can be equipped with the norm

$$|\mu| = \|\mu\|_I + \|\mu'\|_I$$

inspired from distribution theory (Chap. V, § 10), or maybe it is the other way round; $C^{1/2}(I, E)$ then becomes a complete space, i.e. a Banach space: the proof is the same as in Chapter VIII. If $C^{1/2}(I; G)$ is the set of all admissible paths in the given open subset G of E, then $C^{1/2}(I; G)$ is clearly an open subset of $C^{1/2}(I, E)$, i.e. for any $\mu \in C(I; G)$, any path $\nu \in C(I, E)$ with sufficiently small $|\mu - n|$, or merely sufficiently small $\|\mu - \nu\|_I$, is also in $C^{1/2}(I; G)$.

The function $F(\mu) = \int \omega[\mu(t); \mu'(t)]dt$, defined on the open set $C^{1/2}(I, G)$, is everywhere continuous on it. For this, let μ, ν be two elements in this open set. To find an upper bound for $|F(\nu) - F(\mu)|$ for given μ and ν near μ, it is necessary to search for a uniform upper bound of

$$\omega[\nu(t); \nu'(t)] - \omega[\mu(t); \mu'(t)] = p_i[\nu(t)] D\nu^i(t) - p_i[\mu(t)] D\mu^i(t)$$

if $\omega = p_i(x)dx^i$. The problem is similar to that of the proof of the continuity of a product: the uniform continuity of the $p_i(x)$ in a compact neighbourhood

---

[28] To understand the rest of the book, reading this n° is not essential.

of $\mu(I)$ shows that $|p_i[\nu(t)] - p_i[\mu(t)]|$ is everywhere $< r$ if $\|\nu - \mu\|_I < r'$, the differences $|D\nu^i(t) - D\mu^i(t)|$ being bounded above by $\|\nu' - \mu'\|_I$. Usual calculations then lead to the result.

Having done this, note that the notion of a differential introduced in n° 2, (i) for functions defined on an open subset of a Cartesian space generalizes automatically to the case of a numerical-valued function F defined on an open subset of a Banach space B or even with values in another Banach space H; F will be said to be differentiable at a point x if there is a continuous linear map[29] u from B to H such that, for sufficiently small $h \in B$,

$$(7.6) \qquad F(x + h) = F(x) + u(h) + o(h)$$

with, as usual, $\lim \|o(h)\|/\|h\| = 0$ as $\|h\|$ tends to 0; then set $dF(x; h) = u(h)$. With this definition and taking $s = 0$ in formula (4'), it is tempting to write that

$$(7.7) \qquad dF(\mu; \nu) = \omega\,[\mu(1); \nu(1)] - \omega\,[\mu(0); \nu(0)] \ .$$

This expression is indeed linear in ν for given μ. To justify it, it still remains to be shown that

$$F(\mu + \nu) = F(\mu) + dF(\mu; \nu) + o(\nu)$$

as the norm of the path $\nu \in C^{1/2}(I, E)$ tends to 0. Now, (4') and the FT show that, if ω is closed, then

$$(7.8) \qquad F(\mu + \nu) - F(\mu) =$$
$$= \int_0^1 \left\{ \omega\,[\mu(1) + s\nu(1); \nu(1)] - \omega\,[\mu(0) + s\nu(0); \nu(0)] \right\} ds\,,$$

at least if $|\nu|$ is sufficiently small so that $\mu + s\nu \in C^{1/2}(I, G)$ for any $s \in I$. But as $\omega(x; h)$ is a C^1 function of (x, h) and is linear in h, there is an equality of the form

$$|\omega(x + k; h) - \omega(x; h)| \leq M\|k\|.\|h\|$$

which holds for given x, for any h and sufficient small $\|k\|$. The error made by replacing s by 0 in the functions integrated in (8) is, up to a constant factor, bounded above by $\|\nu(1)\|^2$ in the first case and $\|\nu(0)\|^2$ in the second and hence, up to a constant, the total error made is inferior to $|\nu|^2$. But replacing s by 0 in the integrals (8), the right hand side is replaced by expression (7).

---

[29] as already remarked, this is a superfluous assumption. On analysis in Banach spaces, see Dieudonné, *Eléments d'Analyse*, vol. 1, VIII, or Serge Lang, *Analysis I*, or Henri Cartan, *Calcul différentiel*, etc.

So finally,

$$(7.9) \quad F(\mu + \nu) - F(\mu) = \omega \left[\mu(1); \nu(1) \right] - \omega \left[\mu(0); \nu(0) \right] + O \left(|\nu|^2 \right).$$

This result is more than sufficient to justify (7).

But (7) is based on formula (4'), i.e. supposes that ω is closed. If it is not the case, it is necessary to return to the formula (4) in its entirety. Taking $s = 0$ in it, it can be deduced that probably

$$(7.10) \quad dF(\mu; \nu) = \omega \left[\mu(1); \nu(1) \right] - \omega \left[\mu(0); \nu(0) \right] + \int_I d\omega \left[\mu(t); \nu(t), \mu'(t) \right] dt.$$

Like in the case of the simpler formula (7), we leave it to the reader to check this formula.

§ 3. Integration of Differential Forms

8 – Exterior Derivative of a Form of Degree 1

(i) *Physicists' Vector Analysis.* The problem of primitives occurs in physics and in mechanics, but, apart from some theoreticians, physicists and mechanical engineers, being quite conservative revolutionaries, prefer the terminology inherited from the 19th century which they pass on from generation to generation. For them, a differential form $Pdx + Qdy$ with two variables is a *vector field*, namely the function whose value at (x, y) is the vector of the coordinate plane $P(x, y)$ and $Q(x, y)$; usually its origin is taken to be the point (x, y) rather the origin of the coordinates. This explains the use of the word "field", in the same way as we talk about a wheat field. The expression $D_1Q - D_2P$ – a numerical or "scalar"-valued function – is the *rotational* of this vector field and for them, the primitive F, when it exists, is the *potential* from which the given vector field is derived. They also express the relation $dF = D_1Fdx + D_2Fdy$ by saying that the vector field with coordinates D_1F and D_2F is the *gradient* of the function F.

There is above all a vocabulary for 3-dimensional physics. Note first that the physical space of the Creator is not $\mathbb{R}^3$; it can be identified with it only if an origin O, a unit length and for what follows, a rectangular coordinate system Ox, Oy and Oz, are chosen. As a real valued linear functional $h \mapsto c_ih^i$ on a Cartesian space has exactly as many coordinates (its coefficients c_i) as a vector with respect to a basis of it, these two types of objects often tend to be confused; the differential $dF = Pdx + Qdy + Rdz$ of a function is, therefore, identified with a vector field

$$\operatorname{grad} F : (x, y, z) \longmapsto \Big(P(x, y, z), Q(x, y, z), R(x, y, z) \Big),$$

generally written $Pi + Qj + Rk$, where the letters i, j and k, topped with arrows, a delight for typesetters, denote the "unit vectors", i.e. basis vector, of the rectangular coordinate system chosen.

Physicists exploit the fact that *der Herr*, as Einstein used to call him, has once for all defined the "scalar product" of two vectors of the physical space, provided, however, that a unit of length as well as an origin O are chosen as above in the space in order to transform it into a vector space. If the scalar product of two vectors is written

$$(h|k) = h^1k^1 + h^2k^2 + h^3k^3$$

in *rectangular* coordinates, – mechanical engineers and physicists often write it $h.k$ –, any linear functional $h \mapsto c_1h^1 + c_2h^2 + c_3h^3$ can then be written $h \mapsto (h|c)$, with a vector $c = (c_1, c_2, c_3)$ which it is entirely determined by; the linear functional in question can then be identified[30] with the vector c.

---

[30] In mathematics, identifying two objects means that no difference is made between them. It is often convenient to avoid useless complications (for example,

But as already mentioned in n° 1, (ii), this identification has no intrinsic or absolute meaning, whether in a general vector space lacking a distinguished scalar product, or in an Euclidean space, when non-rectangular coordinate systems are used.

Moreover, defining the "gradient" of a function F as we have done, i.e. by differentiating $F(x+th)$ at $t = 0$, is perfectly natural even and especially from a physics viewpoint. For example what is a "temperature gradient" but a measure of the rate of change of temperature when going from a point x to a point $x + th$ in the direction defined by a given vector h? There are no coordinates in this definition, and a "rate of change" has always been a derivative. In this case, the physical reason for writing $dF(x; h)$ as $(\operatorname{grad} F(x)|h)$ is to highlight the direction, that of the vector $\operatorname{grad} F(x)$, in which the temperature change in the neighbourhood of x is fastest.

Remaining in dimension 3, there are three independent conditions (5.14'); as 3 is the unique number for which $n(n-1)/2 = n$, the Creator must have invented this miracle on the eve of the Big Bang to mystify his creatures into believing that they would thus more easily understand his Complete Works than if he had, for example, chosen a four-dimensional space. This leads physicists to wrongly associate to the vector field (P, Q, R), the vector field $(D_2 R - D_3 Q, D_3 P - D_1 R, D_1 Q - D_2 P)$, that they again call the rotational of the given vector field; its cancelation is a necessary (and sufficient in a simply connected domain) condition for the vector field (P, Q, R) to come from a potential (Theorem 3). But for $n = 4$, the Relativity space that the mystified eventually discovered, there are already 6 conditions (5.14') and it is no longer a question of rotationals in the sense of a vector field. For $n > 4$ this is even less so.

(ii) *Differential forms of degree 2.* The proper generalization of physicists' rotational is the exterior derivative of a differential form, which have already cropped up in n° 5 at the end of the previous n° :

$$d\omega(x; h, k) = \omega'(x; h, k) - \omega'(x; k, h) \,.$$

In dimension 3, if $\omega = P dx + Q dy + R dz$, writing (5.15) explicitly, the expression $d\omega(x; h, k)$ is indeed equal to

$$(D_2 R - D_3 Q)\left(h^2 k^3 - h^3 k^2\right) + (D_3 P - D_1 R)\left(h^3 k^1 - h^1 k^3\right)$$
$$+ (D_1 Q - D_2 P)\left(h^1 k^2 - h^2 k^1\right) ,$$

which gives rise to the components of the physicists' rotational.

As a function of h and k for a given x, $d\omega(x; h, k)$ is an alternating bilinear form in h, k. Its generalization is a function $\omega(x; h, k)$ of $x \in G$ and of two

---

in Chapter II rational numbers were identified to real ones) when the objects identified satisfy for example the same computation rules. But this is highly questionable when for example it suppose a particular choice of a coordinate system.

varying vectors $h, k \in E$, called a *differential form of degree* 2 and class C^p on an open subset G of a Cartesian space E, which, for given x, is an alternating bilinear form[31]

$$\omega(x) : (h, k) \longmapsto \omega(x; h, k)$$

in h, k and which, for given h, k is a C^p function of x. The case $p = 2$ is sufficient for what follows.

Let $B(h, k)$ be an alternating bilinear form (a purely algebraic notion) and (a_i) a basis for E. Since $h = h^i a_i$ and $k = k^j a_j$,

$$B(h, k) = h^i B(a_i, k) = h^i k^j B(a_i, a_j) = b_{ij} h^i k^j$$

with antisymmetric coefficients

$$b_{ij} = B(a_i, a_j) = -b_{ji},$$

that are, therefore zero for $i = j$. This can be also written

$$B(h, k) = b_{ij} h^i k^j = \sum_{i<j} b_{ij} \left(h^i k^j - h^j k^i \right) = \frac{1}{2} b_{ij} \left(h^i k^j - h^j k^i \right).$$

The factor $1/2$ corrects the fact that each term is written twice. In this form, the antisymmetry is highlighted.

Applying this calculation to $B = \omega(x)$, we, therefore, get the relation

$$(8.1) \qquad \omega(x; h, k) = p_{ij}(x) h^i k^j = \sum_{i<j} p_{ij}(x) \left(h^i k^j - h^j k^i \right) =$$

$$= \frac{1}{2} p_{ij}(x) \left(h^i k^j - h^j k^i \right)$$

with coefficients

$$(8.2) \qquad\qquad p_{ij}(x) = \omega(x; a_i, a_j) = -p_{ji}(x)$$

depending on the basis (a_i) used. Note that in (1), we have been forced to abandon Einstein's convention in the first sum. When $\omega = d(p_i dx^i)$, $p_{ij} = D_i p_j - D_j p_i$.

In degree 2, expressions similar to $p_i(x) dx^i$ can be used. For this, given two linear functionals $u(h) = u_i h^i$ and $v(h) = v_i h^i$ on the vector space considered, the alternating bilinear form

$$(8.3) \qquad u \wedge v : (h, k) \longmapsto u(h)v(k) - u(k)v(h) = u_i v_j \left(h^i k^j - h^j k^i \right)$$

$$= (u_i v_j - u_j v_i) h^i k^j$$

is called the *exterior product* of u and v. This is a purely algebraic notion.

---

[31] See for example *Cours d'Algèbre* by the author, §§ 21 to 24. A differential form of degree 2 is, therefore, an "antisymmetric" tensor field of type $(0, 2)$.

Obviously,

(8.4) $$u \wedge v = -v \wedge u$$

and the product is an alternating bilinear form of u and v; in particular, $u \wedge u = 0$ always holds.

Applying this definition to the forms $u = dx^i$ and $v = dx^j$, given by $u(h) = h^i$ and $v(h) = h^j$, we get the form

$$dx^i \wedge dx^j : (h, k) \longmapsto dx^i(h)dx^j(k) - dx^i(k)dx^j(h) = h^i k^j - h^j k^i .$$

As a result, (1) can also be written

$$\omega(x; h, k) = \sum_{i<j} p_{ij}(x).dx^i \wedge dx^j(h, k)$$

where $dx^i \wedge dx^j(h, k)$ denotes the value of the alternating bilinear form $dx^i \wedge dx^j$ at (h, k); hence the shorthand expression

(8.5) $$\omega = \sum_{i<j} p_{ij} dx^i \wedge dx^j = \frac{1}{2} p_{ij} dx^i \wedge dx^j$$

involving both the product of the bilinear form $dx^i \wedge dx^j$ and the scalar function p_{ij}. In dimension 3, setting x, y, z for the three coordinates, we always write

(8.5') $$\omega = p\, dy \wedge dz + q\, dz \wedge dx + r\, dx \wedge dy .$$

The exterior product of two forms ω and ϖ of degree 1 is similarly defined: it is the form $x \mapsto \omega(x) \wedge \varpi(x)$ of degree 2. If $\omega = p_i dx^i$ and $\varpi = q_i dx^i$, clearly

$$\omega \wedge \varpi = p_i q_j dx^i \wedge dx^j = \frac{1}{2}\left(p_i q_j - p_j q_i\right) dx^i \wedge dx^j .$$

For example, if f and g are two functions on an open subset of $\mathbb{R}^2$, and if s and t denote the standard coordinates, then[32]

(8.6) $$df \wedge dg = (D_1 f.D_2 g - D_2 f.D_1 g)\, dt \wedge dt , = \frac{D(f, g)}{D(s, t)} dt \wedge dt ,$$

an expression involving the Jacobian of the map $(s, t) \mapsto \big(f(s, t), g(s, t)\big)$.

With these conventions, the exterior differential of a form $\omega = p_i dx^i$ of degree 1 can be calculated by writing that

---

[32] Direct calculation: write that $df \wedge dg = (D_1 f ds + D_2 f dt) \wedge (D_1 g ds + D_2 g dt)$, merely develop and take into account the relations $ds \wedge ds = dt \wedge dt = 0$, $dt \wedge ds = -ds \wedge dt$.

$$dw(x; h, k) = dp_i(x; h)k^i - dp_i(x; k)h^i =$$
$$= dp_i(x; h)dx^i(k) - dp_i(x; k)dx^i(h) =$$
$$= dp_i(x) \wedge dx^i(h, k),$$

a relation involving the value at (h, k) of the exterior product of the linear functionals $h \mapsto dp_i(x; h)$ and $h \mapsto dx^i(h) = h^i$.

(8.7) $$\omega = p_i dx^i \implies dw = dp_i \wedge dx^i$$

for short.

Exercise 1. Using the definition of dw, show that

(8.8) $$d(f\omega) = df \wedge \omega + f d\omega$$

if f is a function and ω a form of degree 1. Deduce (7).

In dimension 3 and in rectangular coordinates, but only in this case, (5') can be identified to the vector field $H(x)$ with coordinates p, q, r, though the vector with coordinates

$$h^2 k^3 - h^3 k^2, h^3 k^1 - h^1 k^3, h^1 k^2 - h^2 k^1$$

is called the *vector product* (or, which is better, exterior product) of the vectors h and k, and is written $h \times k$ or $h \wedge k$. Then

$$\omega(x; h, k) = (\overrightarrow{H(x)}|h \wedge k),$$

the scalar product of the vectors $H(x)$ and $h \wedge k$. This does not substitute for the theory of alternating bilinear forms.

(iii) *Forms of degree p.* All this can be generalized[33] and differential forms of arbitrary degree p as well as an operation d taking a form of degree p to a form of degree $p + 1$ can be defined. A form of degree p is a tensor field of type $(0, p)$, i.e. a function

$$\omega(x; h_1, \ldots, h_p)$$

multilinear in h_i for given x, which is also *alternating*, i.e. multiplied by -1 when two variables h are permuted. It can easily be shown that[34] $\omega = 0$ if p is greater than the dimension n of E and that when $p = n$,

$$\omega(x; h_1, \ldots, h_n) = p(x) \det(h_1, \ldots, h_n),$$

---

[33] See for example Henri Cartan, *Calcul différentiel.*

[34] If $p > n$, the vectors $h_1, \ldots, h_p$ are never linearly independent; hence one of them can be expressed as the linear combination of the others. Substituting in a p-linear alternating form, we get a linear combination of its values for vectors that are not pairwise distinct, hence that are zero because of the antisymmetry of the form considered. See *Cours d'algèbre,* § 23, also for determinant theory.

where $p(x)$ is a numerical function and $\det(h_1, \ldots, h_n)$ is the determinant of the vectors h_i (i.e. of the matrix of their coordinates) with respect to a basis of E.

Exercise 2. How does the coefficient p change when the basis of E with respect to which the determinant of n vectors is defined is changed?

In the general case, choosing a basis for the space E, ω can be written as

$$\omega\left(x; h_1, \ldots, h_p\right) = a_{i_1 \ldots i_p}(x) h_1^{i_1} \ldots h_p^{i_p} =$$

$$= \frac{1}{p!} a_{i_1 \ldots i_p}(x) \det{}^{i_1 \ldots i_p}\left(h_1, \ldots, h_p\right)$$

with antisymmetric coefficients

$$a_{i_1 \ldots i_p}(x) = \omega\left(x; e_{i_1}, \ldots, e_{i_p}\right).$$

These are the values of $\omega(x)$ at the canonical basis vectors, and the upper indices bearing upon the determinants indicate the p rows that need to be extracted from the $p \times n$ matrix for the coordinates of the h_i. The term $1/p!$ could be avoided by summing over the systems of strictly increasing indices. All this, except the variable x which does not play any role in this context, expresses the standard formulas of multilinear algebra and of the theory of determinants. It will henceforth be hardly needed since the the way forward is sufficiently indicated by forms of degree 1,2, and 3.

In the general case, there is also an exterior differentiation operation taking forms of degree p to forms of degree $p + 1$. For example to understand what physicists call the "divergence" of a vector field, it is necessary to know how to associate a form $d\omega$ of degree 3 to a form

$$\omega = \frac{1}{2} p_{ij} dx^i \wedge dx^j = p_{23} dx^2 \wedge dx^3 + p_{31} dx^3 \wedge dx^1 + p_{12} dx^1 \wedge dx^2$$

of degree 2 on $\mathbb{R}^3$ (or on an arbitrary Cartesian space); its value at some point $x \in G$ is an alternating *trilinear* form, i.e. an antisymmetric function of three variable vectors $h, k, l \in E$. To define it, we first need to introduce a covariant derivative

$$(8.9) \qquad \omega'(x; h, k, l) = \frac{d}{dt} \omega(x + th; k, l) \text{ for } t = 0$$

$$= \frac{1}{2} dp_{ij}(x; h) \left(k^i l^i - k^j l^j\right),$$

a linear expression in h, k, l and alternating in k, l. Then set

$$(8.10) \quad d\omega(x; h, k, l) = \omega'(x; h, k, l) - \omega'(x; k, h, l) + \omega'(x; l, h, k) =$$

$$= \omega'(x; h, k, l) + \omega'(x; k, l, h) + \omega'(x; l, h, k)$$

$$= \frac{1}{2} D_i p_{jk}(x) \det{}^{ijk}(h, k, l),$$

where $\det^{ijk}(h,k,l)$ is the determinant of order 3 consisting of the coordinates with indices i,j,k of the vectors h,k,l; we could get rid of the factor $1/2$ by summing only over the i,j,k such that $j \leq k$. We do the bare minimum to transform ω' into an *alternating* form. In particular, considering the form

$$\omega = pdy \wedge dz + qdz \wedge dx + rdx \wedge dy$$

of degree 2 on $\mathbb{R}^3$, easily leads to

$$d\omega(x; h,k,l) = (D_1p + D_2q + D_3r) . \det(h,k,l),$$

where $\det(h,k,l)$ is the determinant of the vectors h,k,l with respect to the canonical basis. For physicists, the function $D_1p + D_2q + D_3r$ is the *divergence* of the vector field (p,q,r); the determinant is the *scalar triple product* of the vectors h,k and l, and is written

$$(h,k,l) = (h|k \wedge l) = h^1\left(k^2l^3 - k^3l^2\right) + h^2\left(k^3l^1 - k^1l^3\right) + h^3\left(k^1l^2 - k^2l^1\right),$$

where $k \wedge l$ is the vector or exterior product of k and l.

In the general case, start with a function $\omega(x; h_1, \ldots, h_p)$ which, for given x, is alternating multilinear with respect to the variables $h_i E$ and compute its covariant derivative

$$(8.11) \qquad \omega'(x; h_0, \ldots, h_p) = \frac{d}{dt}\omega(x + th_0; h_1, \ldots, h_p) \quad \text{for } t = 0,$$

then set

$$(8.12) \qquad d\omega(x; h_0, \ldots, h_p) = \sum_{0 \leq i \leq p} (-1)^i \omega'\left(x; h_0, \ldots \widehat{h_i}, \ldots, h_p\right),$$

where the accent over the letter h_i indicates it is *omitted*. the result is easily seen to be multilinear and alternating in $h_0, \ldots, h_p$; any differential form that can be written as $d\omega$ is said to be *exact*.

Exterior differentiation satisfies some classical properties; proofs can be found everywhere, for example in Cartan, *Calcul différentiel*, but as the best way for understanding them is to recover them oneself, I will only state them as exercises.

Exercise 3. For any form ω of degree p, $dd\omega = 0$, in other words: any exact differential form is closed. Corollary: the divergence of a rotational is always zero.

Exercise 4. Let ω be a closed form ($d\omega = 0$) of degree 2 on a star domain with respect to 0. Define a form ϖ of degree 1 by setting

$$(8.13) \qquad\qquad \varpi(x; h) = \int \omega(tx; x, h)tdt,$$

where integration is over $[0,1]$. Show that $d\varpi = \omega$. [Imitate calculation (5.6)]. For a form of degree $p+1$, set (Poincaré theorem)

$$(8.14) \qquad \varpi\left(x; h_1, \ldots, h_p\right) = \int \omega\left(tx; x, h_1, \ldots, h_p\right) t^p dt.$$

These calculations show that *locally*, any differential closed form of degree p is exact, i.e. is the exterior derivative of a form ϖ of degree $p-1$; the latter is not unique: a closed form, i.e. as we are using local arguments, an exact differential can always be added to it.

This argument still holds globally if G is, for example, star shaped, and hence convex. But the general case is far more difficult to deal with, even in a simply connected domain if forms of degree ≥ 2 are considered; the problem is directly connected to the topology of G (De Rham cohomology); the following simple argument may give a vague idea.

G can always be written as the finite or infinite union of open *convex* non-empty subsets U_i; the intersections

$$U_{ij} = U_i \cap U_j, \quad U_{ijk} = U_i \cap U_j \cap U_k, \quad \text{etc.},$$

remain convex (and possibly empty). Then take for example a closed form ω of degree 3 on G, de degré 3. There are forms ω_i of degree 2 on U_i such that

$$\omega = d\omega_i \text{ on } U_i.$$

As $d\omega_i = d\omega_j$ on U_{ij}, there are forms ω_{ij} of degree 1 on the non-empty U_{ij} such that

$$\omega_i - \omega_j = d\omega_{ij} \text{ on } U_{ij}.$$

Then $d(\omega_{jk} - \omega_{ik} + \omega_{ij}) = 0$ on each U_{ijk}, and so there are forms ω_{ijk} of degree 0 (i.e functions) on the non-empty U_{ijk} such that

$$\omega_{jk} - \omega_{ik} + \omega_{ij} = d\omega_{ijk} \text{ on } U_{ijk}.$$

So

$$d(\omega_{jkh} - \omega_{ikh} + \omega_{ijh} - \omega_{ijk}) = 0 \text{ on } U_{ijkh},$$

and as the non-empty U_{ijkh} are convex, the relations

$$\omega_{jkh} - \omega_{ikh} + \omega_{ijh} - \omega_{ijk} = c_{ijkh}$$

follow, where the c_{ijkh} are constants associated to the non-empty U_{ijkh} and satisfying

$$c_{jkhl} - c_{ikhl} + c_{ijhl} - c_{ijkl} + c_{ijkh} = 0$$

whenever U_{ijkhl} is *non-empty*. Thus the scheme of *non-empty* pairwise, three-wise, etc. intersections of the open subsets U_i – he "simplicial structure" of the cover considered – is involved, and this takes us into the realm of cohomology... See Andre Weil, *Sur les théorèmes de de Rham* (Comm. Math. Helvetici, 1952, or *Œuvres*).

Corollaries: *In dimension 3, any vector field with zero divergence is locally the rotational of a vector field, unique up to a gradient, and every function is locally the divergence of a vector field, which is unique up to a rotational.* For example in the second case, if the given function f is assumed to be on a star domain with respect to 0, then we find a vector field (p^1, p^2, p^3) with divergence f by applying formula (14) for $p = 2$ to the differential form $\omega = f(x)dx^1 \wedge dx^2 \wedge dx^3$, for which

$$\omega(x; h, k, l) = f(x) \det(h, k, l) ;$$

since $\det(h, k, l)$ is the "scalar triple product" $(h, k, l) = (h|k \wedge l)$ of physicists,

$$\varpi(x; h, k) = \int f(tx) (x|h \wedge k) t^2 dt = (x|h \wedge k) \int f(tx) t^2 dt ,$$

follows. This means that the vector field we sought is

$$p^i(x) = F(x)x^i \quad \text{with} \quad F(x) = \int f(tx) t^2 dt ,$$

where integration is over $(0, 1)$.

Exercise 5. Defined the external product of two alternating multilinear forms f and g of degrees p and q by anti-symmetrizing their tensor product:

$$(8.15) \qquad f \wedge g (h_1, \ldots, h_{p+q}) = \frac{1}{p!q!} \sum \varepsilon(s) f \left(h_{s(1)}, \ldots, h_{s(p)} \right)$$
$$\times g \left(h_{s(p+1)}, \ldots, h_{s(p+q)} \right) ,$$

where summation is over all permutations s of $\{1, \ldots, p+q\}$ and where $\varepsilon(s)$ denotes the signature of s; the factor $1/p!q!$ could be omitted by summing only over the permutations such that $s(1) < \ldots < s(p)$ and $s(p+1) < \ldots < s(p+q)$. Show that

$$(8.16) \qquad\qquad g \wedge f = (-1)^{pq} f \wedge g$$

and – this is the hardest part – that the exterior product is associative.

Exercise 6. The exterior product of two differential forms is defined by the previous exercise. Show that

$$(8.17) \qquad\qquad d(\omega \wedge \varpi) = d\omega \wedge \varpi + (-1)^p \omega \wedge d\varpi$$

if ω is of degree p.

The notion of inverse image formulated in n° 6 for forms of degree 1 can be trivially made to apply to the general case: given a map $f : U \longrightarrow V$ and a form ω of degree p on V, the inverse image of ω under f is the form

$$(8.18) \qquad \omega \circ f : (x, h_1, \ldots, h_p) \longmapsto \omega \left[f(x); f'(x)h_1, \ldots, f'(x)h_p \right] \ ;$$

as in the case of the differentiation of a composite functions and of forms of degree 1. This is the only definition conceivable given the data available; it can be generalized to any tensor field T of type $(p, 0)$ since antisymmetry obvious plays no role in the definition. The reader will easily prove the following formulas:

$$(8.19) \qquad \omega \circ (g \circ f) = (\omega \circ g) \circ f \,,$$

$$(8.20) \qquad (\omega \wedge \varpi) \circ f = (\omega \circ f) \wedge (\varpi \circ f) \,,$$

$$(8.21) \qquad d(\omega \circ f) = d\omega \circ f \,.$$

They are just trivial consequences of multivariate chain rule: apply definitions. They can even be generalized to tensor fields, provided exterior differentiation is replaced by covariant differentiation (3.10).

9 – Extended Integrals over a 2-Dimensional Path

Physicists say that if we consider a surface bounded by a regular curve in an electromagnetic field, the flow of the field through the surface is equal to the circulation of the electric current vector around its boundary. This fundamental law, experimentally discovered by Ampère and Faraday in geometrically trivial cases – for example, a circular plane surface –, was formulated mathematically by Maxwell about 1870 by taking account of the fact that the "magnetic field" vector is the rotational of the "electric" vector. It is based on a precise mathematical result, Stokes' formula which gives a relationship between curvilinear integrals and "surface" integrals over R^3.

With their practice of mathematical conjuring tricks and their possession of what a recent author[35] calls – with admiration? irony? – *a powerful weapon: a striking intuition, literally based on centuries of collective experience*, physicists give almost instantaneous proofs of this;[36] these astound mathematicians who, having stopped arguing as they did hundred fifty years

---

[35] Michel Talagrand, *Verres de spin et optimisation combinatoire* (talk in the N. Bourbaki Seminar, n° 859, Mars 1999, p. 8)

[36] In Paris, physicists have even been seen to write Taylor's formula, Maxwell's equations, Stokes' formula, and chemists to calculate eigenfunctions of Schrodinger's operator for the hydrogen atom, in front of first year students not having yet understood or even learnt what a partial derivative was, and to reproach mathematicians who argue correctly and in the natural pedagogical order of making students "lose their time". Apparently, a number of physicists do not understand that mathematicians give as much importance to rigour as they give to experiments. Besides, it is interesting to observe that the hostility with which many physicists regard abstract or modern mathematics does not extend to modern physics, some sectors of which, like quantum mechanics invented in the same period, are all the same quite abstract and "modern" too. This evolution can also be understood by comparing chemistry and biology classes of the 1930s with those taught today, especially in high schools.

ago, laboriously reach the result. In fact, Stokes' formula is one the hardest to prove rigorously in an elementary manner; and even using the highest level of mathematics, no one knows how to exactly define the right category for "surfaces" (of arbitrary dimension) to apply it to; the problem is including integration domains whose "edges" are sufficiently regular for integration to be possible on it, while including sufficiently general singularities so as not to exclude important practical or theoretical cases. For example, the border of a polyhedra is not a "smooth" surface; it has sharp corners and edges. Excluding such a simple example from the application domain of Stokes' formula would run counter to common sense, but proving it in a sufficiently general framework so as to include polyhedrons (in other words, "simplicial complexes" from algebraic topology) poses substantial difficulties since at the same time the case of perfectly smooth surfaces has to be covered. Physicists reply that the surface of a tetrahedron is in reality made of four perfectly smooth triangles and that the edges do not count in the integration, or that "corners may be cut" without significantly changing the result. True, but they obviously do not have to prove it. It is for good reason that this problem is at the origin of the Bourbaki group; in the 1930s, when its program was to write a usable treatise for university teaching, a mathematician of the level of Andre Weil asked Henri Cartan if he knew a good method for *proving* the formula, it being agreed that all mathematicians – I used to do in in 1947 – have always been able to present the type of "proof" that satisfy physicists.

(i) *The exterior derivative as an infinitesimal integral.* Let us again consider a differential form $p_i(x)h^i = \omega(x; h)$ of degree 1 and class C^1 on an open subset G of a Cartesian space E. Take a point $x \in G$, and fix two vectors h and k and, for given $s, t \in \mathbb{R}$, consider the plane parallelogram $P(x, sh, tk) = P$, i.e. the set of points of the form $x + uh + vk$, where u (resp. v) vary between 0 and s (resp. t); assume s and t to be sufficiently small so that P can be contained in G. When we follow the sides of this parallelogram in the direction indicated in fig. 2, the border of P is transformed into a closed integration path written ∂P, the *boundary* of P. Let us calculate the integral $I(x; sh, tk)$ of ω along this path. On the side connecting x to $x + sh$, the parametric representation $u \mapsto x + uh$ can be used, which gives a contribution equal to

$$\int_0^s \omega(x + uh; h)du.$$

Contributions from the other sides are calculated in a similar fashion. $I(x; sh, tk)$ is thus seen to be equal to

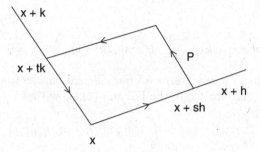

Fig. 9.2.

$$\int_0^s \omega(x + uh; h)du + \int_0^t \omega(x + sh + vk; k)dv + \int_s^0 \omega(x + uh + tk; h)du +$$

$$+ \int_t^0 \omega(x + vk; k)dv,$$

and so

$$(9.1) \qquad I(x; sh, tk) = \int_0^t [\omega(x + sh + vk; k) - \omega(x + vk; k)]\, dv -$$

$$- \int_0^s [\omega(x + uh + tk; h) - \omega(x + uh; h)]\, du.$$

As the functions of (s, t, u, v) being integrated are C^1, it is for instance possible to differentiate under the $\int$ sign with respect to s. By the FT, the derivative of the second integral is $\omega(x + sh + tk; h) - \omega(x + sh; h)$. To differentiate the first one, the second term in the substraction can be omitted since it does not depend on s; to differentiate the first one, use the definition

$$(9.2) \qquad \omega'(x + sh; h, k) = \frac{d}{ds}\omega(x + sh; k)$$

of the covariant derivative and so finally,

$$(9.3) \qquad \frac{d}{ds}I(x; sh, tk) = \int_0^t \omega'(x + sh + vk; h, k)dv -$$

$$- [\omega(x + sh + tk; h) - \omega(x + sh; h)].$$

Now differentiate with respect to t; by the FT, the derivative of the integral is $\omega'(x + sh + tk; h, k)$; by (2), that of the expression between [ ] is equal to $\omega'(x + sh + tk; k, h)$ since $\omega(x + sh; h)$ does not depend on t. It follows that

$$(9.4) \qquad \frac{d^2}{dtds}I(x; sh, tk) = \omega'(x + sh + tk; h, k) - \omega'(x + sh + tk; k, h) =$$

$$= d\omega(x + sh + tk; h, k).$$

In particular,

$$dw(x; h, k) = \frac{d^2}{dt\,ds} I(x; sh, tk) \quad \text{for } s = t = 0.$$

Having done this, we can "retrace" our calculations; as the right hand side of (3) is clearly zero for $t = 0$, the FT and (4) show that

$$\frac{d}{ds} I(x; sh, tk) = \int_0^t dw(x + sh + vk; h, k).dv\,;$$

and as $I(x; sh, tk) = 0$ for $s = 0$, it also follows that

(9.5) $$I(x; sh, tk) = \int_0^s du \int_0^t dw(x + uh + vk; h, k)dv\,.$$

Hence for $s = t = 1$,

(9.6) $$\int_{\partial P(x;h,k)} w = \iint_{I^2} dw(x + uh + vk; h, k)dudv\,,$$

which is an extended double integral over the square $I^2 = I \times I$ in the plane; this supposes that h and k are sufficiently small so that the surface of the parallelogram $P(x; h, k)$ can be contained in the open subset on which w is defined.

The simplest case can be obtained by assuming that $w = pdx + qdy$ is a form on $\mathbb{R}^2$, and so

(9.7) $$dw = (D_1 q - D_2 p)dx \wedge dy = p_{12}dx \wedge dy\,.$$

(4) can then be written

(9.8) $$\int_{\partial P} pdx + qdy = \iint_{I^2} p_{12}(x + uh + vk)\left(h^1 k^2 - h^2 k^1\right) dudv\,,$$

where P is a parallelogram with initial point x generated by the vectors h and k. In particular, if $x = 0$ and if h and k are the unit vectors of the coordinate axes, then we get the *Green-Riemann formula*, unless it be the Gauss formula,

(9.9) $$\int_{\partial I^2} pdx + qdy = \iint_{I^2} (D_1 q - D_2 p)\,dxdy$$

for the square I^2, provided the boundary ∂I^2 of I^2 is given the usual positive orientation. Cauchy proved it directly:

$$\iint_{I^2} D_1 q\,dxdy = \int_0^1 dy \int_0^1 D_1 q(x, y)dx = \int_0^1 [q(1, y) - q(0, y)]\,dy =$$

$$= \int_{\partial I^2} qdy$$

since the integral qdy over the horizontal sides of I^2 is obviously zero. He used this calculation to show that the integral of a holomorphic function along the boundary of a square is zero, the relation $D_1 q - D_2 p = 0$ being just its holomorphy condition in this case. Gauss, Green, Cauchy, Stokes, Riemann: many fathers for essentially the same result. There is also an Ostrogradsky in dimension three.

(ii) *Stokes' formula for a 2-dimensional path.* Replacing the map $(s,t) \mapsto x + sh + tk$ by a C^2 map $\sigma : I \times I \longrightarrow G$ on $I \times I$ in the sense of §1, n° 2, (iv) leads to a generalization: σ is of class C^2 on the open square and its derivatives of order ≤ 2 can be extended by continuity to the closed square. As already seen, σ defines two families of paths in G:

$$(9.10) \qquad \mu_s : t \longmapsto \sigma(s,t)$$

and

$$(9.11) \qquad \nu_t : s \longmapsto \sigma(s,t).$$

Given a differential form ω of class C^1 and degree 1 on the open subset G, set

$$(9.12) \qquad F(s) = F(\mu_s) = \int_{\mu_s} = \int_0^1 \omega\left[\sigma(s,t); D_2\sigma(s,t)\right] dt.$$

Let us compute the derivative of $F(s)$ by direct calculations that will produce a less primitive version of Stokes' formula than (6); it will be seen further down that that the same final result can be obtained more quickly by using Green-Riemann, but it is necessary to accustom the reader to using the multivariate chain rule...

The $\int$ sign will always denote an extended integral over $I = [0,1]$.

We start, in telegraphic style, from the formula

$$(9.13) \qquad F'(s) = \int D_1\left[\omega(\sigma, D_2\sigma)\right] dt.$$

Therefore,

$$(9.14) \qquad D_1\left[\omega\left(\sigma; D_2\sigma\right)\right] = \frac{d}{ds}\left\{\omega\left[x(s); h(s)\right]\right\}$$

needs to be computed, where $x(s) = \sigma(s,t)$, $h(s) = D_2\sigma(s,t)$ for fixed t. The multivariate chain rule shows that

$$D_1\left\{\omega\left[x(s), h(s)\right]\right\} = \omega'\left[x(s); D_1x(s), h(s)\right] + \omega\left[x(s); D_1h(s)\right],$$

and so

$$(9.15) \qquad D_1\left[\omega(\sigma; D_2\sigma)\right] = \omega'\left(\sigma; D_1\sigma, D_2\sigma\right) + \omega\left(\sigma; D_1D_2\sigma\right).$$

But the proof of (15) also shows that

$$D_2 \left[\omega(\sigma; D_1\sigma) \right] = \omega' \left(\sigma; D_2\sigma, D_1\sigma \right) + \omega \left(\sigma; D_2 D_1 \sigma \right) = $$
$$= \omega' \left(\sigma; D_2\sigma, D_1\sigma \right) + \omega \left(\sigma; D_1 D_2 \sigma \right),$$

and so, by (15) and the definition of $d\omega$,

(9.16) $D_1 \left[\omega \left(\sigma; D_2\sigma \right) \right] = D_2 \left[\omega \left(\sigma; D_1\sigma \right) \right] + d\omega \left(\sigma; D_1\sigma; D_2\sigma \right).$

Hence, by (15), it follows that

(9.17) $F'(s) = \int D_2 \left[\omega \left(\sigma; D_1\sigma \right) \right] dt + \int d\omega \left(\sigma; D_1\sigma, D_2\sigma \right) dt.$

As $D_2 = d/dt$, the first integral is the change of $\omega(\sigma; D_1\sigma)$ between $t = 0$ and $t = 1$. Integrating $F'(s)$ over $(0, 1)$, we then get (FT)

(9.18) $$F(\mu_1) - F(\mu_0) = \int \omega \left[\sigma \left(s, 1 \right); D_1\sigma \left(s, 1 \right) \right] ds - $$
$$- \int \omega \left[\sigma \left(s, 0 \right); D_1\sigma \left(s, 0 \right) \right] ds + $$
$$+ \iint_{I^2} d\omega \left(\sigma; D_1\sigma; D_2\sigma \right) . ds \, dt.$$

Using the paths μ_s and ν_t defined above, (18) can be written as

(9.19) $$F \left(\mu_1 \right) - F \left(\mu_0 \right) = F \left(\nu_1 \right) - F \left(\nu_0 \right) + \iint d\omega(\ldots).$$

If the four integration paths occurring in (19) are concatenated into a single path $\partial\sigma : [0, 4] \longrightarrow G$ given by the formulas

$$\partial\sigma(t) = \begin{array}{lll} \sigma(0, t) & = \mu_0(t) & (0 \leq t \leq 1) \\ \sigma(t-1, 1) & = \nu_1(t-1) & (1 \leq t \leq 2) \\ \sigma(1, 3-t) & = \mu_1(3-t) & (2 \leq t \leq 3) \\ \sigma(4-t, 0) & = \nu_0(4-t) & (3 \leq t \leq 4), \end{array}$$

relation (19) becomes

(9.20) $$\int_{\partial\sigma} \omega = \iint_{\sigma} d\omega$$

provided, generally speaking, that we set

(9.21) $$\iint_{\sigma} \varpi = \iint_{I^2} \varpi \left[\sigma(s, t); D_1\sigma(s, t), D_2\sigma(s, t) \right] ds \, dt$$

for any sufficiently differentiable *2-dimensional path* $\sigma : I^2 \longrightarrow G$ and any differential form of degree 2 on G; the similarity with the definition of a

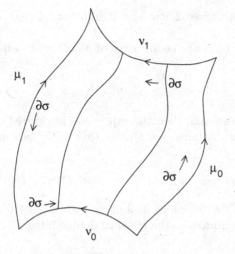

Fig. 9.3.

curvilinear integral is clear, and the reader will surely think of generalizing it to forms of arbitrary degree.

When the initial form ω is closed, the double integral vanishes from (20), which, therefore, indicates that the integral of ω along the closed path $\partial\sigma$ is zero. We thereby recover the invariance of the integral under homotopy (modulo obvious conditions: fixed endpoints, or contour remaining closed), but by imposing differentiability conditions on σ that are too strong.

Exercise. Generalize the proof to linear homotopies between paths of class $C^{1/2}$ [imitate the calculations of Chapter VIII, n° 3, (iii)].

Formula (20) is one of the possible versions of Stokes' formula in dimension two in a case which, compared to the physicists' traditional version where integration is over an excellent perfectly smooth surface, can seemingly present pathological aspects: the image of the square I^2 under σ can have all sorts of singularities – sharp corners, edges, pleats,[37] etc. – if the rank (n° 2, (v)) of σ is not assumed to be everywhere equal to 2, i.e. maximum. Moreover, as σ is not assumed to be injective, even if σ has maximum rank everywhere, the image of I^2 may resemble a paper sheet or a tube with multiple crossings, analogous to a curve in dimension 2 with multiple points.

(iii) *Integral of an inverse image.* The expression integrated on the right hand side is clearly the inverse image $\varpi \circ \sigma$ of ϖ under σ, defined at the end of n° 8. Hence, by definition,

(9.21')
$$\iint_\sigma \varpi = \iint_{I^2} \varpi \circ \sigma$$

---

[37] Consider the map $(s,t) \mapsto (s^2, t^2)$ de $J \times J$ on the plane, where $J = [-1, +1]$, or the map $(s,t) \mapsto (\sin^2 s, t)$ on the strip $0 \leq t \leq 1$.

for any form ϖ of degree 2 and any 2-dimensional path $\sigma : I^2 \longrightarrow X$ of class C^1.

Denoting by $\sigma^i(s,t)$ the coordinates of $\sigma(s,t)$ with respect to a basis for E, if

$$\varpi = p_{ij}dx^i \wedge dx^j ,$$

with respect to this basis, then the expression integrated can be calculated by replacing the dx^i by the $d\sigma^i = D_1\sigma^i(s,t)ds + D_2\sigma^i(s,t)dt$ and so $dx^i \wedge dx^j$ by

$$J^{ij}(s,t)ds \wedge dt \quad \text{where} \quad J^{ij} = D_1\sigma^i.D_2\sigma^j - D_1\sigma^j.D_2\sigma^i$$

is the Jacobian of the functions σ^i and σ^j with respect to s,t. So the right hand side of (21') reduces to the classical double integral

$$(9.22) \qquad \iint p_{ij}\,[\sigma(s,t)]\,J^{ij}(s,t).dsdt = \iint r(s,t)dsdt .$$

The computation is immediate as long as the mechanism of exterior products and inverse images has been understood.

Version (20) of the result obtained suggests a far quicker proof, which amount to applying Gauss' formula (9) to the square $I^2 = K$. Indeed, the left hand side is by definition a curvilinear integral: the integral over I of the inverse image $w \circ \sigma$. But denoting by ∂K the path with initial point 0 consisting in following the border of K counterclockwise, clearly

$$\partial\sigma = \sigma \circ \partial K ,$$

and so $w \circ \partial\sigma = w \circ (\sigma \circ \partial K) = (w \circ \sigma) \circ \partial K$ by (8.19). The left hand side of (20) is, therefore, the integral of $w \circ \sigma$ along the path ∂K. On the other hand, by (8.21),

$$dw \circ \sigma = d(w \circ \sigma) .$$

Setting $w \circ \sigma = \theta$ to be a form of degree 1 on K, by (21'), relation (20) means that

$$\int_{\partial K} \theta = \iint_K d\theta ,$$

which reduces to (9) as expected.

The preceding calculation can be generalized. Consider two Cartesian spaces E and F, two open sets $U \subset E$ and $V \subset F$, a map $f : U \longrightarrow V$ and a form ϖ of degree 2 on V. Let $\sigma : I^2 \longrightarrow U$ be a 2-dimensional path in U. This gives a path $f \circ \sigma : I^2 \longrightarrow V$ in V. Having said this,

$$(9.23) \qquad \iint_\sigma \varpi \circ f = \iint_{f \circ \sigma} \varpi .$$

Indeed the two sides are, by definition, the integrals over I^2 of the forms $(\varpi \circ f) \circ \sigma$ and $\varpi \circ (\sigma \circ f)$. It, therefore, suffices to use (8.19) in order to obtain (23).

(iv) *A planar example.* Consider the simplest case: $E = \mathbb{R}^2$. There are two real C^1 functions f_0 and f_1 on I. Set

$$\mu_0(t) = (t, f_0(t)) \,, \quad \mu_1(t) = (t, f_1(t)) \,,$$

so that these two paths amount to following the graphs of f_0 and f_1. Under the corresponding linear homotopy

$$(9.24) \qquad \sigma(s,t) = (1 - s)\mu_0(t) + s\mu_1(t) =$$
$$= (t, (1 - s)f_0(t) + sf_1(t))$$

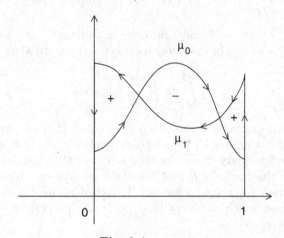

Fig. 9.4.

the image of $I \times I$ is the subset of $\mathbb{R}^2$ bounded by these graphs and by the verticals with coordinates 0 and 1 with respect to the x-axis, the boundary $\partial\sigma$ of σ together with the direction followed being indicated in the above figure. If

$$\omega = pdx + qdy$$

is a form of degree 1, its integral can be computed along $\partial\sigma$ by choosing parameters y along the vertical paths and x along the graphs of f_0 and f_1; thus

$$\int_{\partial\sigma} \omega = \int_0^1 \{p\,[x, f_0(x)] + q\,[x, f_0(x)]\,f_0'(x)\}\,dx + \int_{f_0(1)}^{f_1(1)} q(1,y)dy +$$
$$+ \int_1^0 \{p\,[x, f_1(x)] + q\,[x, f_1(x)]\,f_0'(x)\}\,dx + \int_{f_1(0)}^{f_0(0)} q(0,y)dy \,.$$

These four simple *oriented* integrals correspond to the obvious four paths making up $\partial\sigma$. As for the double integral of

$$d\omega = (D_1q - D_2p)\, dx \wedge dy = r(x,y)dx \wedge dy$$

occurring in (20), it is calculated by replacing (x,y) by $\sigma(s,t)$ and $dx \wedge dy$ by $J_\sigma(s,t)dsdt$, where

$$J_\sigma(s,t) = \begin{vmatrix} 0 & f_1(t) - f_0(t) \\ 1 & ? \end{vmatrix} = f_0(t) - f_1(t)$$

is the Jacobian of map (24). Here, integral (21') can, therefore, be written

$$\iint_{I^2} r\left[t, (1-s)f_0(t) + sf_1(t)\right] \cdot \left[f_0(t) - f_1(t)\right] . dsdt .$$

Replacing t by x and making a change of variable $y = (1-s)f_0(x) + sf_1(x)$ in the integration with respect to s, we get the integral

$$\int_0^1 dx \int_{f_0(x)}^{f_1(x)} r(x,y)dy ,$$

where the integral with respect to y is *oriented*. If the Lebesgue-Fubini formula (33.7) of Chapter V, §9 is naively applied, the result seems to be just the extended ordinary double integral of r over the compact subset A of $\mathbb{R}^2$ bounded by the graphs of f_0 and f_1 and the verticals $x = 0$ and $x = 1$. This is the case only if $f_0(t) \leq f_1(t)$ for all t. In fact, denoting by A_+ (resp. A_-) the subset of A on which the Jacobian $f_1(x) - f_0(x)$ is positive (resp. negative), (20) can be written as

$$(9.25) \qquad \int_{\partial\sigma} pdx + qdy = \iint_{A_+} (D_1q - D_2p)\, dxdy -$$

$$- \iint_{A_-} (D_1q - D_2p)\, dxdy .$$

This corresponds to the fact the the path $\partial\sigma$, which is the image under σ of the border of the square I^2, is made up of several closed simple curves, some oriented counterclockwise, others clockwise. If $f_1(t) \geq f_0(t)$ everywhere, we recover the Green-Riemann formula in a slightly more general case, but as simple to prove directly with the help of a small Cauchy calculation and of the most elementary version of Lebesgue-Fubini.

(v) *Classical version.* In classical analysis, there was a "surface" S – a sphere, a torus, a somewhat deformed rectangle, etc. – and a vector field (P, Q, R) in the usual three dimensional Euclidean space; the aim was to define the extended integral $\iint Pdydz + Qdzdx + Rdxdy$ over S. The general method consisted in using a "bi-univocal parametric representation", i.e. bijective, of S, given by functions

(9.26) $$x = \varphi(s,t), \quad y = \psi(s,t), \quad z = \theta(s,t)$$

of two real "parameters" s, t varying in a subset A of $\mathbb{R}^2$ bounded by one or many simple curves such as those encountered in Cauchy's residue formula; this amounts to setting

$$\sigma(s,t) = (\varphi(s,t), \psi(s,t), \theta(s,t)) .$$

Having said this, $dydz, dzdx, dxdy$ had to be replaced by

(9.27) $$\frac{D(\psi,\theta)}{D(s,t)}dsdt, \quad \frac{D(\theta,\varphi)}{D(s,t)}dsdt, \quad \frac{D(\varphi,\psi)}{D(s,t)}dsdt$$

and x, y, z by their expressions in terms of s and t in the function P, Q, R; an expression of the form $r(s,t)dsdt$ was thus obtained and integrated over the domain of variation A of (s,t). To make sure that S was indeed a "smooth" surface, without sharp corners, edges or any other singularities, it was assumed that the above three Jacobians were never simultaneously zero, in other words, that the tangent linear map to σ at (s,t) was everywhere injective (the necessity of this assumption regarding submanifolds of a Cartesian space will be explained in n° 13, (iv)) or else that σ was of rank 2; the image of $\mathbb{R}^2$ under the tangent map $\sigma'(s,t)$ was then the "tangent plane" the the surface S at the point $\sigma(s,t)$.

Clearly, the expression $Pdydz + Qdzdx + Rdxdy$ to be integrated is just the differential form .

$$\omega = Pdy \wedge dz + Qdz \wedge dx + Rdx \wedge dy$$

on $\mathbb{R}^3$ and, if K is the unit square I^2, then the surface integral to be computed is just the integral (21') of ω extended to the path σ.

The integral thus defined also had to be shown to depend solely on the given vector field and surface S, and not on the chosen parametric representation; as will be seen in the next n°, this was ensured by a change of variable formula in ordinary double integrals, the only difficulty residing in its proof.

As for (20), it is the traditional Stokes formula; it was applied by assuming that the vector field (P,Q,R) to be integrated was the rotational of a vector field H. The integral of H along the curve C bounding S gave the left hand side of (20). The expression to be integrated was often described as the scalar product of H and of an infinitesimal vector with components dx, dy, dz, written dM, where the letter M denotes a variable point of the curve. Physicists wrote the double integral in a similar way by regarding expressions (27) as the components of the metaphysical vector dS normal to S (i.e. orthogonal to the tangent plane to S) at the point considered and with length "the element of infinitesimal surface" of S, whatever that meant, so that (20) finally became

$$\int_C H.dM = \iint_S \text{rot}\, H.dS \,.$$

After much thought, it was possible to understand that dS was the "vector product" of the vectors with initial point $M(s,t)$ and terminal points the points $M(s+ds,t)$ and $M(s,t+dt)$ of the surface, in other words, in our notation, the vectors $D_1\sigma(s,t)ds$ and $D_2\sigma(s,t)dt$ generating the tangent plane to S at the point M considered. As the classical "vector product" $h \wedge k$ of two vectors is orthogonal to them and its length is the area of the parallelogram given by h and k, saying that the components of the vector dS of physicists are the expressions (27) amounts to choosing, at each point $M \in S$, an orientation of the normal to S at M or, equivalently, a "positive direction of rotation" in the tangent plane to the surface S, namely that which takes the vector $D_1\sigma(s,t)$ to the vector $D_2\sigma(s,t)$; this choice determines the orientation of the normal: its unit vector together with the previous two vectors must form a "direct" trihedron.

Since the left hand side supposes the border curve C to be oriented and changes sign if the orientation is reversed, the surface integral inevitably involved a question of orientation. This difficulty was overcome by orienting coherently the normals to S and by orienting C accordingly. The "coherence" of the orientations of the normals to S meant (more accurately, presupposed) that the unit vector of the oriented normal at a point $M \in S$ had to a *continuous* function of M. As will be shown in n° 13, (iv), any smooth surface admits a parametric representation (26) in the neighbourhood of one of its points, and even, up to a permutation of its canonical coordinates, an equation $z = f(x,y)$; so its normals can be oriented *in the neighbourhood* of each of its points. When a global orientation can be found, the surface S' is said to be *orientable*. The well-known "Möbius strip" is not so (it is the image of a square under a map σ of rank 2 everywhere, but which is not injective since, to obtain a closed strip, the images of two of the opposite sides need to be identical) and in this case there are no Stokes formula in the traditional sense.

The orientation of C was then chosen by a simple rule. In good cases – physicists do not consider any other –, the curve C is indeed, as seen above, the image under σ of the boundary of the set $A \subset \mathbb{R}^2$ in which (s,t) varies. If S is oriented in this manner – i.e. by using the map σ to transfer to S the positive rotational direction in $\mathbb{R}^2$ –, then C must be oriented by transferring the traditional "positive" orientation of the boundary A by using $\partial\sigma$. It was then explained that if you follow the curve in the chosen direction while remaining constantly upright on the tangent plane to S so that the normal vector having your feet as initial and coming out of your head be oriented like the normal to the surface, then looking straight in front of you, you should see the surface on your left. Maxwell's immortal corkscrew rule was also available.

All this mess made mathematicians with "modernist" tendencies laugh or repelled them, as the case may be, especially those inspired by Elie Cartan.

Nevertheless, it should be mentioned that he spent far more efforts to give the theory a formal concise and esthetic aspect than to justify its formulas: a quite impossible task before the modern development of topology and of the theory of differentiable manifolds and which continues to raise serious problems today.[38]

10 – Change of Variables in a Multiple Integral

Thus in the classical version, a "surface integral" needed to be shown to be solely dependent on the geometric surface and not on its parametric representation (9.26). It is exact – up to sign as in dimension one where a question of orientation arises – for injective maps σ of maximum rank everywhere, but even in this case the answer is obviously not easily available: the corresponding statement in dimension one already requiring the change of variable formula in a simple integral, the same will necessarily hold in dimension two. Indeed, suppose that σ is replaced by $\sigma \circ \varphi$ where φ is a diffeomorphism from I^2 to I^2, which leaves the image of the square invariant. The integral over I^2 of

$$\omega \circ \sigma = r(s,t)ds \wedge dt$$

is then replaced by that of

$$\omega \circ (\sigma \circ \varphi) = (\omega \circ \sigma) \circ \varphi = r\left[\varphi(s,t)\right] J_\varphi(s,t)ds \wedge dt .$$

To solve the problem, it is thus necessary to already know that

$$(10.1) \qquad \iint_{I^2} r(s,t)dsdt = \iint_{I^2} r\left[\varphi(s,t)\right] J_\varphi(s,t)dsdt$$

holds *up to sign* (orientation!); it is a particular case of the change of variable formula in multiple integrals which will be proved in this n°. Indeed, if φ is a diffeomorphism, its Jacobian does not vanish in K, hence its sign remains constant there. If the function r is positive, then so is the second integral. Hence the sign to be used is necessarily that of $J_\phi(s,t)$, which will be written $\mathrm{sgn}(\varphi)$. In other words, the correct formula is

$$(10.2) \qquad \iint_K r\left[\varphi(s,t)\right] J_\varphi(s,t)dsdt = \mathrm{sgn}(\varphi) \iint_K r(x,y)dxdy$$

or, equivalently,

$$(10.3) \qquad \iint_K r\left[\varphi(s,t)\right] \cdot |J_\varphi(s,t)|\, dsdt = \iint_K r(x,y)dxdy .$$

---

[38] Henri Cartan, *Calcul différentiel*, still need nineteen pages to prove Stokes' formula and that of change of variable in the most simple case of a compact subset of $\mathbb{R}^2$ bounded by a reasonable curve.

This formula can be made to apply to much more general situations:

Theorem 4. *Let $U \subset \mathbb{R}^n$ be an open bounded set, A its closure and φ a diffeomorphism of class C^1 from A to a compact subset $B \subset \mathbb{R}^n$. Suppose that the borders of A and B have measure zero. Then, for any integrable*[39] *function f on B,*

$$(10.4) \qquad \int_{\varphi(A)} f(x)dx = \int_A f[\varphi(t)] \cdot |J_\varphi(t)| \, dt$$

denoting by $x = (x^1, \ldots, x^n)$ or $t = (t^1, \ldots, t^n)$ the integration variable, by $dm(x) = dx = dx^1 \ldots dx^n$ or dt the Lebesgue measure on $\mathbb{R}^n$ and writing $\int$ for a multiple integral. This is similar to what happens in a *non-oriented* simple integral in which a strictly monotone change of variable $x = \varphi(t)$ is being carried out: the factor $\varphi'(t)$ involved in the oriented integrals has to be replaced by its absolute value:

$$\int_{\varphi(I)} f(x)dx = \int_I f[\varphi(t)] \cdot |\varphi'(t)| \, dt$$

since otherwise the sides could have opposite signs.

We will prove the formula for continuous functions. By n° 10 and 11 of Chapter V, § 2 it can then be immediately generalized to lsc (resp. usc) functions; for this, simply observe that if there is an increasing (resp. decreasing) philtre Φ for continuous functions, then the functions $f[\varphi(s,t)]|J_\varphi(s,t)|$,where $f \in \Phi$, form an increasing (resp. decreasing) philtre of continuous functions on A. Hence it is possible to pass to the limit under the $\int$ sign on both sides of this formula. Lebesgue's theorems lead to the general case in a few lines. In particular, it applies if f is the characteristic function of an open or closed subset of $\varphi(A)$, hence of the form $\varphi(M)$ where $M \subset A$ is an open or closed subset in A. We find the measure of $\varphi(M)$ on the left hand side; on the right, we integrate the characteristic function of M, whence

$$(10.4') \qquad m[\varphi(M)] = \int_M |J_\varphi(t)| \, dt \, .$$

(i) *Case where φ is linear.* This is the simplest case and we will give a proof using classical properties[40] of the group $GL_n(\mathbb{R}) = G$ of $n \times n$ real

---

[39] In the sense of Lebesgue.

[40] As shown by K. Iwasawa about 1950 in an article rightly famous, in particular lemma d below if it is properly interpreted, these properties generalize to all semisimple Lie groups, a class of groups singled out by Élie Cartan whose extraordinary properties continue to be the subject of numerous studies mixing algebraic geometry, number theory, generalizations of the theory of modular functions, non-commutative harmonic analysis (there is a quite sophisticated version of Fourier transforms for these groups), PDEs, etc. Apart from "classical" groups such as the matrix group equipped with a symmetric or alternating bi-

invertible matrices and, contrary to some shorter ones, requiring little integral calculations.

Set $E = \mathbb{R}^n$ and let $L(E)$ be the set of continuous functions with compact support in E. We will not differentiate between a matrix $g \in G$ and the linear map $x \mapsto g(x)$ corresponding to it on E and for which g is the matrix with respect to the canonical basis (e_i): the formulas

$$(10.5) \qquad g(e_i) = g_i^j e_j, \quad g(x)^i = g_j^i x^j,$$

where the $g(x)^i$ are the canonical coordinates of the vector $g(x)$, in conformity with tensor conventions, then follow. The Jacobian of $y = g(x)$ with respect to x is then equal to $\det(g)$, so that it suffices to prove that for all $f \in L(E)$,

$$(10.6) \qquad \int f[g(x)]\, dx = |\det(g)|^{-1} \int f(x) dx,$$

where integration is over all of E.

Lemma a. *Let μ be a Radon measure on $\mathbb{R}^n$. Assume μ to be invariant under translations. Then μ is proportional to the Lebesgue measure m.*

Note first that, for any $f, g \in L(E)$, the function $(x, y) \mapsto f(x)g(y - x)$ has compact support in $E \times E$, since if f and g are zero outside the compact sets M and N, then it can be $\neq 0$ only if (x, y) belongs to the compact set $M \times (M + N)$. The most elementary Lebesgue-Fubini theorem can then be applied to this functions, and in the following, formal calculations are possible. Having said this,

$$m(f)\mu(g) = \iint f(x)g(y)dm(x)d\mu(y) = \iint f(x)g(y - x)dm(x)d\mu(y)$$

by the change of variable $y \mapsto y - x$ in the integration with respect to μ; the change of variable $x \mapsto x + y$ in the integration with respect to m then gives

$$m(f)\mu(g) = \iint f(x + y)g(-x)dm(x)d\mu(y) =$$
$$= \int g(-x)dm(x) \int f(x + y)d\mu(y) =$$
$$= \mu(f) \int g(-x)dm(x).$$

---

linear form, this category contains "exceptional" groups whose construction is far less obvious.

For a more elementary but far less instructive proof, see for example Rudin, *Real and Complex Analysis*, end of Chapter 8, where a proof of the complete result analogous to ours can also be found, as well as more subtle results from the theory of integration. Dieudonné, *Eléments d'analyse* (vol. 3, XVI.22) uses the fact that, locally, all diffeomorphisms decompose into simpler maps (changing one variable at a time) for which the formula is more or less obvious, thereby avoiding all approximation calculations presented in parts (ii) and (iii) below.

Choosing g so that $\int g(-x)dm(x) = 1$,

$$\mu(f) = \mu(g)m(f)$$

for all f, qed.

Lemma b. *For every $g \in G$, there is a number $\Delta(g) > 0$ such that*

(10.7) $$\int f[g(x)]\,dm(x) = \Delta(g)\int f(x)dm(x)$$

for all $f \in L(E)$.

Formula $\mu(f) = \int f[g(x)]dm(x)$ defines a Radon measure on E and μ is clearly invariant, hence the formula. $\Delta(g) > 0$ because if f is positive, then so are the two integrals.

As can now be seen, if g is the diagonal matrix $(t_1, \ldots, t_n)$, in which case the left hand side of (7) is the integral of $f(t_1 x^1, \ldots, t_n x^n)$, then the change of $x^i \mapsto t_i x^i$ shows that $\Delta(g) = |t_1 \ldots t_n|^{-1} = |\det(g)|^{-1}$.

Lemma c. *The map Δ is a continuous homomorphism from G to the multiplicative group $\mathbb{R}_+^*$.*

The relation $\Delta(gh) = \Delta(g)\Delta(h)$ can be obtained by apply lemma b twice. To show that Δ is continuous, observe fist that, like $g(x)$, $f[g(x)]$ is a continuous function of the couple $(g, x) \in G \times E$; it is, moreover, zero outside $g^{-1}(M)$, where $M \subset E$.is the compact support of f; but when g varies in a compact subset of G, the set $g^{-1}(M)$ remains in the image K of the compact set $N \times M$ under the map $(g, x) \mapsto g^{-1}(x)$. As $g \mapsto g^{-1}$, and so $(g, x) \mapsto g^{-1}(x)$ is also continuous, K is compact. Hence if g remains in a fixed compact subset N of G, as can be seen, the integral of lemma b in fact generalizes to a fixed compact subset of E. Continuity with respect to the parameter $g \in N$ is then clear (Chapter V, §2, n° 9, Theorem 9). Observing that G being an open subset of $M_n(\mathbb{R})$, all $g \in G$ have a compact neighbourhood N, leads to the conclusion.

The rest of the proof consists in determining all the continuous homomorphisms from G to $\mathbb{R}_+^*$: they are all of the form $g \mapsto |\det(g)|^s$, with $s \in \mathbb{R}$.

Lemma d. *For every matrix $g \in G$, there are orthogonal matrices u and v and a diagonal matrix $t = (t_1, \ldots, t_n)$, with $t_i > 0$, such that $g = utv$.*

Denote the usual scalar product on $\mathbb{R}^n$ by $(x|y)$ and write g' for the transpose of a matrix g. It is characterized by the identity

$$(g(x)|y) = (x|g'(y)) .$$

The orthogonal subgroup $K = O_n(\mathbb{R})$ of G is the set of matrices such that $g'g = 1$, i.e. such that $\|g(x)\| = \|x\|$ for all x; it is obviously closed and bounded in $M_n(\mathbb{R})$, and so compact. On the other hand, there are symmetric matrices in G, and even in $M_n(\mathbb{R})$, i.e. such that $h' = h$; for such matrices, we know there is an orthonormal basis (a_i) in $\mathbb{R}^n$ and scalars $t_i \in \mathbb{R}$ such that $h(a_i) = t_i a_i$ for all i (diagonalization), and conversely. Writing t for the diagonal matrix $(t_1, \ldots, t_n)$ and u for the orthogonal matrix transforming the canonical basis (e_i) into (a_i),

$$hu\,(e_i) = h\,(a_i) = t_i u\,(e_i) = u\,(t_i e_i) = ut\,(e_i)$$

for all i (no summation over i, obviously!), and so $hu = ut$, i.e.

$$h = utu^{-1}.$$

This result holds for all $g \in G$ and $h = g'g$. As

$$(g'g(x)|x) = (g(x)|g(x)) > 0$$

for all $x \neq 0$, $t_i > 0$ in this case. Then write $h^{1/2}$ for the operator given by $h^{1/2}(a_i) = (t_i)^{1/2} a_i$; it is symmetric and

$$(g(x)|g(y)) = (g'g(x)|y) = \left(h^{1/2}h^{1/2}(x)|y\right) = \left(h^{1/2}(x)|h^{1/2}(y)\right)$$

for all x and y, and so $(gh^{-1/2}(x)|gh^{-1/2}(y)) = (x|y)$. The orthogonality of $gh^{-1/2} = w$ follows. But the argument showing that $h = utu^{-1}$ shows as well that $h^{1/2} = ut^{1/2}u^{-1}$, and so, finally,

$$g = wh^{1/2} = wut^{1/2}u^{-1} = vt^{1/2}u^{-1},$$

qed.

Lemma e. *Let K be a compact topological group and Δ a continuous homomorphism from G to the multiplicative group $\mathbb{C}^*$. Then $|\Delta(k)| = 1$ for all $k \in K$.*

The image of K under Δ is indeed a compact subgroup H of $\mathbb{R}^*$. For any $t \in H$, the set of the $t^n (n \in \mathbb{Z})$ must, therefore, be bounded, and so $|t| = 1$. Corollary: $|\det(u)| = 1$ for all $u \in O_n(\mathbb{R})$.

Lemma f. *Any continuous homomorphism Δ from $\mathbb{R}_+^*$ to $\mathbb{R}_+^*$ is of the form $\Delta(t) = t^s$ for some $s \in \mathbb{R}$.*

This is the characterization of power functions: Chapter IV, n° 6, Theorem 4.

We can now return to the calculation of the factor $\Delta(g)$ of lemma b. Writing $g = utv$, by lemma e, $\Delta(g) = \Delta(u)\Delta(t)\Delta(v) = \Delta(t)$. If t is the diagonal matrix $(1, \ldots, 1, t, 1, \ldots, 1)$, where $t > 0$ is in the ith place, $\Delta(t)$ is a power function of t by lemma f. As any positive diagonal matrix is a product of like matrices, we get a formula of type

$$\Delta(t) = t_1^{s_1} \ldots t_n^{s_n}$$

with the $s_i \in \mathbb{R}$ a priori arbitrary.

But let us consider the linear operator w_σ which, for a given permutation σ of $\{1,\ldots,n\}$, transforms e_i into $e_{\sigma(i)}$. Then $w_\sigma \in K$ and $w_\sigma t w_\sigma^{-1} = (t_{\sigma(1)},\ldots,t_{\sigma(n)})$. As $\Delta(w_\sigma t w_\sigma^{-1}) = \Delta(t)$, $s_1 = \ldots = s_n = s$ by lemma e, and so

$$(10.8) \qquad\qquad \Delta(t) = \det(t)^s .$$

For arbitrary $g \in G$, formula $g = utv$ then shows that

$$(10.9) \qquad\qquad \Delta(g) = |\det(g)|^s$$

with an absolute value since $\Delta(g)$ must be > 0. This is the general form of continuous homomorphisms from $GL_n(\mathbb{R})$ to $\mathbb{R}^*_+$.

To finish the proof of the change of variable formula, it remains to observe that, if $g(x) = tx$ where $t \in \mathbb{R}$ is non-zero, then the factor $\Delta(g)$ of lemma b is clearly equal to $|t|^{-n}$, and so $s = -1$.

Exercise. Show that any $g \in GL_n(\mathbb{R})$ can be written as $g = khu$ where k is orthogonal, h positive diagonal and u triangular with diagonal $(1,\ldots,1)$; verify (6) for $g = u$ by changing variables in the simple integrals and deduce (6) for g. (The decomposition $g = khu$ means that any basis (a_i) of $\mathbb{R}^n$ can orthonormalized by a *triangular* linear map applied to the a_i : Gram-Schmidt orthogonalization process.

Exercise. Set $G = GL_n(\mathbb{C})$. Show that any continuous homomorphism from G to $\mathbb{C}^*$ (resp. $\mathbb{R}^*$) is of the form $g \mapsto |\det(g)|^s \det(g)^p$ with $s \in \mathbb{C}$ and $p \in \mathbb{Z}$ [resp. $s \in \mathbb{R}$, $p \in \{1,-1\}$].

(ii) *Approximation Lemmas.* We now return to the general case of the theorem. To prove it, we need some preliminary results justifying what physicists take to be obvious, that in the neighbourhood of a point a, a diffeomorphism is approximately linear, and "so" multiplies the volumes by the absolute value of the determinant of its differential map. It then suffices to add the results and one easily obtains the general formula...

In what follows, the length or the norm $|u|$ of a vector $u \in \mathbb{R}^n$ is defined by

$$(10.10) \qquad\qquad |u| = \sup \left(|u^1|,\ldots,|u^n|\right) ;$$

for this "cubic" norm, an open ball centered at a and of radius r is the set $|u^i - a^i| < r$. To avoid confusion, it will be call the *cube* centered at a and of radius radius r, and will be written $U(a,r)$ or $K(a,r)$ according to whether the cube is open or closed. Because of this norm, a reasonable set can be approximately decomposed into parallelepipeds whose pairwise intersections are faces that are negligible in the integration. This operation cannot be performed using real Euclidean balls. As in any normed vector space, the norm of a linear map A is defined by

$$\|A\| = \sup |Ax|/|x| .$$

The crucial point is the following lemma, in which $K(r) = K(0, r)$.

Lemma 1. *Let U be an open subset of $\mathbb{R}^n$ containing 0 and $\psi : U \longrightarrow \mathbb{R}^n$ a C^1 map such that $\psi(0) = 0$, $\psi'(0) = 1$. Given a number q such that $0 < q < 1$, let r be a number > 0 such that $K(r) \subset U$ and*

$$(10.11) \qquad |x| < r \Longrightarrow \|\psi'(x) - 1\| < q .$$

Then

$$(10.12) \qquad K\left[(1 - q)r\right] \subset \psi\left[K(r)\right] \subset K\left[(1 + q)r\right] .$$

The derivative $\psi'(x)$ being a continuous function of x equal to 1 at $x = 0$, the existence of r for given $q > 0$ is obvious. Having said that, suppose $x \in K(r)$, so that $tx \in K(r)$ for $0 \leq t \leq 1$. The derivative of $t \mapsto \psi(tx)$ is $\psi'(tx)x$; as $\psi(0) = 0$, $\psi(x) = \int \psi'(tx)x dt$, and so

$$\psi(x) - x = \int \left[\psi'(tx) - 1\right] x . dt ,$$

where integration is over $(0, 1)$. Since, by (11),

$$|\psi'(tx)x - x| \leq \|\psi'(tx) - 1\| . |x| \leq \|\psi'(tx) - 1\| r \leq qr$$

$|\psi(x) - x| \leq qr$ and so

$$|\psi(x)| \leq |x| + r \leq (1 + q)r ,$$

which proves the right half of (12).

To prove the other less easy inequality, imitating the proof of the local inversion theorem· is a possibility. It all amounts to showing that, for all $\zeta \in K[(1 - q)r]$, there exists $z \in K(r)$ such that $\psi(z) = \zeta$. For this, set $\psi(z) = z + p(z)$. So $p'(z) = \psi'(z) - 1$; by (11),

$$(10.13) \qquad |z| \leq r \Longrightarrow |p(z)| \leq q|z| .$$

Then construct a sequence of points

$$z_1 = \zeta, \quad z_2 = \zeta - p(z_1) , \quad z_3 = \zeta - p(z_2) , \quad \ldots$$

as in Chap. III, § 5, Theorem 24, whose proof we follow (except that, lack of foresight, $q = 1/2$ was chosen in Chapter III). We must make sure that the construction continues without obstruction, i.e. that $z_1 \in K(r)$ – obvious – and that $z_n \in K(r)$ implies $z_{n+1} \in K(r)$. Now, by (13),

$$|z_{n+1}| \leq |\zeta| + |p(z_n)| \leq (1 - q)r + q|z_n| \leq r .$$

Having done this, (13) shows that

$$|z_{n+1} - z_n| = |p(z_n) - p(z_{n-1})| \leq q|z_n - z_{n-1}|$$

with $q < 1$. Hence $\lim z_n = z \in K(r)$ exists, with obviously $\psi(z) = \zeta$, qed.

Let us now suppose that the assumptions of theorem 4 hold.

Lemma 2. *For all $q > 0$, there exists $r > 0$ such that, for $x, y \in A$,*

$$(10.14) \qquad |x - y| \leq r \implies \|\varphi'(x) - \varphi'(y)\| \leq q/\|\varphi'(x)^{-1}\| \,.$$

The map $x \mapsto \varphi'(x)$ being continuous on A and $\varphi'(x)$ being invertible for all $x \in A$, the map $x \mapsto \varphi'(x)^{-1}$ is also continuous on A: indeed Cramer's formulas[41] tell us how to calculate the entries of the matrix of $\varphi'(x)^{-1}$ from those of the matrix of $\varphi'(x)$. The norm $\|\varphi'(x)^{-1}\|$ is, therefore, also a continuous function on A and, A being compact, it is bounded on A. Setting $\sup \|\varphi'(x)^{-1}\| = 1/M$, $M \leq 1/\|\varphi'(x)^{-1}\|$ for all $x \in A$ and (14) will hold if

$$(10.15) \qquad |x - y| \leq r \implies \|\varphi'(x) - \varphi'(y)\| \leq Mq \,.$$

But $x \mapsto \varphi'(x)$ is uniformly continuous since A is compact. So for all $q > 0$, there exists r satisfying (15), qed.

In the following statement, $m(X)$ denotes the Lebesgue measure of a measurable set $X \subset \mathbb{R}^n$, as it happens a compact set.

Lemma 3. *For all q such that $0 < q < 1$, there exists $r > 0$ satisfying the following property: for all $a \in U$ such that $K(a, r) = K \subset U$,*

$$(10.16) \qquad |m[\varphi(K)] - |J_\varphi(a)|.m(K)| \leq q.m(K) \,.$$

Take a point $a \in U$ and replace φ by

$$\varphi_a : x \longmapsto \varphi'(a)^{-1}[\varphi(a + x) - \varphi(a)] \,,$$

---

[41] There is an easier argument in the case of an arbitrary Banach space E; it is based on the fact that, for any linear operator T with norm < 1, the operator $1 - T$ has an inverse, namely $\sum T^n$ (the series converges absolutely since $\|T^n\| \leq q^n$, where $q = \|T\|$). Let A and X be two continuous linear operators on E; set $X = A - Y$ and suppose that A is invertible. Then, $X = A(1 - A^{-1}Y)$, so that X is invertible if $\|A^{-1}Y\| = q < 1$; as $\|A^{-1}Y\| < \|A^{-1}\|.\|Y\|$, this is the case if $\|Y\| < 1/\|A^{-1}\|$, i.e. if X is sufficiently near A. Then, $X^{-1} = (1 - A^{-1}Y)^{-1}A^{-1} = \sum(A^{-1}Y)^n A^{-1}$, and so

$$\|X^{-1}\| < \|A^{-1}\|/(1 - q) < 2\|A^{-1}\|$$

if $\|Y\| \leq 1/2\|A^{-1}\|$. As $X^{-1} - A^{-1} = A^{-1}(A - X)X^{-1} = A^{-1}YX^{-1}$,

$$\|X^{-1} - A^{-1}\| < \|A^{-1}\|.\|X - A\|.\|X^{-1}\| < 2\|A^{-1}\|^2.\|X - A\|$$

follows. Hence $X \mapsto X^{-1}$ is continuous at A.

where $\varphi'(a)$ is the tangent linear map to φ at a. Then $\varphi_a(0) = 0$ and

$$\varphi'_a(x) = \varphi'(a)^{-1}\varphi'(a+x)\,,$$

and so $\varphi'_a(0) = id$. Choose a number $q' \in\,]0,1[$ such that

(10.17) $$1 - q \le (1 - q')^n \le (1 + q')^n \le 1 + q$$

– the end of the proof will explain this bizarre condition – and apply lemma 1 to φ_a by replacing q by q' in it. To be able to apply it, is suffices that φ_a be defined for $|x| < r$, i.e. that $K(a, r) \subset U$, and that

(10.18) $$|x| \le r \Longrightarrow \|\varphi'_a(x) - 1\| \le q'\,.$$

But

$$\begin{aligned}
\|\varphi'_a(x) - 1\| &= \|\varphi'(a)^{-1}\varphi'(a+x) - 1\| = \\
&= \|\varphi'(a)^{-1}\left[\varphi'(a+x) - \varphi'(a)\right]\| \le \\
&\le \|\varphi'(a)^{-1}\|.\|\varphi'(a+x) - \varphi'(a)\|\,.
\end{aligned}$$

Condition (18) will, therefore, hold if, for $x, y \in U$,

(10.19) $$|x - y| \le r \Longrightarrow \|\varphi'(y) - \varphi'(x)\| \le q'/\|\varphi'(x)^{-1}\|\,.$$

Lemma 2 shows that, for all $q' > 0$, there exists r satisfying this condition. Hence r can indeed be chosen so that (18) holds for all $a \in U$ such that $K(a, r) \subset U$.

Having done this, consider these points $a \in U$. Lemma 1 applied to φ_a shows that

$$K\left(r - q'r\right) \subset \varphi_a\left[K(r)\right] \subset K\left(r + q'r\right)\,,$$

where $K(r) = K(0, r)$. Applying $\varphi'(a)$ to the terms of this relation, φ_a is replaced by the map $x \mapsto \varphi(a+x) - \varphi(a)$; the image of $K(r)$ under this map is[42] $\varphi[a + K(r)] - \varphi(a) = \varphi[K(a, r)] - \varphi(a)$; hence, setting $K = K(a, r)$ as above,

$$\varphi'(a)\left[K\left(r - q'r\right)\right] \subset \varphi(K) - \varphi(a) \subset \varphi'(a)\left[K\left(r + q'r\right)\right]\,,$$

and so, applying the translation by the vector $\varphi(a)$,

(10.20) $$\varphi'(a)\left[K\left(r - q'r\right)\right] + \varphi(a) \subset \varphi(K) \subset \varphi'(a)\left[K\left(r + q'r\right)\right] + \varphi(a)\,.$$

But since $\varphi'(a)$ is linear, formula (6) shows that, for all $r > 0$,

$$m\left\{\varphi'(a)\left[K(r)\right]\right\} = |J_\varphi(a)|\, m\left[K(r)\right]\,.$$

---

[42] For a set $E \subset \mathbb{R}^n$ and some $b \in \mathbb{R}^n$, the notation $E + b$ denotes the image of E under the translation $u \mapsto u + b$. In particular, $K(r) + a = K(a, r)$.

The measure of $K(r)$ being proportional to that of r^n, (20) shows that

(10.21) $$(1 - q')^n \le m\left[\varphi(K)\right]/|J_\varphi(a)|m(K) \le (1 + q')^n \,.$$

However, condition (17) has been imposed on q'; it necessarily follows that

$$1 - q \le m\left[\varphi(K)\right]/|J_\varphi(a)|m(K) \le 1 + q \,,$$

and hence (16) holds, qed.

Lemma 4. *Let U be an open subset and φ a C^1 map defined on U. Then $\varphi(M)$ has measure zero for any compact set[43] of measure zero $M \subset U$.*

Let $d > 0$ be the distance from the compact subspace M to the border of U and M' the set of $x \in U$ such that $d(x, M) \le d/2$. It is also a compact set contained in U. Let k be an integer > 0 such that $1/2^k < d/2$ and let us take a grid of $\mathbb{R}^n$ by hyperplanes defined by a single equation $x^i = p/2^k$, with $p \in \mathbb{Z}$ and $i \in \{1, \dots, n\}$. $\mathbb{R}^n$ can thereby be decomposed into cubes of the form $K(a, 1/2^{k+1})$ whose pairwise intersections are at most faces of dimension $\le n - 1$. Finitely many of these cubes intersect M non-trivially since M is bounded; they are all contained in M' since their diameter d_k is $< d/2$; and finally, they cover M. Let M_k be their union. $\varphi(M) \subset \varphi(M_k) = \bigcup \varphi(K)$, where K varies in the set of cubes $K(a, 1/2^{k+1})$ comprising M_k.

To find an upper bound for the measures of these $\varphi(K)$, observe that, these cubes being convex, by the (FT),

$$\varphi(x) - \varphi(y) = \int_0^1 \varphi'\left[tx + (1 - t)y\right](x - y)dt$$

and so

$$|\varphi(x) - \varphi(y)| \le |x - y|.\|\varphi'\|_K \le |x - y|.\|\varphi'\|_{M'} \,,$$

for all $x, y \in K$. In conclusion, since all cubes K considered have the same diameter d_k, $\varphi(K)$ is contained in a cube of diameter $\le cd_k$ where $c = \|\varphi'\|_{M'}$, a uniform norm on M'. The measure of a cube of diameter d being proportional to d^n, it follows that

$$m\left[\varphi(K)\right] \le cm(K) \,.$$

However, $m(M_k) = \sum m(K)$, where summation is over the cubes comprising M_k, because their pairwise intersections have measure zero since they are contained in the hyperplanes of $\mathbb{R}^n$. Also,

---

[43] There is a far stronger result: if U is an open subset of $\mathbb{R}^n$, any C^1 map $\varphi : U \longrightarrow \mathbb{R}^n$ transforms sets contained in U into sets of measure zero, without any compactness assumption. See Dieudonné, *Eléments d'analyse*, XVI.22, exercises 1 and 2. More complete results can be found in Rudin, *Real and Complex Analysis*, chap. 8.

(10.22) $m\left[\varphi(M)\right] \leq m\left[\varphi\left(M_k\right)\right] \leq \sum m\left[\varphi(K)\right] \leq \sum cm(K) = cm\left(M_k\right)$.

The Lemma will, therefore, follow once we have shown that $\lim m(M_k) = 0$.

But let us compare the sets M_k and M_{k+1}. To obtain the first one, take a grid of $\mathbb{R}^n$ by the hyperplanes $x^i = p/2^k$, whereas the second ones is obtained by using the hyperplanes $x^i = q/2^{k+1}$. Clearly, any cube K of the second grid is contained in at least one cube K' of the first; if K meets M, the same holds for K'. Hence $M_{k+1} \subset M_k$, which gives a decreasing sequence of compact sets contained in M. By definition, any point of M_k belongs to a cube of radius $1/2^k$ meeting M, and so is at a distance $\leq 1/2^k$ from M. Any point common to all the M_k is, therefore, at distance zero from M, in other words belongs to M since M is *closed*.

However, when there is a decreasing sequence of closed sets[44] M_k with intersection M, we know that $m(M) = \lim m(M_k)$; this was shown in Chapter V, §2, end of n° 11, for an increasing sequence of open sets, but, as mentioned then, the result and the proof remain the same for a decreasing sequence of closed sets. Since here $m(M) = 0$, relation (22) shows that $m[\varphi(M)] = 0$, qed.

(iii) *Change of variable formula.* The general formula

$$\int_{\varphi(A)} f(x)dx = \int_A f\left[\varphi(t)\right].|J_\varphi(t)|dt$$

can now be proved by replacing $A = \bar{U}$ with simpler sets, unions of cubes, and then passing to the limit.

Choose a number $r > 0$. For any integer $k > 0$, let us once again take a grid of $\mathbb{R}^n$ by the hyperplanes $x^i = p/2^k$. Denote the finitely many cubes in this grid contained in U by $K_1, \ldots, K_N$ and let $A_k \subset U$ be their union. The pairwise intersections of these cubes being compact and having measure zero, the same holds (lemma 4) for their images; hence

(10.23) $\displaystyle\int_{A_k} f\left[\varphi(t)\right].|J_\varphi(t)|dt = \sum \int_{K_i} f\left[\varphi(t)\right].|J_\varphi(t)|dm(t)$.

The function integrated in (23) being uniformly continuous on the compact set A, for any $r > 0$, k may be assumed to be sufficiently large for it to be constant, up to r, in each cube K_i. Setting a_i to be the centre of K_i and $b_i = \varphi(a_i)$, the general term of the right hand side of (23) is seen to be equal to $f(b_i).|J_\varphi(a_i)|m(K_i)$, up to $m(K_i)r$.

Now, setting $D_i = \varphi(K_i)$, $B_k = \varphi(A_k) = \bigcup D_i \subset B = \varphi(A)$, lemma 4 shows that the pairwise intersections of the $\varphi(A_k)$ have measure zero. So once again,

(10.23') $\displaystyle\int_{B_k} f(x)dx = \sum \int_{D_i} f(x)dx$.

---

[44] more generally of "measurable" sets as Lebesgue's complete theory will show.

As in (23), the uniform continuity of f shows that, for k sufficiently large, the general term of the right hand side of (23') is equal to $f(b_i)m(D_i)$ up to $m(D_i)r$. Since $m(A_k) = \sum m(K_i)$ and $m(B_k) = \sum m(D_i)$, the left hand sides of (23) and (23') are respectively equal to

(10.24) $\qquad\qquad \sum f(b_i).|J_\varphi(a_i)|m(K_i) \qquad\qquad$ up to $m(A_k)r$,

(10.24') $\qquad\qquad\qquad \sum f(b_i)m(D_i) \qquad\qquad$ up to $m(B_k)r$.

But lemma 3 applies to the K_i – they are contained in U – provided k sufficiently large. Then

(10.25) $\qquad\qquad m(D_i) = |J_\varphi(a_i)|m(K_i)$ up to $m(K_i)r$,

so that replacing sum (24) by sum (24') the error made is bounded above by

$$\sum |f(b_i)|m(K_i) \leq \|f\|_A \sum m(K_i) = \|f\|_A m(A_k).$$

In view of the errors made while replacing the right hand sides of (23) and (23') by the "Riemann" sums (24) and (24') , the absolute value of the difference between these right hand sides is bounded above by

$$m(A_k)r + m(B_k)r + \|f\|_A m(A_k)r.$$

But clearly, $m(A_k) \leq m(A)$ and $m(B_k) \leq m(B)$; the error found is, therefore, $< cr$, where $c = (1 + \|f\|_A)m(A) + m(B)$ does not depend on r.

This argument shows that, for all $r > 0$, the inequality

(10.26) $\qquad\qquad \left| \int_{A_k} f[\varphi(t)].|J_\varphi(t)|\, dt - \int_{B_k} f(x)dx \right| \leq cr$

holds for all sufficiently large k. Next consider what happens when k is replaced by $k+1$. Each cube from the first grid is the the union of cubes from the second; if a cube from the first one is contained in U, those from the second one comprising it are necessarily so. Hence $A_k \subset A_{k+1}$, and so this time we get an *increasing* sequence of compact (and so "measurable") sets contained in U. Their union is equal to U since any $a \in U$ is at a distance > 0 from the border of U and so is contained in one of the cubes of A_k for sufficiently large k.

If our familiarity with the theory of integration goes slightly further than Chap. V, §2, n° 11, we can conclude that the extended integrals over A_k and B_k in (26) converge to the extended integrals over U and V as $k \longrightarrow +\infty$. $r > 0$ being arbitrary, it follows that

$$\int_U f[\varphi(t)].|J_\varphi(t)|dt = \int_V f(x)dx.$$

To finish the proof of theorem 4, it remains to observe that if the borders $A - U$ and $B - V$ have measure zero,[45] the previous integrals remain invariant if U and V are replaced by A and B. This indicates that the essential result is in fact the previous formula, which does not assume anything about the borders of U and V.

(iv) *Stokes' formula for a p-dimensional path.* The definition of the extended integral of a form of degree 2 over a 2-dimensional path σ can be generalized in an obvious manner: if ω is a form of degree p on an open subset G of a Cartesian space E and if $\sigma : I^p \longrightarrow G$ is a a *p-dimensional path* (or *singular cube*) assumed to be of class at least C^1 so that the partial derivatives of σ extend by continuity to the border of I^p, then, by definition, set

$$(10.27) \qquad \int_\sigma \omega = \int_{I^p} \omega\left[\sigma(t); D_1\sigma(t), \ldots, D_p\sigma(t)\right] dt,$$

where $t = (t^1, \ldots, t^p)$ and where $dt = dt^1 \ldots dt^p$ is the usual Lebesgue measure. This obviously amounts to setting

$$\omega \circ \sigma = r(t) dt^1 \wedge \ldots \wedge dt^p \quad \text{and} \quad \int_\sigma \omega = \int_{I^p} r(t) dt$$

as in degree 2. If σ is replaced by $\sigma \circ \varphi$, where $\varphi : I^p \longrightarrow I^p$ is a diffeomorphism, then, by the associativity formula (8.19), $\omega \circ \sigma = \varpi$ is replaced by

$$\omega \circ (\sigma \circ \varphi) = (\omega \circ \sigma) \circ \varphi = \varpi \circ \varphi.$$

So $r(s)$ is replaced by $r[\varphi(t)] J_\varphi(t)$. Formula (2), seen to be equivalent to Theorem 4, then shows that

$$(10.28) \qquad \int_{\sigma \circ \varphi} \omega = \operatorname{sgn}(\varphi) \int_\sigma \omega,$$

where $\operatorname{sgn}(\varphi)$ is the, necessarily constant, sign of the Jacobian of φ.

This formula resembles the definition of the integral of a form ω along a path σ such as the integral over the cube of its inverse image under σ, but they should not be confused: (28) is a theorem and not a definition. More generally, consider the open subsets U and V of two Cartesian spaces, a C^1 map $f : U \longrightarrow V$ and a path p-dimensional σ in U. This gives a path $f \circ \sigma$ in V. Then

$$(10.29) \qquad \int_{f \circ \sigma} \omega = \int_\sigma \omega \circ f$$

---

[45] By lemma 4, this is always the case when φ can be extended to a C^1 function defined on an open set containing A.

for any differential form ω of degree p on V, which is even more like (28). But (29) means that integrals over the cube of $\omega \circ (f \circ \sigma)$ and $(\omega \circ f) \circ \sigma$ are equal; these two forms being identical by (8.19), there is nothing to show. Relation (29) is, therefore, almost a tautology or, equivalently, a direct consequence of the multivariate chain rule formula (2.13).

To take an example, let us return to the computations of n° 9, (ii) related to the effect of a homotopy σ on the integral of a form ω of degree 1 along a path; writing the final formula as

$$(10.30) \qquad \int_{\partial \sigma} \omega = \int_\sigma d\omega ,$$

it reduces to Gauss' formula for the square I^2.

An analogue of the Green-Riemann formula for the cube exists in arbitrary dimension; using Cauchy's simple calculation, the proof is the same as in dimension 2. The only difficulty concerns the definition of the extended integral of a form ω of degree p over ∂K for $K = I^{p+1}$; it is the sum of extended integrals over the faces of the cube, but the signs $+$ and $-$ placed before these integrals need to be determined. However, denoting the canonical coordinates of $\mathbb{R}^{p+1}$ by $t^0, \ldots, t^p$ gives formulas of type

$$\omega = \sum p_i(t) dt^0 \wedge \ldots \widehat{dt^i} \ldots \wedge dt^p ,$$
$$d\omega = \sum (-1)^i D_i p_i(t) dt^0 \wedge \ldots \wedge dt^p ,$$

so that the integral of $d\omega$ is the sum of the integrals over K of the functions $(-1)^i D_i p_i(t)$. To calculate them, first integrate with respect to the corresponding t^is. This gives (FT)

$$(-1)^i \int_{I^p} p_1 \left(t^0, \ldots, 1, \ldots, t^p \right) dt^0 \ldots dt^i \ldots dt^p +$$
$$+ (-1)^{i+1} \int_{I^p} p_1 \left(t^0, \ldots, 0, \ldots, t^p \right) dt^0 \ldots dt^i \ldots dt^p$$

involving the extended integrals of ω over the faces of the cube I^{p+1}. More precisely, write F_i^+ for the face $t_i = 1$ of the cube and F_i^- for the face $t_i = 0$, and define the extended integrals of ω over these faces by using the parametric representation

$$(10.31) \qquad \varphi_i^+ : (t_0, \ldots, \widehat{t_i}, \ldots, t_p) \longmapsto (t_0, \ldots, 1, \ldots, t_p)$$

in the first case and the the analogous formula in the second. If F is a face of the cube, set

$$(10.32) \qquad \varepsilon(F) = \begin{array}{ll} (-1)^i & \text{if } F = F_i^+ , \\ (-1)^{i+1} & \text{if } F = F_i^- . \end{array}$$

The Green-Riemann formula can then be written

$$(10.33) \qquad \int_K d\omega = \sum \varepsilon(F) \int_F \omega,$$

where, as mentioned above, the extended integrals of ω over the faces F of $K = I^{p+1}$ are defined by using the parametric representations (31), which amounts to regarding these faces as p-dimensional singular cubes.

To go from here to a formula of Stokes type for a form ω of degree p and an arbitrary path of dimension $p+1$ in an open subset G of a Cartesian space, we need to define what will be meant by the extended integral of ω over $\partial\sigma$; denoting the parametric representation (31) of the face F by φ_F, it will be the expression

$$\int_{\partial\sigma} \omega = \sum \varepsilon(F) \int_{\sigma\circ\varphi_F} \omega = \sum \varepsilon(F) \int_{\varphi_F} \omega\circ\sigma.$$

Using formula (33) for the cube, Stokes' formula for σ,

$$\int_{\partial\sigma} \omega = \int_\sigma d\omega$$

then becomes trivial.

The presence of the signs $\varepsilon(F)$ will become clear in n° 16 where a slightly different version of Stokes' formula will be proved; as in dimension 1 or 2, it corresponds to the necessity of choosing an "orientation" for each face F of K.

Exercise. Let $\sigma_0, \sigma_1 : I^p \longrightarrow G$ be two p-dimensional paths in an open subset G of a Cartesian space, coinciding on the border of I^p (the analogue of two one-dimensional paths with the same endpoints). They will be said to be fixed-border-homotopic if there is a path $\sigma : I \times I^p \longrightarrow G$ satisfying the following conditions: (i) $\sigma(0,t) = \sigma_0(t)$, $\sigma(1,t) = \sigma_1(t)$ for all $t \in I^p$, (ii) for every point t in the border of I^p, $\sigma(s,t)$ is independent of s. Show that, if ω is a closed form of degree p on G, then the extended integrals of ω over σ_0 and σ_1 are equal. Similarly, generalize the invariance under homotopy of the integral over a closed path.

§ 4. Differential Manifolds

This § gives a very summary treatment of the simplest aspects of the theory of differential manifolds. Even just considering its most basic notions, it has several other aspects. There are many excellent presentations of the subject and far more complete than ours.[46]

11 – What is a Manifold?

(i) *The sphere in* $\mathbb{R}^3$. To understand this problem, consider the unit sphere X having equation $x^2 + y^2 + z^2 = 1$ in $\mathbb{R}^3$. What would be a reasonable way to define differential functions on an open subset of X?

A first condition they need to satisfy is to be continuous and defined by properties of a local nature. Then consider a point $(a, b, c) \in X$ and a function f defined and continuous in the neighbourhood of this point. Suppose that $c > 0$. The upper hemisphere $H_+ : z > 0$ of X is an open subset of X and also the graph of a C^∞ function

$$z = \left(1 - x^2 - y^2\right)^{1/2},$$

defined on the open subset $x^2 + y^2 < 1$ of $\mathbb{R}^2$. Setting

$$\varphi(x, y, z) = (x, y),$$

define a homeomorphism from H_+ onto an open subset of $\mathbb{R}^2$ transforming f into a function of (x, y), defined and continuous in the neighbourhood of the point $\varphi(a, b, c) \in \mathbb{R}^2$. It is then natural to say that f is of class C^r in the neighbourhood of (a, b, c) if, as a function of (x, y), it is of class C^r in the classical sense. We adopt the same convention if $c < 0$, i.e. if we are in the lower hemisphere H_- of X, the graph of the function

$$z = -\left(1 - x^2 - y^2\right)^{1/2}.$$

If $c = 0$, perhaps $b < 0$; then replace H_+ by the hemisphere $y < 0$, which is the graph of the equation

$$y = -\left(1 - x^2 - z^2\right)^{1/2}.$$

Formula $\varphi(x, y, z) = (x, z)$ again defines a homeomorphism from this hemisphere onto an open subset of the plane and differentiable function in the

---

[46] Marcel Berger and Bernard Gostiaux, *Géométrie différentielle* (A. Colin, 1972), Paul Malliavin, *Géométrie différentielle intrinsèque* (Hermann, 1972), Pham Mau Quan, *Introduction à la géométrie des variétés différentiables* (Dunod, 1969), Frank W. Warner, *Foundations of Differential Manifolds and Lie Groups* (Scott, Foresman, 1971), Michael Spivak, *A Comprehensive Introduction to Differential Geometry* (Publish or Perish, Inc, 5 vol.), Shlomo Sternberg, *Lectures on Differential Geometry* (Prentice-Hall, 1964), Serge Lang, *Fundamentals of Differential Geometry* (Springer, 1999).

neighbourhood of (a, b, c) are, by definition, those that can be expressed in a differentiable manner using its coordinates x, z. Etc. Thus, X can be written as the union of six open subsets. On each of them there is a special homeomorphism onto an open subset of $\mathbb{R}^2$. Each of these homeomorphisms makes it possible to give a reasonable definition of C^r ($r \leq \infty$) functions on the corresponding open subset of X.

This definition would, however, be insignificant if it gave incompatible definitions of differentiability on the intersections of these open subsets; it is not at all so. For example, take an open set U, where both $\{z > 0\}$ and $\{y < 0\}$ hold; using the hemisphere $z > 0$, the relation $f \in C^r(U)$ means that f is a C^r function of (x, y); using the hemisphere $y < 0$, it means that f is a C^r function of (x, z). Hence it suffices to show that (x, y) is a C^∞ function of (x, z) on the open subset $\{z > 0\} \cap \{y < 0\}$ of X, and conversely. This is obvious since

$$y = - \left(1 - x^2 - z^2\right)^{1/2} \quad \text{and} \quad z = \left(1 - x^2 - y^2\right)^{1/2}$$

with $1 - x^2 - z^2 > 0$ and $1 - x^2 - y^2 > 0$.

This leads to a coherent definition of functions of class C^r on a random open subset of the sphere.

It can also be formulated more directly. First, given a C^r function of (x, y, z) on an open subset V of $\mathbb{R}^3$, it restriction to $U = V \cap X$ is of class C^r in the previous sense. Conversely, consider a C^r function f on an open subset U of X and take a neighbourhood of a point $(a, b, c) \in U$. If, for example $c > 0$, the definition shows that in the neighbourhood of (a, b, c), f is a C^r function of (x, y) defined in the neighbourhood of a point (a, b) of $\mathbb{R}^2$. The composition of f and the projection $(x, y, z) \mapsto (x, y)$ of $\mathbb{R}^3$ onto $\mathbb{R}^2$ is a C^r function of (x, y, z) on a vertical cylinder having an open subset of the plane (x, y) as base. The restriction of this function to X is equal to the given function f on a neighbourhood of (a, b, c). In conclusion, *a fuction f defined on an open subset U of X is of class C^r if and only if, in the neighbourhood of every point of U, it has a C^r extension on an open subset of $\mathbb{R}^3$.*

(ii) *The notion of a manifold of class C^r and dimension d* is obtained by generalizing the construction of C^r functions on the sphere.

To start with, a manifold X is a *separated* topological space. Hence there is a category of sets in X called open and satisfying the two obvious conditions (any union of open sets is open, the intersection of a finite number of open sets is open), as well as Hausdorff's condition: if a and b are two distinct points of X, there are open disjoint subsets U and V containing a and b. Topological spaces are the natural realm of the notion of continuity: a map $f : X \longrightarrow Y$ is continuous if and only if the inverse image $f^{-1}(V)$ of any open subset V of Y is an open subset of X.

A differential manifold X must also be a *locally Cartesian* topological space: for each $a \in X$, there must be an open neighbourhood U homeomorphic to an open subset of some space $\mathbb{R}^d$, where d is a given integer,[47] the *dimension* of X. Such a homeomorphism $\varphi = (\varphi^1, \ldots, \varphi^d)$, where the $\varphi^i(x)$ are the canonical coordinates of $\varphi(x)$, is, by definition, a *local topological chart* (U, φ) of X which makes it possible to identify points $x \in U$ by using d real scalars $\xi^i = \varphi^i(x)$, its *coordinates* in the chart considered.[48] The integer d is uniquely determined thanks to a well-known theorem (J. L. E. Brouwer) according to which an open subset of $\mathbb{R}^p$ can be homeomorphic to an open subset of $\mathbb{R}^q$ only if $p = q$. Peano's curve ($p = 1, q = 2$) is not a homeomorphism.

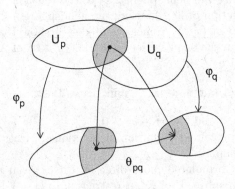

Fig. 11.5.

This definition supplies the topological or C^0 manifolds: only continuous functions can be reasonably defined on them. To turn X into a manifold of class C^r, the charts admitted need to be selected. Like in the case of the terrestrial sphere, a possible method is to take an *atlas* of class C^r of X, i.e. a finite or infinite family of topological charts (U_p, φ_p) covering X and pairwise C^r-*compatible*: for all p and q, there is a C^r map θ_{pq} (in the usual sense) from the open set $\varphi_p(U_{pq})$ to the open set $\varphi_q(U_{pq})$ taking $\varphi_p(x)$ to $\varphi_q(x)$ in $U_p \cap U_q = U_{pq}$:

$$(11.1) \qquad \varphi_q(x) = \theta_{pq}\left[\varphi_p(x)\right].$$

Changing the roles of p and q, it follows that θ_{pq} and θ_{qp} are mutually inverse and hence are diffeomorphisms.

For $s \leq r$, a function f defined on an open subset U of X will then be said to be of class C^s if, for all p, $f(x)$ is a function of class C^s from $\varphi_p(x)$ to the open set $U \cap U_p$.

---

[47] Some authors allow the dimension to depend on the point a. As it is locally constant, and so constant in each connected component of X, this generalization is of little interest.

[48] Hence, like any open subset of a Cartesian space, a manifold is locally compact. Apart from some rare exceptions, all manifolds encountered are unions of countably many compact and metrizable sets.

Manifolds have been defined using atlases, but the only thing that really matters in a manifold X is the set $C^r(U)$ associated to all open subsets $U \subset X$ of numerical functions defined and of class C^r on U. As in the case of the sphere, these functions are characterized by a local property: if a function f is defined on an union U of open subsets U_i, then $f \in C^r(U)$ if and only if the restriction of f to each U_i belongs to $C^r(U_i)$. Apart from its atlas, there are many other useful local charts in a manifold X defined in this manner, namely the *charts of class C^r* that, for short, will be henceforth almost always called charts, or local charts, without further precision when there is no ambiguity. These are the topological charts (U, φ) such that, for any open subset $U' \subset U$, the homeomorphism φ transforms $C^r(U')$ into the set of C^r functions (in the usual sense) on the open subset $\varphi(U')$ of $\mathbb{R}^d$. In other words, a function defined on U' is of class C^r if and only if it is a C^r function in the classical sense of the coordinates $\xi^i = \varphi^i(x)$. In particular, the coordinate functions $\varphi^i(x)$ must be of class C^r on U, but choosing randomly d functions in $C^r(U)$ is obviously not sufficient to obtain such a chart. Loosely speaking, an open subset U of X will be said to be *open Cartesian* if it is the domain of a chart (U, φ).

If (U, φ) and (V, ψ) are charts of class C^r of X, the C^r functions on $U \cap V$ must be the same, whether they be given by $\xi = \varphi(x)$ or $\eta = \psi(x)$. As in the above case of an atlas, this leads to the conclusion that, that the relations $\eta = \theta(\xi)$, $\xi = \rho(\eta)$ must hold in $U \cap V$, i.e.

$$(11.2) \qquad\qquad \varphi = \rho \circ \psi, \quad \psi = \theta \circ \varphi,$$

where $\theta : \varphi(U \cap V) \longrightarrow \psi(U \cap V)$ and $\rho : \psi(U \cap V) \longrightarrow \varphi(U \cap V)$ are of class C^r, in other words are mutually inverse diffeomorphisms. The (formulas of) *change* of *chart* (or coordinate) will be denoted ρ and θ. I do so because this was the notation used by the inventors of *absolute differential calculus* (n° 3, (ii)), who without knowing it were working in the theory of manifolds.

(iii) *Some Examples.* Take E to be a d-dimensional Cartesian space and choose a basis (a_i) for E. The point $\varphi(x) = \xi^i e_i \in \mathbb{R}^d$ can be associated to each $x = \xi^i a_i \in E$. Its *canonical* coordinates are those of x with respect to the basis of E. This gives a topological chart (E, φ) for E which, on its own, is an atlas of E and so turns E into a manifold of class C^∞. The ξ^i undergo a linear transformation under a change of basis in E. Clearly, the manifold structure thus defined on E does not depend on the choice of the basis (a_i) and for any open subset $U \subset E$, $C^r(U)$ is for all r a set of C^r functions on U in the classical sense.

In the case of the unit sphere X discussed above, there are, among others, two charts, a priori of class C^0, using the stereographic projection from the north or the south pole of X; they map the open complement U of this pole in X to $\mathbb{R}^2$. If we consider the projection from the north pole, then the function φ is easily seen to be given by

$$\varphi(x, y, z) = [x/(1 - z), y/(1 - z)]$$

with $x^2 + y^2 + z^2 = 1$. But considering any of the six local charts defined above, the corresponding coordinates are two of the variables x, y, z, the third one being, as we have seen, a C^∞ function of the other two. The chart φ is, therefore, C^∞- compatible with the atlas initially used since $z \neq 1$ outside the north pole and conversely. Hence the differentiable structure of S could have been defined by using these two stereographic projections.

To find an example of an "abstract" manifold, i.e. which is not given as a subset of a Cartesian space, consider the *projective space* $X = P_n(\mathbb{R})$; by definition, it is the set of 1-dimensional vector subspaces (lines with initial point the origin) of $\mathbb{R}^{n+1}$. It would amount to the same to establish the equivalence relation "x and y are proportional" on the set of *non-zero* vectors of $\mathbb{R}^{n+1}$ and to say that $P_n(\mathbb{R})$ is the quotient space of $\mathbb{R}^{n+1} - \{0\}$ modulo this relation would come to the same. An element of $P_n(\mathbb{R})$ is, therefore, characterized by $n+1$ numbers $x^0, \ldots, x^n$, not all of which are zero, defined up to a factor; these are its *homogeneous coordinates*. Any function f defined on a subset of $P_n(\mathbb{R})$ can be identified with a *homogeneous* function of these coordinates, i.e. such that

$$f\left(tx^0, \ldots, tx^n\right) = f\left(x^0, \ldots, x^n\right) \text{ for all } t \neq 0.$$

If $p(x)$ denotes the image of $x \in \mathbb{R}^{n+1} - \{0\}$ in $X = P_n(\mathbb{R})$, i.e. the subspace D generated by x, then a topology can be defined on X by requiring $U \subset X$ to be open if and only $p^{-1}(U)$ is open in $\mathbb{R}^{n+1} - \{0\}$ or, equivalently, if the union of the lines $D \in U$ is an open subset of $\mathbb{R}^{n+1} - \{0\}$. In particular, X is the union of the $n+1$ open sets $U_i (0 \leq i \leq n)$ that are images under p of the open subsets of $\mathbb{R}^{n+1}$ defined by $x^i \neq 0$. As $p(x) = p(x/x^i)$, $U_i = p(E_i)$ also holds, where E_i is the hyperplane with equation $\xi^i = 1$. U_i is, therefore, the set of lines D having non-trivial intersection with E_i and as such a line is determined by its (unique) intersection point with E_i, the map $p : E_i \longrightarrow U_i$ is bijective. This leads to an inverse map $q_i : U_i \longrightarrow E_i$ which takes every line $D \in U_i$ onto its intersection point with E_i. For all $D \in U_i$,

$$q_i(D) = \left(\xi^0, \ldots, 1, \ldots, \xi^n\right)$$

with well-defined $\xi^p = x^p/x^i$, so that the formula

$$\varphi_i(D) = \left(\xi^0, \ldots, \xi^{i-1}, \xi^{i+1}, \ldots, \xi^n\right)$$

gives a bijection from U_i onto $\mathbb{R}^n$, which together with its inverse is obviously continuous. Hence the couples (U_i, φ_i) thus defined are charts of X of class C^0 covering $P_n(\mathbb{R})$. In fact, they are pairwise C^∞-compatibles. Indeed, for $D \in U_i \cap U_j$, there are relations of the form

$$q_i(D) = \left(\xi^0, \ldots, \xi^{i-1}, 1, \xi^{i+1}, \ldots, \xi^n\right),$$
$$q_j(D) = \left(\eta^0, \ldots, \eta^{j-1}, 1, \eta^{j+1}, \ldots, \eta^n\right);$$

these points of $\mathbb{R}^{n+1}$ being on D, $\xi^k = \eta^k/\eta^i$ and $\eta^k = \xi^k/\xi^j$ for all k. As $U_i \cap U_j$ corresponds to $y \in E_j$ such that $\eta^i \neq 0$, the coordinates $\varphi^i(D)$

in $U_i \cap U_j$ are C^∞ (or even rational) functions of the coordinates $\varphi^j(D)$ and conversely. The result follows, and this atlas defines a C^∞ manifold structure on $P_n(\mathbb{R})$. C^r functions on an open subset $U \subset P_n(\mathbb{R})$ are just the homogeneous functions of class C^r (in the usual sense) on the open set $p^{-1}(U)$.

Exercise. Check Hausdorff's axiom.

Replacing the 1-dimensional subspaces in the previous construction by p-dimensional ones for given p, it is possible to generalize. But it is slightly less simple and we leave it to the reader to give detailed proofs. Let E be an n-dimensional Cartesian space and $\Omega \subset E^p$ the set of sequences $x = (x_1, \ldots, x_p)$ of p linearly independent vectors in E; it is an open subset of E^p. Any p-dimensional subspace H of E is generated by the "components" x_i of some $x \in \Omega$ and $x, y \in \Omega$ generate the same subspaces if and only if there is a matrix $(g_i^j) \in GL_p(\mathbb{R})$ such that $y_i = g_i^j x_j$. This is an equivalence relation, so that the set $X = G_p(E)$ of p-dimensional subspaces of E is the quotient of Ω modulo this relation. If $p : \Omega \longrightarrow X$ denotes the obvious map, we then get a topology on X as in the case $p = 1 : U \subset X$ is open if and only if so is $p^{-1}(U)$ in Ω (or E^p). Having done this, if $U \subset X$ is open, $C^\infty(U)$ is, by definition, the set of functions f for which $f \circ p$ is C^∞ on the open subset $p^{-1}(U)$ of Ω. There are verifications to be done and charts to be found; we proceed as follows.

For this, choose a $(n-p)$-dimensional vector subspace F in E and let X_F denote the set of $H \in X$ such that $H \cap F = \{0\}$ or, what amounts to the same for reasons of dimension, such that $E = F \oplus H$, a direct sum. It is not hard to see that X_F is open in X. Then choose p vectors $a_i \in E$ generating a subspace H_0 such that $E = F \oplus H_0$. If $H \in X_F$, the relation $E = F \oplus H$ shows that, for all i, there is a unique $x_i \in F$ such that $a_i - x_i \in H$. Setting $\varphi(H) = (x_1, \ldots, x_p)$ for all $H \in X_F$, define a bijection from X_F onto F^p – a linear algebra exercise – and hence a chart (X_F, φ) for X_F, a priori purely set-theoretical.

Couples (X_F, φ) depending on the choice of F and of vectors a_i are in fact topological charts pairwise C^∞-compatibles and they define the $p(n-p)$-dimensional manifold structure of the *grassmannian* $X = G_p(E)$.

The latter is *compact*. To see this, choose a Hilbert or Euclidean scalar product $(x|y)$ on E and note that in all subspaces H of E, there are bases $x = (x_1, \ldots, x_p)$, orthonormal with respect to it: $(x_i|x_j) = 1$ or 0. As these relations are conserved when passing to the limit and prove that the x_i are linearly independent, the set $\Omega_0 \subset \Omega$ of these systems is closed in E^p. It is also compact since it is bounded. As $p : \Omega \longrightarrow X$ is continuous and maps Ω_0 onto X, the compactness of X is obvious.[49]

Real or complex (replace $\mathbb{R}$ by $\mathbb{C}$ in what precedes) Grassmannians were invented in the 19th century in order to generalize of projective geometry and its real or imaginary "points at infinity". In former times, they used to

---

[49] For further information and other methods, see Dieudonné, XVI.11.

delight undergraduates, including the present author during his youth. We thus learnt that all circles in the plane, irrespective of their centre, go through the same two points at infinite, the "cyclic points" with homogeneous coordinates $(1, i, 0)$ and $(1, -i, 0)$. These manifolds play an important role in algebraic geometry and their topological properties have been extensively studied.

(iv) *Differentiable maps.* In the same way that topological spaces are adapted to the general notion of a continuous map, manifolds are adapted to the notion of a differentiable map. It is all based on the following remark: given two manifolds X and Y and a map $f : X \longrightarrow Y$ taking the domain U of a chart (U, φ) to X in the domain V of a chart (V, ψ) of Y, there is a unique map $F : \varphi(U) \longrightarrow \psi(V)$ such that

$$\psi \circ f = F \circ \varphi$$

on U; it is the map which, for any $x \in U$, makes it possible to calculate the coordinates η^j of the point $y = f(x)$ in the chart (V, ψ) in terms of the coordinates ξ^i of x in the chart (U, φ). We will sometimes say that F *expresses* f in the charts considered. Note that, if f is continuous at a point a of X, then for all charts (V, ψ) of Y at $b = f(a)$, there exists a chart (U, φ) of X at a such that $f(U) \subset V$, since $f^{-1}(V)$ is a neighbourhood of a, and so contains an open set containing a, which contains the domain of a chart of X at a. Because of this trivial observation, it is possible to generalize the definitions and results relating to Cartesian spaces to maps from a manifold to another, provided we check they have are well-defined independently of the charts used, a fact generally immediate.

For example, given two manifolds X and Y of class at least C^r, f will be said to be *of class C^r* if f is continuous and if, for any charts (U, φ) and (V, ψ) such that $f(U) \subset V$, the function F expressing f in these charts is of class C^r in the usual sense. This means that, for any open subset W of Y and any function $g \in C^r(W)$, the composite function $g \circ f$, defined on the open subset $f^{-1}(W)$, is of class C^r on it. If, moreover, f is a homeomorphism and if f^{-1} is of class C^r, f is said to be a *diffeomorphism* of class C^r from X to Y. The reader will easily check that the map $X \longrightarrow Z$ obtained by composing two maps $X \longrightarrow Y$ and $Y \longrightarrow Z$ of class C^r is also of class C^r.

Maps of class C^r are also called *homomorphisms* of manifolds, in line with homomorphisms of groups, rings or vector spaces, etc. in algebra. The Grothendieck school censored the controversial prefix "homo" and invented the term *morphism*, which the Greeks would have probably considered doubly barbarian.[50]

This definition makes the characterization of open Cartesian subsets of X possible. First, any open subset U of X is itself a manifold since there are C^r functions on any smaller open subset. In particular, any open subset U of a

---

[50] Barbarian: a foreigner, with respect to the Greeks and Romans (Littré).

space $\mathbb{R}^d$ is a manifold if C^r functions are defined on it. Having said that, U is an open Cartesian subset of a manifold X if and only if, as a manifold, U is diffeomorphic to an open subset of $\mathbb{R}^d$; if, moreover, (U, φ) is a chart, then φ is a diffeomorphism from U onto the open subspace $\varphi(U)$ and conversely. Proofs reduce to rewording exercises.

Many other such trivialities can be found in detailed presentations, in particular in volume 3 of Dieudonné's *Eléments d'analyse* which, fortunately, gives frequent illustrations in the form of examples or exercises that are a great deal harder than the soporific but crucial definitions, scholia and sorites[51] that we end up learning through repeated use.

In particular, it is possible to define the notion of a *product manifold*: take two manifolds X and Y of class C^r, and as (U, φ) and (V, ψ) are charts of X and Y with values in $\mathbb{R}^p$ and $\mathbb{R}^q$, consider the map $(x, y) \mapsto (\varphi(x), \psi(y))$ from $U \times V$ to $\mathbb{R}^{p+q}$. Making these charts vary gives an atlas of class C^r for $X \times Y$, whence a manifold structure on the topological space[52] $X \times Y$. You will have no difficulty in showing that a map $z \mapsto (f(z), g(z))$ from a manifold Z to $X \times Y$ is of class C^r if and only if so are $f : Z \longrightarrow X$ and $g : Z \longrightarrow Y$, or that the projections $X \times Y \longrightarrow X$ are $X \times Y \longrightarrow Y$ are of class C^r. And many more wonders... One may laugh, but this is what transformed the loose theory available in the 1930s to a perfectly clear and precise mechanism, whose concepts often suffice to indicate the notions that should be introduced and the theorems that should be proved, at least at an elementary level: they only need to well-defined.

12 – Tangent vectors and Differentials

(i) *Vectors and tangent vector spaces.* In what way can the calculations of the preceding §§ be generalized to a d-dimensional manifold X, for example the notion of a differential form? We instantaneously encounter a fundamental difficulty: it is possible to talk about "vectors" and "linear forms" in a Cartesian space, but there are no vectors in a "curved" space, not even in a sphere in $\mathbb{R}^3$. To bypass this obstacle, associate to each $a \in X$ an "abstract" vector space having the same dimension d as X, called the *tangent vector space to X at a* and which I will denote $X'(a)$, other authors adopting other conventions, for example $T_a(X)$ which I will sometimes use.

---

[51] *Scholium*: In philology, a grammatical or critical note explaining classical texts. In geom. A remark on several propositions made in order to show their link, restriction or extension. *Sorite*: Sort of argument, in which a series of propositions is so arranged that the second one must explain the predicate of the first one, the third one the predicate of the second one, and so one, until the conclusion wanted is reached. *Predicate*: In log. and gram. What is denied or affirmed of the subject of a proposition. In the proposition: All men are mortals, *mortals* is the predicate. (Littré).

[52] If X and Y are two topological spaces, ordaining a subset of $X \times Y$ to be open if and only if it is a union of sets $U \times V$, where U and V are open in X and Y defines a topology on $X \times Y$.

To arrive at a definition of $X'(a)$, what is more generally called a *tensor of type* (p, q) *at the point* a can first be defined by drawing on the Italians. A priori, we do not know what the concrete nature of such an object is, but we suspect that it must have "components" in each local chart (U, φ) at a; for example, if $(p, q) = (2, 1)$, these components should be numbers $T_{ij}^k(\varphi)$ depending on three indices and on the chart considered. Having admitted this, we stipulate that these components should transform according to formula (3.9) of §1 when (U, φ) is replaced by another chart (V, ψ) at a: if coordinates $\xi = \varphi(x)$ and $\eta = \psi(x)$ of a variable point $x \in U \cap V$ are connected by

$$\eta = \theta(\xi), \quad \xi = \rho(\eta),$$

where θ maps $\varphi(U \cap V)$ diffeomorphically onto $\psi(V \cap U)$ and conversely, then this gives formulas

$$d\eta^\alpha = \theta_i^\alpha(\xi) d\xi^i, \quad d\xi^i = \rho_\alpha^i(\eta) d\eta^\alpha$$

with partial derivatives

(12.1) $$\theta_i^\alpha(\xi) = d\eta^\alpha / d\xi^i, \quad \rho_\alpha^i(\eta) = d\xi^i / d\eta^\alpha.$$

Having said this, the numbers $T_{\alpha\beta}^\gamma(\psi)$ corresponding to the chart (V, ψ) must satisfy the relation

(12.2) $$T_{\alpha\beta}^\gamma(\psi) = \rho_\alpha^i(\eta) \rho_\beta^j(\eta) \theta_k^\gamma(\xi) T_{ij}^k(\varphi),$$

where the coefficients are calculated at the points $\xi = \varphi(a)$ and $\eta = \psi(a)$. At this stage of the definition, we are reduced to "absolute differential calculus": we do not know over what we are calculating, but we calculate. We continue doing so every day in our times, and not only in physics. . .

Then the tangent vectors to X at a are, by definition, the tensors of type $(0, 1)$ at α. We thus get some $h \in X'(a)$ by taking, in each local chart (U, φ) at a, numbers $h^i(\varphi)$ that are subject to the equivalent relations

(12.3) $$h^\alpha(\psi) = \theta_i^\alpha(\xi) h^i(\varphi), \quad h^i(\varphi) = \rho_\alpha^i(\eta) h^\alpha(\psi)$$

for any local chart φ and ψ at α. However, the $\theta_i^\alpha(\xi)$ are the entries of the Jacobian matrix with respect to the canonical basis for $\mathbb{R}^d$ at the point $\varphi(a)$, of the chart change[53]

$$\theta : \varphi(x) \longmapsto \psi(x).$$

Hence, setting

(12.4) $$h(\varphi) = h^i(\varphi) e_i,$$

---

[53] The notation below replaces the formula $\theta[\varphi(x)] = \psi(x)$.

where (e_i) is the canonical basis of $\mathbb{R}^d$,

(12.5) $$h(\psi) = \theta'(\xi)h(\varphi).$$

takes $h(\varphi)$ to $h(\psi)$. This is the formula that we will always use. The coordinates of the infinitesimal vector connecting the points x and $x + dx$ of X in the charts φ and ψ being $d\xi$ and $d\eta = \theta'(\xi)d\xi$, formula (5) says that *the components of a tangent vector must transform as those of dx*, despite the fact that the expression $x + dx$ is not well-defined in a curved space.

To construct a tangent vector it then suffices to choose $h(\varphi)$ arbitrarily in *a* particular chart at a and to define $h(\psi)$ in the others by (5); it is nonetheless necessary to check that (5) still holds when passing from a chart φ_2 to a chart φ_3. But if the Jacobian matrices at a of the chart changes $\varphi(x) \mapsto \varphi_2(x)$, $\varphi_2(x) \mapsto \varphi_3(x)$ and $\varphi(x) \mapsto \varphi_3(x)$ are temporarily denoted by $M_{12}(a)$, $M_{23}(a)$ and $M_{13}(a)$, then by construction,

$$h(\varphi_2) = M_{12}(a)h(\varphi), \quad h(\varphi_3) = M_{13}(a)h(\varphi);$$

it therefore suffices to show that

(12.6) $$M_{13}(a) = M_{23}(a)M_{12}(a);$$

up to notation, this is the multivariate chain rule.

The set $X'(a)$ of tangent vectors to X at a being thus defined, it can be turned into a vector space by, for example, defining the sum $h = h' + h''$ of two tangent vectors by $h(\varphi) = h'(\varphi) + h''(\varphi)$: we do the necessary for the map $h \mapsto h(\varphi)$ from $X'(a)$ to $\mathbb{R}^d$ to be linear in any chart valid in the neighbourhood of a. As it is bijective,

$$\dim X'(a) = \dim X.$$

Then, as in the case of a Cartesian space [§ 1, n° 3, (i)], a basis $(a_i(\xi))$ for $X'(x)$ can be associated to any local chart (U, φ) and any $x \in U$,[54] $\xi = \varphi(x)$: the basis which corresponds under $h \mapsto h(\varphi)$ to the canonical basis (e_i) of $\mathbb{R}^d$. As $h(\varphi) = h^i(\varphi)e_i$ for any $h \in X'(x)$,

(12.7) $$h = h^i(\varphi)a_i(\xi)$$

for all $h \in X'(x)$, so that the $h^i(\varphi)$ are now the coordinates of h with respect to the basis $(a_i(\xi))$ for $X'(x)$. Élie Cartan, who knew all this intuitively, took advantage of it to define Leibniz's infinitesimal vector dx: letting $\xi^i + d\xi^i$ be the coordinates of a point "infinitely near" the point x with coordinates ξ^i, set

---

[54] This notation has the inconvenience of not specifying the chart φ used, but the notation $a_i(\varphi, a)$ is too cumbersome. The person who will succeed in introducing in differential geometry a notation system perfectly coherent and comprehensible in all cases is probably not born yet. See the notation index in Dieudonné.

(12.7') $$dx = a_i(\xi)d\xi^i .$$

If the chart is changed, the $a_i(\xi)$ and $d\xi^i$ are inversely transformed, so that the vector dx has an "absolute" meaning, at least metaphysically. Élie Cartan also used to speak of the point $x + dx$ of X but, as mentioned above, this reaches the limit of acceptable misnomers when X is not contained in a Cartesian space, and even in this case.

The construction of $X'(a)$ makes it possible to reduce the definition of tensors at a given above to that of §1, n° 1, (ii). First, *covectors at the point a* can be defined either as elements of the dual $X'(a)^*$ of $X'(a)$, or as tensors of type $(1,0)$ having in each (U,φ) at a components $u_i(\varphi)$ whose transformation is given by the relation

$$u_\alpha(\psi) = \rho^i_\alpha(\eta)u_i(\varphi).$$

Indeed, comparing with the transformation formula (3) shows that the number $u(h) = u_i(\varphi)h^i(\varphi)$ is independent of the chart φ, and so defines a linear functional u on $X'(a)$, and (4) then shows that

$$u_i(\varphi) = u\,[a_i(\xi)] .$$

Now, if there is a tensor T of type $(2,1)$ for example, the transformation formula (2) shows that, for $h, k \in X'(a)$ and $u \in X'(a)^*$, the expression

$$T(h,k\,;u) = T^k_{ij}(\varphi)h^i(\varphi)k^j(\varphi)u_k(\varphi)$$

is independent of φ, and so has an "absolute" meaning. Hence, tensors at a defined in the manner of the Italians of 1900 are merely tensors on the vector space $X'(a)$ in the sense of §1, n° 1, (ii). We now feel like we know what we are talking about, but we have in effect just expressed the founders' concepts in modern algebraic language.

(ii) *Tangent vector to a curve.* A simple method for constructing a vector $h \in X'(a)$ is to use a path or a curve $\mu : I \longrightarrow X$, where $I \subset \mathbb{R}$ is an open interval containing 0, with $\mu(0) = a$. Assuming that the $\mathbb{R}^d$-valued function $\varphi[\mu(t)]$ is differentiable at $t = 0$ in a local chart (U,φ) at a, it is possible to set

(12.8) $$h(\varphi) = \lim_{t=0} \frac{\varphi\,[\mu(t)] - \varphi\,[\mu(0)]}{t} = D\,\{\varphi\,[\mu(t)]\} \text{ for } t = 0 ,$$

where $D = d/dt$. The multivariate chain rule immediately shows that condition (5) holds. This gives some $h \in X'(a)$, which we write $\mu'(0)$, an expression that should not be confused with $\lim[\mu(t) - \mu(0)]/t$, as this is *not well-defined* except when X is contained in a Cartesian space, and as in this case, it is not, strictly speaking, a tangent vector to X, in the more abstract sense adopted here; we will return to this later. It would be natural to say that $\mu'(0)$ is the *tangent vector to μ at the point a* – mechanical engineers would talk of a velocity vector –, provided, once again, of not letting classical, but misleading images mystify us.

Any tangent vector at $a \in X$ can be obtained in this manner: if h corresponds to the vector $h(\varphi) \in \mathbb{R}^d$ in the chart (U, φ), it suffices to choose the map

$$\varphi \left[\mu(t) \right] = \varphi(a) + t h(\varphi)$$

as μ. In particular, the vectors $a_i(\xi)$ of the basis for $X'(a)$ correspond to the trajectories $t \mapsto \varphi(a) + t e_i$ in $\mathbb{R}^d$, where (e_i) is the canonical basis. Denoting the inverse map of φ by $f : \varphi(U) \longrightarrow U$ so that, for all $\xi \in \varphi(U)$, $f(\xi)$ is the point of X whose coordinates in the chart considered are precisely the canonical coordinates ξ^i of the point ξ, the corresponding curve μ is obviously the map

$$t \longmapsto f \left(\xi^1, \ldots, \xi^i + t, \ldots, \xi^d \right) ,$$

where $\xi = \varphi(a)$. The basis for $X'(a)$ associated to the chart considered is obtained by calculating the tangent vectors to these supposed "curvilinear coordinate axes" at $t = 0$.

Suppose, for example, that X is an n-dimensional *Cartesian space* E, and hence isomorphic but not identical to $\mathbb{R}^n$. The simplest charts for E are obtained by choosing a basis (a_i) for E and by associating to each $x = \xi^i a_i \in E$ the point $\varphi(x) = \xi^i e_i$ of $\mathbb{R}^n$. If (b_α) is another basis for E, we get the formulas $b_\alpha = g_\alpha^i a_i$ with an invertible matrix (g_α^i); to calculate $\psi(x)$ for the new chart, set $x = \eta^\alpha b_\alpha$, and so, by definition, $\psi(x) = \eta^\alpha e_\alpha$. Then, as $\xi^i = g_\alpha^i \eta^\alpha$, formula (1) can be applied with $\rho_\alpha^i(\eta) = g_\alpha^i$. Then consider a vector $h \in E'(x)$, where x is an arbitrary point of E. It has corresponding vectors

$$h(\varphi) = h^i(\varphi) e_i , \quad h(\psi) = h^\alpha(\psi) e_\alpha$$

in the charts (E, φ) and (E, ψ) thus constructed. The general formula (3) shows that $h^i(\varphi) = g_\alpha^i h^\alpha(\psi)$, which means that

$$h^i(\varphi) a_i = h^\alpha(\psi) g_\alpha^i a_i = h^\alpha(\psi) b_\alpha .$$

So $E'(x)$ can be *canonically* identified with the vector space E itself, by the map $h \mapsto h^i(\varphi) a_i$ which, in conformity with Italian mechanics, does not depend on the chosen basis.

Conversely, the simplest among the curves μ running through some $x \in E$ are the lines $t \mapsto x + th$, where $h \in E$ is given. Hence, each $h \in E$ canonically defines an element of $E'(x)$. It would be extremely surprising if the map $E \longrightarrow E'(x)$ thus defined was not the inverse of the one obtained by using the bases for E; we leave it to the reader to check this. It is for good reason that Leibniz, a philosopher mathematician, believed in the existence of a pre-established harmony[55] governing Creation and hence differential geometry.

(iii) *Differential of a map.* In the general case, the construction of $X'(a)$ makes it possible to define the differential $df(a)$ of a numerical function differentiable at a. For this, choose a chart (U, φ) at a, write that $f(x) = F[\varphi(x)]$ where the expression F of f in the chart considered is differentiable at $\xi = \varphi(a)$, and, for all $h \in X'(a)$, set

(12.9) $$df(a; h) = dF[\xi; h(\varphi)] = D_i F(\xi) h^i(\varphi),$$

where $D_i = d/d\xi^i$. The multivariate chain rule immediately shows that, under chart change, the $D_i F(\xi)$ and $h^i(\varphi)$ are inversely transformed, in other words, that the $D_i F(\xi)$ are transformed like the components of a tensor of type $(0, 1)$; so the left hand side does not depend on the chosen chart, which legitimizes definition (9). In the particular case of the coordinate function $x \mapsto \varphi^i(x)$, the function F is $(\xi^1, \ldots, \xi^d) \mapsto \xi^i$, and so

(12.9') $$d\varphi^i(a; h) = h^i(\varphi).$$

Like in a Cartesian space,

(12.9'') $$df(a) = D_i f(a).d\xi^i$$

is a shorthand version of formula (5), where $D_i f$ denotes the derivatives of f considered as a function of local coordinates $\xi^i = \varphi^i(x)$ and the differential $d\varphi^i(a; h)$ is shortened to $d\xi^i$. This gives a linear functional on $X'(a)$, i.e. a covector $df(a)$ at a. It is obvious that

$$p = fg \implies dp(a; h) = df(a; h)g(a) + f(a)dg(a; h)$$

for all $h \in X'(a)$.

More generally, if f is a map from a d-dimensional manifold X to a n-dimensional manifold Y, for all $a \in X$, a *tangent linear map*

(12.10) $$f'(a) : X'(a) \longrightarrow Y'(b),$$

where $b = f(a)$, may be defined, obviously by making a differentiability assumption about f. As always, the method is imposed by the data of the situation. Indeed, the aim is to associate to each vector $h \in X'(a)$ a vector $k \in Y'(b)$. For this, choose charts (U, φ) and (V, ψ) of X and Y, with $a \in U$, $b \in V$, $f(U) \subset V$, $\varphi(a) = \xi$ and $\psi(b) = \eta$. The construction of tangent spaces then furnishes us with bijective maps from $X'(a)$ and $Y'(b)$ to $\mathbb{R}^d$ and $\mathbb{R}^n$ and

---

[55] Pre-established harmony is a theory of Leibnitz according to which the spiritual and the physical world are like two perfect, but independent, clocks, always indicating the same time. Littré.

with a vector $h(\varphi) \in \mathbb{R}^d$. To define k, a vector $k(\psi) \in \mathbb{R}^n$ needs to be derived. Hence we only lack a map from $\mathbb{R}^d$ to $\mathbb{R}^n$, which must be linear in order for the map $X'(a) \longrightarrow Y'(b)$ sought to be so. But in the charts considered, f can be expressed by a C^r map

$$F : \varphi(U) \longrightarrow \psi(V)$$

such that $F(\xi) = \eta$; it has a tangent linear map $F'(\xi) : \mathbb{R}^d \longrightarrow \mathbb{R}^n$ at ξ. Hence

(12.11) $$k(\psi) = F'(\xi)h(\varphi) \quad \text{where} \quad \xi = \varphi(a) \ .$$

is the unique conceivable solution of the problem; or in coordinates,

(12.11') $$k^p(\psi) = D_i F^p(\xi).h^i(\varphi) \ .$$

Nonetheless, as always, some verifications need to be made to show the "absolute" character of this construction using charts. The simplest is to observe that, in relation to charts, the derivatives $D_i F^p$ behave in (11') like a covector at a does in relation to the index i and like a vector at $b = f(a)$ in relation to the index p; as the formula respects Einstein's conventions, it has an absolute character...

Another way of defining $f'(a)$ is to use the construction of tangent vectors by curves, presented in (ii). If $h \in X'(a)$ is a vector $\mu'(0)$ of a curve μ drawn in X and such that $\mu(0) = a$, then the image of μ under f is a curve $\nu(t) = f[\mu(t)]$ drawn in Y and such that $\nu(0) = b$. There is vector $\nu'(0) = k \in Y'(b)$ corresponding to the latter; it is just $f'(a)h$. Indeed, in the situation used above to define $f'(a)h$, by (8), the vector h is represented in the chart (U, φ) by the derivative vector $h(\varphi)$ at $t = 0$ of the function $\varphi \circ \mu$; the vector $k \in Y'(b)$ is likewise represented in the chart (V, ψ) by the vector $k(\psi)$, the derivative at $t = 0$ of the function $\psi \circ \nu$; as

$$\psi \circ \nu = \psi \circ (f \circ \mu) = (\psi \circ f) \circ \mu = (F \circ \varphi) \circ \mu = F \circ (\varphi \circ \mu),$$

the classic multivariate chain rule shows that the derivative at $t = 0$ of $\psi \circ \nu$ and of $\varphi \circ \mu$ are connected by the formula $k(\psi) = F'(\xi)h(\varphi)$, which reduces to definition (11).

Consider, for example, the *case of a Cartesian space* $Y = E$. At the end of (ii) above, we saw that tangent spaces to E can be canonically identified with E; $f'(a)$ can, therefore, be interpreted as a linear map from $X'(a)$ to E. To write it explicitly, start with some $h \in X'(a)$ defined by a curve $\mu(t)$ such that $\mu(0) = a$; so the image $f'(a)h \in E'(b)$, where $b = f(a)$, is defined by the curve $t \mapsto f[\mu(t)] = \nu(t)$ in E. However, under the above identification of $E'(b)$ with E, $f'(x)h$ becomes an ordinary vector $\nu'(0) = \lim[\nu(t) - \nu(0)]/t$; hence, if, as appropriate, no difference is made between the "abstract" vector $f'(a)h$ and the "concrete" vector corresponding to it in E, it can be computed by the relation

$$(12.12) \qquad f'(a)h = \frac{d}{dt} f\left[\mu(t)\right] \quad \text{for} \quad t = 0 .$$

Even more particularly, suppose that $E = \mathbb{R}^d$, where $d = \dim(X)$, consider a chart (U, φ) of X in the neighbourhood of a and take $f = \varphi$, so that $f'(a)$ is bijective. This leads to the derivative at the origin of the function $\varphi[\mu(t)]$; but by (9), it is precisely the element $h(\varphi)$ of $\mathbb{R}^d$ which defined h in the chart (U, φ). Hence

$$(12.13) \qquad \varphi'(a)h = h(\varphi)$$

for all $h \in X'(a)$ and all charts (U, φ) in the neighbourhood of a.

This is not surprising. Indeed, the aim is to find a preferably natural way to transform every $h \in X'(a)$ into a vector of $\mathbb{R}^d$ by using φ. Now, the definition of tangent vectors itself provides us with such a vector, namely $h(\varphi)$. What other possibilities are there? The preestablished harmony in these domains could have saved us a proof...

The multivariate chain rule can be expressed in the language of manifolds. For this, consider homomorphisms $f : X \longrightarrow Y$, $g : Y \longrightarrow Z$ and $p = g \circ f : X \longrightarrow Z$; the linear maps $f'(a) : X'(a) \longrightarrow Y'(b)$, where $b = f(a)$, and $g'(b) : Y'(b) \longrightarrow Z'(c)$, where $c = g(b) = p(a)$, are available at $a \in X$. Since the aim is to find a linear map $p'(a) : X'(a) \longrightarrow Z'(c)$ that can de deduced naturally from the data, it leads us to believe it is

$$(12.14) \qquad p'(a) = g'(b) \circ f'(a) .$$

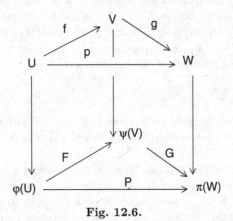

Fig. 12.6.

The reader unhappy with this philosophical and theological argument can always read Dieudonné (vol. 3, p. 24): "It is an immediate consequence of the definitions of the theorem (8.2.1) on composite functions." In fact, the full demostration would consist in using the charts (U, φ), (V, ψ) and (W, π)

of X, Y and Z at a, b and c, in drawing the diagram of the nine maps involved in the question: f, g and p, φ, ψ and π, and the maps F, G and P expressing f, g and p in the charts considered, and in applying repeatedly the definitions and the multivariate chain rule: an excellent exercise to understand the mechanism of manifolds.

Because of the definition of $f'(a)$, as in the classic situation ($\S 1$, n° 2, (v)), it is possible to define the *rank of f at a*: it is the dimension of the image subspace $Y'(b)$ of $X'(a)$ under $f'(a)$, i.e. the rank of the linear map $f'(a)$; if F expresses f in the local charts at a and $b = f(a)$, the rank of f at a is clearly equal to that of F at the point ξ corresponding to a. It cannot exceed $\dim(X)$, nor $\dim(Y)$; if it is equal to $\dim(X)$, then $f'(a)$ is injective and we have an *immersion*; if it is equal to $\dim(Y)$, $f'(a)$ is surjective and we have a *submersion*. In the neighbourhood of a point a, the rank of f is at least equal to its rank at a, so that the rank of an immersion or of a submersion is constant in the neighbourhood of a. Maps with this latter property are the *subimmersions*. They are characterzed by Theorem 1 of $\S 1$, n° 3, (i), whose generalization to manifolds is obvious.

(iv) *Partial differentials.* The tangent space $Z'(a)$ at a point $c = (a, b)$ of a product manifold $Z = X \times Y$ is easily determined. Indeed, in this case, there are projections $pr_1 : Z \longrightarrow X$ and $pr_2 : Z \longrightarrow Y$ given by $(x, y) \mapsto x$ and $(x, y) \mapsto y$; so their tangent maps define maps $Z'(c) \longrightarrow X'(a)$ and $Z'(c) \longrightarrow Y'(b)$, whence a map $Z'(c) \longrightarrow X'(a) \times Y'(b)$. As local charts reduce the case to one where X and Y are open subsets of Cartesian spaces, this map is clearly linear and bijective. We make no distinctions between $Z'(c)$ and $X'(a) \times Y'(b)$. If $h \in Z'(c)$ is defined by a curve

$$t \longmapsto \mu(t) = (\mu_1(t), \mu_2(t)) \,,$$

its images in $X'(a)$ and $Y'(b)$ are defined by the curves μ_1 and μ_2.

Now, let X, Y, Z be three manifolds and $f : X \times Y \longrightarrow Z$ a homomorphism. We compute its tangent map at (a, b). If $c = f(a, b)$, it maps $X'(a) \times Y'(b)$ to $Z'(c)$, and if $h \in X'(a)$, $k \in Y'(b)$ are defined by the paths $\gamma(t)$ and $\delta(t)$ with initial points a and b, the image of (h, k) is defined by the path $t \mapsto f[\gamma(t), \delta(t)]$. As $(h, k) = (h, 0) + (0, k)$, it suffices to add the images of $(h, 0)$ and $(0, k)$. The first one is defined by the path $t \mapsto f[\gamma(t), b]$. Thus the tangent map to $x \mapsto f(x, b)$, which will be denoted $f'_X(a, b)$ or $d_1 f(a, b)$, as well as the tangent map $f'_Y(a, b)$ or $d_2 f(a, b)$ to $y \mapsto f(a, y)$ need to be considered. This implies that $f'(a, b)$ is the map

$$(12.15) \qquad f'(a, b) : (h, k) \longmapsto f'_X(a, b)h + f'_Y(a, b)k \,.$$

The relation with the formulas of n° 2, (iii), in particular (2.24), is clear; besides, thanks to local charts, (15) reduces to (2.24).

(v) *The manifold of tangent vectors.* Denote by X' or $T(X)$ the set of all tangent vectors to X, i.e of couples (x, h) with $x \in X$ and $h \in X'(x)$. By associating to each $h \in X'(x)$ its "initial point" x, we get a canonical map $p : T(X) \longrightarrow X$. The set $T(X)$ can easily be made into a manifold. First, if (U, φ) is a chart of X, the inverse image $p^{-1}(U)$ is the set $T(U)$ of tangent vectors to the manifold U; if $n = \dim(X)$, the map $\varphi' : (x, h) \mapsto (\varphi(x), h(\varphi))$ from $T(U)$ to the Cartesian product $\varphi(U) \times \mathbb{R}^n$ is bijective. Formulas for chart change of section (i) of the present n° show that, if (V, ψ) is another chart of X and if the chart change is of class C^r on $U \cap V$, then the image of $(\varphi(x), h(\varphi))$ is obviously $(\psi(x), h(\psi))$ under a C^{r-1} map. This gives a C^{r-1} structure on $T(X) = X'$: the open subsets Ω of $T(X)$ are defined by the condition that, for any chart (U, φ) of X, the image of $\Omega \cap T(U)$ under φ' is an open subset of $\varphi(U) \times \mathbb{R}^n$, so that the couples $(T(U), \varphi')$ become charts of class C^0 for $T(X)$; as they constitute an atlas of class C^{r-1} for $T(X)$, the definition of the *manifold* $T(X)$ follows.

To any homomorphism $f : X \longrightarrow Y$ of manifolds can be associated a homomorphism $f' : X' \longrightarrow Y'$, namely

(12.16) $$f' : (x, h) \longmapsto (f(x), f'(x)h) \ .$$

If there is another homomorphism $g : Y \longrightarrow Z$, then setting $p = gf$, the multivariate chain rule shows that

(12.17) $$p' = g' \circ f' \ .$$

Indeed, f' maps (x, h) to $(f(x), f(x)h)$, which is then mapped by g' to $(g(f(x)), g'(f(x))f'(x)h)$. It, therefore, remains to check that $p(x) = g(f(x))$, which is trivial, and that $p'(x) = g'(f(x)) \circ f'(x)$.

For a product manifold $Z = X \times Y$, there is a canonical isomorphism $Z' = X' \times Y'$ since, for $x \in X$ and $y \in Y$, the tangent space $T_{(x,y)}(Z)$ was identified with $T_x(X) \times T_y(Y)$.

All this is too easy though sometimes convenient, especially in Lie group theory. Explaining the structure of the manifolds $T(T(X)) = T^2(X)$, $T(T(T(X))) = T^3(X)$, etc. is, however, far less simple. Then, any homomorphism $f : X \longrightarrow Y$ has "extensions" $f^{(r)} : T^r(X) \longrightarrow T^r(Y)$ that are homomorphisms and for which formula (17) becomes

$$p^{(r)} = g^{(r)} \circ f^{(r)} \ .$$

Interpreted in classical terms this order r multivariate chain rule is also not obvious. For a start, try to understand the case $r = 2$.

Exercise. Call an arbitrary basis of a space $T_x(X)$ a *frame*. Construct a natural manifold structure on the set of frames of X.

13 – Submanifolds and Subimmersions

The most immediately obvious manifolds are certain subsets X of a Cartesian space E which, for simplicity's sake, we will often suppose to be $\mathbb{R}^n$. There are several equivalent and equally important methods to equip them with a natural differentiable structure;[56] these methods were conceived in the 17th

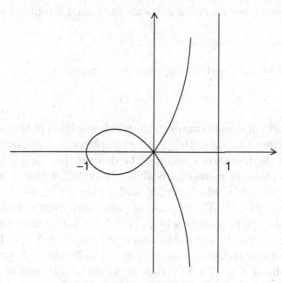

Fig. 13.7.

---

[56] It should not be thought that all possible methods for defining a manifold structure on a given set, however familiar they may be, lead to the same result. First, a method to construct two different, though isomorphic, differentiable structures on a manifold X involves choosing a non-differentiable homeomorphism σ from X onto X and to declare that the differentiable functions for the second structure are those obtained by composing σ with a differentiable function for the first one. This procedure, which can, to start with, be applied in $\mathbb{R}$, being in everyone's reach, consider equivalent two such manifold structures. The real question is whether there are others. Some experts in algebraic topology (M. Kervaire and J. Milnor, Annals of Math., **77**, 1963) have calculated the number ν_n, which happens to be finite, of non-equivalent C^∞ structures that can be defined on the unit sphere of $\mathbb{R}^n$:

n :	≤ 6	7	8	9	10	11	12	13	14	15	16	17
ν_n :	1	28	2	9	6	992	1	3	2	16256	2	16

Others have explicitly described some of these bizarre structures that were unknown. C^∞ structures essentially different from those of everyone else can also be defined on $\mathbb{R}^4$ (but not on $\mathbb{R}^n$ with $n \leq 3$). Finally, there are topological manifolds, i.e of class C^0, on which no C^1 structure can be defined. Naive ideas are sometimes incorrect.

century by using Cartesian coordinates; the theorems used are, except for generality and language, known since the 19th century. Since the 18th, people have studied "curves" and "surfaces" in the plane or the space, sometimes defined by an equation $y = f(x)$ in the plane (the parabola $y = x^2$ for example), sometimes by a parametric representation (the ellipse $x = a.\cos t$, $y = b.\sin t$ for example), sometimes by a relation between coordinates (the sphere $x^2 + y^2 + z^2 = 1$ for example). More complicated curves and surfaces have quickly been considered, for example the planar strophoid with equation

$$(x - 1)y^2 + x^2(x + 1) = 0,$$

which can also be obtained by the parametric representation

$$x = \left(t^2 - 1\right) / \left(t^2 + 1\right), \quad y = tx;$$

there are two arcs of simple curves in the neighbourhood of the origin meeting at 0, with distinct tangents; this is an example of a singularity which does not fall within the framework about to be defined (fig. 7).

The three classical methods recalled above fall within a more general pattern: take two manifolds X and Y and a map $f : X \longrightarrow Y$ and consider either the image $f(X) \subset Y$ (the case of parametric representations), or, for some $b \in Y$, the set of solutions of $f(x) = b$ in X (submanifold defined by relations between coordinates). The graph of a function $X \longrightarrow Y$ falls within this general pattern either by considering it the image of X under the map $x \mapsto (x, f(x))$ from X to $X \times Y$, either as the set of solutions of $y - f(x) = 0$. In this n°, we are going to define submanifolds and then show how some may be obtained by these methods – direct image of a map, inverse image of a point under a map – provided we restrict ourselves to *subimmersions*, i.e. to maps of constant rank.

(i) *Submanifolds.* Let X be a subset of a manifold Y of class C^r and dimension q. For any open subset U of X, let $C^r(U)$ be the set of functions f defined on U and satisfying the following property: for all $a \in U$, there is an open neighbourhood $V(a)$ of a in Y and a C^r function on $V(a)$ which is equal to f on $U \cap V(a)$. The example of the sphere (n° 11, (i)) suggests that X should be called a submanifold of Y if and only if the topological space X has the structure of a manifold of a priori arbitrary dimension p, and for which the C^r functions are precisely those that have been defined.

A first consequence of this definition is that the identity map $x \mapsto x$ from X to Y is then of class C^r: it follows from the definition of these maps. In fact, it is an immersion, and so $p \leq q$ as expected since *there is then a chart* (V, ψ) *of* Y *at any* $a \in X$ – for convenience, it can always be supposed to be cubic – *such that* $\psi(X \cap V)$ *is the face of the cube* $\psi(V)$ *defined by the relations* $\xi^{p+1} = \ldots = \xi^q = 0$, i.e. such that

(13.1) $\psi(V \cap X) = \psi(V) \cap \mathbb{R}^p,$

*as the restrictions φ^i to $V \cap X = U$ of the functions $\psi^1, \ldots, \psi^p$ then define
a chart (U, φ) of X.* The identity map $X \longrightarrow Y$ can, therefore, be expressed
in these charts by

$$(\xi^1, \ldots, \xi^p) \longmapsto (\xi^1, \ldots, \xi^p, 0, \ldots, 0) .$$

This result means that, locally and up to a diffeomorphism, a p-dimensional
submanifold of a q-dimensional manifold resembles a p-dimensional vector
subspace in a q-dimensional vector space. On the other hand, as (V, ψ) is a
cubic chart of Y, relation (1) shows the existence of a C^r map $p : V \longrightarrow V \cap X$
such that $p(x) = x$ for all $x \in V \cap X$: for p, it suffices to choose the map
expressed in (V, ψ) by the projection

$$(\eta^1, \ldots, \eta^q) \longmapsto (\eta^1, \ldots, \eta^p, 0, \ldots, 0) \in \mathbb{R}^q .$$

To prove (1), let us choose arbitrary charts (U, φ) and (V, ψ) of X and Y
at $a \in X$, with $\varphi(a) = \psi(a) = 0$. Since, in the neighbourhood of a, the φ^i are
restrictions to X of C^r functions on an open subset of Y, replacing U and V
by smaller open subsets, we may suppose that $U = X \cap V$ and that there are
$g^i \in C^r(V)$ such that $\varphi^i = g^i$ in U. Denoting by $u^i \in C^r(V')$ the functions
on $V' = \psi(V)$ expressing the g^i, we get

$$\varphi^i(x) = u^i \left[\psi^1(x), \ldots, \psi^q(x) \right] \quad \text{for all } x \in U .$$

Since the restrictions of the ψ^j to U are of class C^r, similarly there are v^j of
class C^r on the open subset $U' = \varphi(U)$ of $\mathbb{R}^p$ such that

$$\psi^j(x) = v^j \left[\varphi^1(x), \ldots, \varphi^p(x) \right] \quad \text{for all } x \in U .$$

As the maps $u = (u^1, \ldots, u^p) : V' \longrightarrow U'$ and $v = (v^1, \ldots, v^q) : U' \longrightarrow V'$
satisfy $u \circ v = id$, $u'(0) \circ v'(0) = 1$, so that the map $v'(0)$ is injective; since v
expresses the map $id : X \longrightarrow Y$ in the charts considered, $id'(a) : X'(a) \longrightarrow$
$Y'(a)$ is also injective, which proves that $id : X \longrightarrow Y$ is an immersion.

It remains to prove that it is possible to choose the chart (V, ψ) in such a
way that (1) holds. This will follow from the standard from of subimmersions
(§ 1, n° 3, (i), Theorem 1). Indeed, this theorem shows that, if there are
manifolds X and Y of dimensions p and q and a map $f : X \longrightarrow Y$ of constant
rank r in the neighbourhood of some $a \in X$, then there exist a chart (U, φ)
of X at a and a chart (V, ψ) of Y at $b = f(a)$ such that $f(U) \subset V$, $\varphi(a) = 0$,
$\psi(b) = 0$ and in which the coordinates $\xi^i = \varphi^i(x)$ of some $x \in U$ are changed
into the coordinates $\eta^j = \psi^j(y)$ of $y = f(x) \in V$ by the formulas

(13.2) $$\eta^1 = \xi^1, \ldots, \eta^r = \xi^r, \eta^{r+1} = \ldots = \eta^q = 0 .$$

If X is a submanifold of Y, this result applies to the immersion $f = id :$
$X \longrightarrow Y$, for which $r = p$. The condition $f(U) \subset V$ is written $U \subset V$, so
that $U = X \cap V$ may be assumed by replacing V by a smaller open set. The

first p relations of (2) then show that the functions $\varphi^i(1 \leq i \leq p)$ are the restrictions to U of the first p functions ψ^j, and the following ones show that $\varphi(U) = \psi(V) \cap \mathbb{R}^p$, which gives (1).

Conversely, if for all $a \in X$ there is a chart (V, ψ) of Y satisfying (1), then X is a submanifold of Y. Indeed, condition (1) shows that for any open subset U of $V \cap X$, the $f \in C^r(U)$, where $C^r(U)$ is defined for all open subsets of X like at the beginning of this section, are the C^r functions of the first p coordinates $\psi^i(x)$ on the open subset $\psi(U)$ of $\mathbb{R}^p$; denoting by $\varphi^i \in C^r(U)$ the restrictions to U of these first p functions ψ^i, we get a chart (U, φ) for X. The charts of X obtained in this manner are pairwise C^r-compatibles since so are the charts of Y used to construct them. Hence the result.

A corollary is that any submanifold X of a manifold Y is *open in its closure* $\bar{X}$, which means that $\bar{X} - X$ is *closed* in Y, or that a sequence of points $a_n \in \bar{X}$ converges to some $a \in X$ only if $a_n \in X$ for large n. Indeed, let (V, ψ) be a chart for Y at a such that $V \cap X$ is defined by relation (1); the a_n are in V for large n and so are the limits of points of $X \cap V$; as (1) shows that $\psi(V \cap X)$ is closed in $\psi(V)$, $\psi(a_n) \in \psi(V \cap X)$, and so $a_n \in V \cap X \subset X$, qed.

It goes without saying that if a chart (V, ψ) of Y at $a \in X$ is chosen randomly, the restrictions of the ψ^j to $U = X \cap V$ do not constitute a chart for X; to start with, there are too many of them. But if X and Y have dimensions p and q, p *functions forming a chart for X can be extracted from these q restrictions*. Indeed, as shown above, there is a chart (U, φ) for X – it may be assumed to be defined on U by choosing V sufficiently small – such that the φ^i are the restrictions to U of the first p coordinate functions of a chart (V, θ) of Y, which can here too be supposed to be defined on V. The θ^k being the C^r functions on V of the ψ^j, the φ^i, are C^r functions on U of the restrictions to U of the ψ^j; but since the φ^i form a chart for X in U, these restrictions are also C^r functions of the φ^i. The $p \times q$ Jacobian matrix of the restrictions of the ψ^j with respect to the φ^i is, therefore, of maximum rank p, and hence extracting a non-zero determinant of order p from it gives the p functions sought.

More generally, if X is a p-dimensional manifold and if, in the neighbourhood of some $a \in X$, there are p C^r functions with non-zero Jacobian in a local (hence in all) chart, these p functions define a chart for X in the neighbourhood of a: this is the local inversion theorem.

Finally, note that, if X is a p-dimensional submanifold of a q-dimensional manifold Y, by the canonical immersion from X to Y, for any $x \in X$, the tangent space $X'(x)$ can be identified with its image in $Y'(x)$ under the linear map $id'(x) : X'(x) \longrightarrow Y'(x)$. In (ii) we will see how to determine this subspace of $Y'(x)$ using local equations of X.

Exercise 1. Let Y be a q-dimensional manifold and X a subset of Y equipped with a manifold structure such that the map $x \mapsto x$ is an immersion. Show that X is a submanifold of Y.

Exercise 2. Let Z be a manifold, Y a submanifold of Z and X a subset of Y. Show that X is a submanifold of Z if and only if its a submanifold of Y.

Exercise 3. Let $f : X \longrightarrow Y$ be a homomorphism mapping a submanifold X' of X to a submanifold Y' of Y. Show that the map $X' \longrightarrow Y'$ induced by f is a homomorphism.

Exercise 4. Let X be the union in $\mathbb{R}^2$ of the coordinate half-axes $\{x \geq 0\}$ and $\{y \geq 0\}$ and P the point $(-1, -1)$. For all $M \in X$, let t the slope of the line PM, so that $M \mapsto t$ is a homeomorphism from X to $\mathbb{R}_+^*$. For any open subset U of X, let $C^\infty(U)$ be the set of functions on U that are C^∞ as functions of t, which give a C^∞ manifold structure on X diffeomorphic (under $M \mapsto t$) to $\mathbb{R}_+^*$. Show that X is not a submanifold of $\mathbb{R}^2$.

Exercise 5. A submanifold is open in its closure. With the help of examples, show that it is not necessarily a submanifold.

(ii) *Submanifolds defined by a subimmersion.* Let X and Y be two manifolds of dimensions p and q and $f : X \longrightarrow Y$ a C^r map. First consider the set Z of solutions of $f(x) = b$ for some given $b \in f(X)$ and suppose that the rank *or* of f is constant in an open subset containing Z (but not necessarily in all of X); Z is then a submanifold of X.

Indeed, for all $a \in X$, there exist charts (U, φ) and (V, ψ) of X and Y at a and b for which $\varphi(a) = 0$, $\psi(b) = 0$, $f(U) \subset V$ and in which the map

$$(13.3) \qquad\qquad F = \psi \circ f \circ \varphi^{-1}.$$

Its rank which is constant in the neighbourhood of 0, is given by

$$(13.4) \qquad F\left(\xi^1, \ldots, \xi^p\right) = \left(\xi^1, \ldots, \xi^r, 0, \ldots, 0\right).$$

$\varphi(U \cap Z)$ is then defined by the relations $\xi^1 = \ldots = \xi^r = 0$, so the result of section (i) implies that Z is a submanifold of dimension $p - r$ of X.

If, moreover, as at the end of section (i), the space $Z'(a)$ is identified with its image in $X'(a)$ under $id'(a)$, where $id : Z \longrightarrow X$, then[57]

$$(13.5) \qquad\qquad Z'(a) = \operatorname{Ker} f'(a).$$

This is the subspace of $h \in X'(a)$ such that $f'(a)h = 0$. Indeed, by definition of Z, the map $f \circ id$ from Z to Y is the constant map $z \mapsto b$. Hence $f'(a) \circ id'(a) = 0$, and so $Z'(a) \subset \operatorname{Ker} f'(a)$. As $f'(a) : X'(a) \longrightarrow Y'(b)$ has rank r and $X'(a)$ dimension p,

$$\dim \operatorname{Ker} f'(a) = p - r = \dim Z'(a).$$

---

[57] If $u : E \longrightarrow F$ is a linear map, $\operatorname{Ker} u$ denotes the set of $h \in E$ such $u(h) = 0$, and $\operatorname{Im} u$ the set of $u(h) \in F$. Then

$$\operatorname{rg}(u) = \dim \operatorname{Im} u = \dim E - \dim \operatorname{Ker} u.$$

(5) follows.

Example: The sphere in $\mathbb{R}^3$. The map from $X = \mathbb{R}^3$ to $Y = \mathbb{R}$ given by $f(x, y, z) = x^2 + y^2 + z^2$ has rank 1 everywhere except at the origin where its three derivatives are zero. Hence, for $R \neq 0$, equation $f = R^2$ defines a 2-dimensional submanifold Z of $\mathbb{R}^3$. At every point (a, b, c) of Z, the subspace $Z'(a, b, c)$ of $X'(a, b, c) = X$ is the set of vectors (dx, dy, dz) orthogonal to (a, b, c) since

$$df\left[(a, b, c); (dx, dy, dz)\right] = 2(a\,dx + b\,dy + c\,dz).$$

Another example: consider the orthogonal group $G = O_n(\mathbb{R}) \subset GL_n(\mathbb{R})$, i.e. the set of $n \times n$ matrices satisfying $g'g = 1$, where g' is the transpose of g. Taking $X = Y = M_n(\mathbb{R})$ and $f(x) = x'x$, a map from X to Y,

$$df(x; h) = h'x + x'h$$

so that, for given $x \in X$, the kernel of $f'(x)$ is the set of $h \in M_n(\mathbb{R})$ such that $h'x + x'h = 0$, i.e. such that $x'h = u$ is an antisymmetric matrix. If x is invertible, the map $u \mapsto x'^{-1}u$ is an isomorphism from the vector space of antisymmetric matrices onto $\operatorname{Ker} f'(x)$; the dimension of $\operatorname{Ker} f'(x)$, which is also the rank of $f'(x)$, is, therefore, constant in the open subset $GL_n(\mathbb{R}) \supset G$ of $M_n(\mathbb{R})$. So the group G is a submanifold of $M_n(\mathbb{R})$, the vector subspace of $M_n(\mathbb{R})$ tangent to it at $g = 1$ being also the kernel of

$$h \longmapsto df(1; h) = h' + h,$$

i.e. the set of antisymmetric matrices.

Exercise 6. Consider $M_n(\mathbb{C})$ as a real vector space. Let $U_n(\mathbb{C})$ be the group of unitary matrices, i.e. satisfying

$$u^*u = 1 \quad \text{where} \quad u^* = \bar{u}^{-1}$$

is the *adjoint* matrix of u (imaginary conjugate of the transpose). Show that $U_n(\mathbb{C})$ is a submanifold of $M_n(\mathbb{C})$.

Let us return to the general case and investigate the image $Z = f(X)$ of $f : X \longrightarrow Y$ by supposing that the rank of f is constant in all of X. For some $b = f(a)$, relations (4) show that the image of $f(U)$ under ψ is defined by $\eta^{r+1} = \ldots = \eta^q = 0$ this time, and so is a submanifold of Y. This would imply that $f(X)$ is a submanifold of Y if $f(U)$ were a neighbourhood of b in $f(X)$; but this condition does not necessarily hold as will be seen in the next section. It is, therefore, prudent to suppose that f is an *open* map from X to $f(X)$, i.e. takes open subsets of X to open subsets of $f(X)$. Then clearly, as a map from the manifold X to the manifold $f(X)$, f is a submersion and the subspace of $Y'(b)$ tangent to $Z = f(X)$ is

(13.6) $$Z'(b) = \operatorname{Im} f'(a).$$

The most important case is that of an immersion or, as used to be said in the past, a submanifold defined by a parametric representation, a variable point of Z depending on the "parameter" $x \in X$. In the best of cases, f is both an immersion from X to Y and a *homeomorphism* from X onto $f(X)$; f is then said to be an *embedding* of X in Y or, in obsolete language, a *parametric eigenspace representation* of Z. To suppose that f is an open and injective immersion would be equivalent since f^{-1} is then continuous. As, for any sufficiently small open subset $U \subset X$, f is a diffeomorphism from U onto the open subset $f(U)$ of the manifold $f(X)$ and as f is a global homeomorphism from X onto $f(X)$, it is in fact a diffeomorphism from X onto $f(X)$.

H. Whitney showed that, given some inoffensive countability assumptions, any n-dimensional manifold admits an embedding into $\mathbb{R}^{2n}$. This well-known theorem is hard to prove – easy theorems rarely become famous – except in the fairly elementary case of compact manifolds;[58] even Dieudonné, who proves it for $\mathbb{R}^{2n+1}$ (XVI.25, exercises 2 and 13) instead of $\mathbb{R}^{2n}$, retreated before the complete result. Doubts may arise as to the practical usefulness of this type of theorem since a useful embedding of a manifold into a Cartesian space is generally one that can be explicitly constructed from the specific data of the situation. If, for example, the universe happened to be a 4-dimensional "curved" manifold, looking for an artificial embedding of it into a 8-dimensional Cartesian space would not be very useful, though with physicists...

---

[58] If X is compact of dimension d, it can be covered by finitely many N charts (U_p, φ_p) such that for all p, $\varphi_p(U_p)$ is the cube $|\xi^i| < 2$ of $\mathbb{R}^d$; denoting by V_p the set of $x \in U_p$ for which $\varphi_p(x)$ is in the cube $|\xi^i| < 1$, the V_p may be assumed to cover X. Now, there is a C^∞ function h on $\mathbb{R}^d$ equal to 1 for $|\xi^i| < 1$ and to 0 for $|\xi^i| > 3/2$ (for $d = 1$, see Chapter V, n° 29; the general case follows in an obvious manner). Replacing the U_p by the V_p and the φ_p by the restrictions to V_p of the functions $h[\varphi_p(x)]$, X can be covered by the N charts (V_p, φ_p) for which the φ_p (extended by 0 outside U_p) are defined and of class C^r on all of X. The map

$$x \longmapsto (\varphi_1(x), \ldots, \varphi_N(x))$$

from X to $\mathbb{R}^d \times \ldots \times \mathbb{R}^d = \mathbb{R}^{Nd}$ then has rank d everywhere, but is not necessarily injective. To obtain a homeomorphism, use a partition of unity, i.e. a family of functions (θ_p) on X satisfying $\sum \theta_p(x) = 1$ for all x and whose supports are contained in the V_p. The map

$$x \longmapsto (\varphi_1(x), \ldots, \varphi_N(x), \theta_1(x), \ldots, \theta_N(x))$$

is then continuous and injective, hence a homeomorphism from X onto its image since X is compact, and has rank d everywhere. This gives an embedding of X into $\mathbb{R}^{N(d+1)}$.

(iii) *One-Parameter Subgroups of a Torus* are the classical examples of immersions that are not necessarily open.[59] To see this, take $X = \mathbb{R}$ and for Y, take the "torus" $\mathbb{T}^2$, where $\mathbb{T}$ is the unit circle of $\mathbb{C}$. It is a compact submanifold of $\mathbb{C}^2$ and, at the same time, is a (multiplicative) group which plays the same role for periodic functions with two variable as $\mathbb{T}$ in Chapter VII and leads to the same theory; the reader will invent it without difficulty and will even be able to generalize it to n variables... As always, setting

$$\mathbf{e}(t) = \exp(2\pi i t),$$

the map

(13.7)
$$f(t) = (\mathbf{e}(at), \mathbf{e}(bt)),$$

where $a, b \in \mathbb{R}$ are given and both non-zero, is both a homomorphism of the additive group $\mathbb{R}$ to the additive group $\mathbb{T}^2$ and an immersion since its derivative

$$f'(t) = (2\pi i a \mathbf{e}(at), 2\pi i b \mathbf{e}(bt))$$

is never zero. To deduce that the "one-parameter subgroup" $Z = f(\mathbb{R})$ is a 1-dimensional submanifold of $\mathbb{T}^2$, we need to make sure that f is an open map from $\mathbb{R}$ onto its image equipped with the topology of $\mathbb{T}^2$. As will be shown, that is the case if and only if the ratio a/b is *rational*. Otherwise, $f(\mathbb{R})$ is everywhere dense in $\mathbb{T}^2$, is neither locally compact nor locally connected with respect to the topology of $\mathbb{T}^2$ and so is not a submanifold of $\mathbb{T}^2$.

Let us first investigate the case when a/b is rational. Multiplying a and b by a convenient real number, a and b can be supposed to be coprime integers. Then there are integers u and v such that $au + bv = 1$ (Bezout's theorem, who before the Revolution was the author of a famous Mathematics Course used in artillery schools that were predecessors of the École Polytechnique). Then map (7) has period 1. Moreover, relation $f(t) = (1,1)$, where this is the identity element of the group $\mathbb{T}^2$, requires $at \in \mathbb{Z}$ and $bt \in \mathbb{Z}$, so that $t = t(au + bv) \in \mathbb{Z}$. As f is a homomorphism from the additive group $\mathbb{R}$ to the multiplicative group $\mathbb{T}^2$, it follows that

(13.8)
$$f(t) = f(t') \Longleftrightarrow t - t' \in \mathbb{Z}.$$

Hence $f(\mathbb{R}) = f(I)$, where $I = [0,1]$, so that $f(\mathbb{R})$ is compact and in particular closed in $\mathbb{T}^2$. But (8) shows that f is obtained by composing the map $\mathbb{R} \longrightarrow \mathbb{R}/\mathbb{Z}$ with the map φ from $\mathbb{R}/\mathbb{Z}$ to $\mathbb{T}^2$, which is obviously continuous and injective. As the space $\mathbb{R}/\mathbb{Z}$ is compact, a generalization of theorem 12 of Chap. III shows that φ is a homeomorphism from $\mathbb{R}/\mathbb{Z}$ onto its image $f(\mathbb{R})$:

---

[59] Since the results of this section are not used in the rest of this Chapter, it can be skipped in the immediate.

any continuous and injective map from a compact space X to a space Y is a homeomorphism from X onto its image. To deduce that the map $f : \mathbb{R} \longrightarrow f(\mathbb{R})$ is open, it is, therefore, sufficient to check that so is the canonical map from $\mathbb{R}$ onto $\mathbb{R}/\mathbb{Z}$, which is clear. Thus the image of $f(\mathbb{R})$ is indeed a (compact) submanifold of $\mathbb{T}^2$ when a/b is rational.

In the general case, let us consider the homomorphism from $\mathbb{R}^2$ onto $\mathbb{T}^2$ given by

(13.9) $$F(x, y) = (\mathbf{e}(x), \mathbf{e}(y))$$

and let D be the line of $\mathbb{R}^2$ generated by the vector $w = (a, b)$. As two points of $\mathbb{R}^2$ have the same image under F if and only if they differ by a point of $\mathbb{Z}^2$, the inverse image of $Z = F(D) = f(\mathbb{R})$ under F is the subgroup $G = D + \mathbb{Z}^2$, i.e. the set of $d + \omega$ where $d \in D$, $\omega \in \mathbb{Z}^2$. We show that when a/b is irrational, G is everywhere dense in $\mathbb{R}^2$.

Let w' be a vector that is not in D; the vector space $\mathbb{R}^2$ is the direct sum of D and the subspace D' generated by w'; hence

$$G = D + G \cap D',$$

so that it all amounts to showing that the subgroup $G' = G \cap D'$ of D' is everywhere dense in D'. The set of $t \in \mathbb{R}$ such that $tw' \in G'$ is obviously a subgroup H of the additive group $\mathbb{R}$, and it all amounts to showing that it is everywhere dense in $\mathbb{R}$.

However, for such a subgroup, there are only four possibilities:[60]

(a) $H = \{0\}$,
(b) $H = \mathbb{R}$,
(c) H is the set $m\mathbb{Z}$ of integer multiples of non-zero $m \in \mathbb{R}$,
(d) H is everywhere dense in $\mathbb{R}$.

Since cases (a) and (b) do not present any problems, first note that H contains numbers $t > 0$ since $t \in H$ implies $-t \in H$. Let $m \geq 0$ be the infimum of these numbers.

If $m = 0$, for $r > 0$, there exists $t \in H$ such that $0 < t < r$. However for all $x \in \mathbb{R}$ and any real number $t > 0$, there exists $q \in \mathbb{Z}$ such that

$$|x - qt| < t.$$

Applying this remark to some $t \in H$ such that $0 < t < r$, this shows that we are in case (d).

If $m > 0$, the same arguments show that, for all $x \in H$, there exists $q \in \mathbb{Z}$ such that $0 \leq x - qm < m$, and so $x - qm = 0$; hence we are in case (c).

---

[60] Other formulations: (i) every *closed* subgroup of $\mathbb{R}$ other than $\mathbb{R}$ is of the form $m\mathbb{Z}$; (ii) every subgroup of $\mathbb{R}$ is either some $\mathbb{Z}m$ (possibly with $m = 0$), or is everywhere dense in $\mathbb{R}$.

We can now return to the torus $\mathbb{T}^2$, to the subgroup H of t such that $tu' \in G' = G \cap D'$ and to the subgroup

$$G = D + G' = D + \mathbb{Z}^2 ,$$

which is the inverse image of $Z = f(D)$ under f.

If $H = \{0\}$, then $G = D$ and so $\mathbb{Z}^2 \subset D$, which is absurd.

If $H = \mathbb{R}$, then $G' = D'$, so that G contains D and D', and so $G = \mathbb{R}^2$, i.e. $\mathbb{R}^2 = D + \mathbb{Z}^2$. The set of lines parallel to D is therefore countable; which is absurd.

If H is in the case (d), $G' = G \cap D'$ is everywhere dense in D', $G = D + G'$ is everywhere dense in $D + D' = \mathbb{R}^2$, and as $f(\mathbb{R}) = F(G)$, $f(\mathbb{R})$ is clearly everywhere dense in $\mathbb{T}^2$. If a/b were rational, the image $f(\mathbb{R}) = F(G)$ would be compact, thus closed, and so $F(G) = \mathbb{T}^2$, which is absurd.

To conclude, let us study $f(\mathbb{R}) = F(D) = Z$ in the neighbourhood of the identity element $(1, 1)$ of $\mathbb{T}^2$. First note that the map F from $\mathbb{R}^2$ onto $\mathbb{T}^2$ given by (9) is already surjective on the closed square

$$K : 0 \leq x \leq 1, \quad 0 \leq y \leq 1$$

and that it is injective on the open square. Drawing the verticals with abscissa p and horizontal ordinates q, where $p, q \in \mathbb{Z}$ gives a grid of $\mathbb{R}^2$ which cuts D into intervals that can be numbered by the $n \in \mathbb{Z}$; each of these intervals can be mapped by an integer translation into a line segment parallel to D contained in K whose endpoints are on each side of K. Figure 8 is obtained from the interval $K \cap D$.

If the slope of D is rational, these segments are periodically reproduced. Indeed we saw at the start of the proof that by assuming a and b to be integers and coprime,[61] relation $f(t) = f(t')$ is equivalent to $t = t' \bmod \mathbb{Z}$; but then the points tw and $t'w$ of D, where $w = (a, b)$, differ from each other by an element of $\mathbb{Z}^2$, so that the intervals of D containing them can be deduced from each other by an integer translation; by bringing them back into K, we thereby get the same line segments, and so the announced periodicity holds. These arguments also prove that all these segments can be obtained by only considering the intersections of the set of points $tw \in D$ with the grid, where $0 \leq t \leq 1$. There are obviously finitely many of them. The union Δ of these line segments constructed in K is, therefore, compact, and hence, as seen earlier, so is the curve $F(\Delta)$ of $\mathbb{T}^2$.

If the slope of D is, on the contrary, irrational, then these segments are pairwise disjoint, for were it not so, there would be numbers t and $t' \neq t$ such that $tw - t'w \in \mathbb{Z}^2$, so that $w = (a, b)$ would be proportional to an integer vector: so the ratio a/b would be rational.[62] As $f(\mathbb{R}) = F(D)$

---

[61] This means that the vector $w = (a, b)$ generating D is a *primitive* element of $\mathbb{Z}^2$. We use this term for any integral vector belonging to a basis of $\mathbb{Z}^2$.

[62] These arguments show that f is injective if and only if $a/b \notin \mathbb{Q}$.

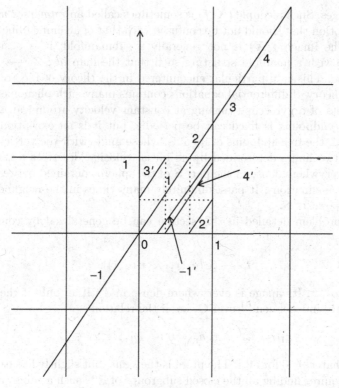

Fig. 13.8.

is everywhere dense in $\mathbb{T}^2$, the union Δ of the segments obtained in K is everywhere dense in K. The map F being, however, a homeomorphism, if it is restricted to a neighbourhood of 0 in D, then the set of points of the trajectory $Z = F(\Delta)$ in the neighbourhood of the point $(1,1)$ of $\mathbb{T}^2$ is seen to be homeomorphic to the intersection of Δ and of a neighbourhood of 0 in K; this intersection is composed of infinitely many pairwise disjoint line segments. In the neighbourhood of the origin in $\mathbb{T}^2$, the trajectory Z can, therefore, be decomposed into an infinitely countable number of excellent pairwise disjoint arcs of curve. Hence if Z is equipped with the topology of $\mathbb{T}^2$, we get a space, which while being connected, is not locally connected,[63] nor even locally compact since the intersection of Δ with a closed neighbourhood of 0 in K is obviously not closed and even less compact. All is now proved.

The non-closed geodesics of the torus are examples of immersions $f :$ $X \longrightarrow Y$ which, while being injective, are not homeomorphisms from X onto

---

[63] A topological space Z is said to be locally connected if, for every $x \in Z$ and every neighbourhood V of x in Z, there is a connected neighbourhood U of a such that $U \subset V$: existence of arbitrarily small connected neighbourhoods.

their images. Such a couple (X, f) is sometimes called an *immersed manifold* in Y, a notion that should not be confused with that of an embedding defined above. The image $f(X)$ is not generally a submanifold; it is a subset Z equipped with a manifold structure such that the map $id : Z \longrightarrow Y$ is an immersion. This is in particular encountered in the theory of Lie groups.

The theory of differential equations contains many such phenomena. The movements of a gyroscope turning at constant velocity around an axis one of whose endpoints is fixed can be periodic, but it is an exceptional case. In general, the free endpoint of its axis, whose angle with the vertical varies between two limits determined by its initial velocity, describes a trajectory that is everywhere dense in the part S of the sphere contained between these two limit inclinations. It passes infinitely many times in the neighbourhood of every point of S.

The problem detailed in this section can be generalized by considering the map

$$t \longmapsto (\mathbf{e}\,(a_1 t), \ldots, \mathbf{e}\,(a_n t))$$

from $\mathbb{R}$ to $\mathbb{T}^n$. Its image is everywhere dense in $\mathbb{T}^n$ if an only if the $a_i \in \mathbb{R}$ are linearly independent[64] over $\mathbb{Q}$, i.e. if the relation

$$x_1 a_1 + \ldots + x_n a_n = 0 \qquad x_1, \ldots, x_n \in \mathbb{Q}$$

implies that $x_i = 0$ for all i. The proof is the same, but slightly less easy since it first requires finding all the closed subgroups of $\mathbb{R}^n$: such a subgroup is the set of vectors which, with respect to a conveniently chosen basis (u_i) for $\mathbb{R}^n$ can be written as $\sum x^i u_i$, where $x^i \in \mathbb{R}$ for $1 \le i \le p$, $x^i \in \mathbb{Z}$ for $p+1 \le i \le q$ and $x^i = 0$ for $i \ge q$. Let us now return to more general manifolds.

(iv) *Submanifolds of a Cartesian space : tangent vectors.* In the case of a p-dimensional submanifold X of a q-dimensional Cartesian space Y, a more concrete description of tangent spaces $X'(a)$ can be given; it has no use – as remarked with reason by Dieudonné (*Eléments d'analyse*, vol. 3, p. 2), it may even lead the reader on a wrong track – but it makes it possible to relate "abstract" tangent spaces $X'(x)$ to "tangent planes" of classical geometry. In what follows, we suppose that $Y = \mathbb{R}^q$, and we will identify $\mathbb{R}^q$ to the Cartesian product $\mathbb{R}^p \times \mathbb{R}^{q-p}$.

First note that, in this case, the 18th century method for representing a plane curve either by an equation $y = f(x)$, or by an equation $x = g(y)$, can be generalized here. Indeed if, as a general rule, we denote by y^j the canonical coordinates of some $y \in Y$, we know that (end of (i)) in the neighbourhood of all $a \in X$, the restrictions x^j of these q functions to X form a system of rank p. Hence, up a permutation of the canonical coordinates, the first p functions x^i may be supposed to be defined by a chart of X in the neighbourhood U of a in X, which means that the projection

---

[64] $\mathbb{R}$ is an infinite-dimensional vector space over $\mathbb{Q}$.

$$y \longmapsto \left(y^1, \ldots, y^p\right)$$

from $\mathbb{R}^q = Y$ onto $\mathbb{R}^p$ induces a diffeomorphism from U onto an open subset U' of $\mathbb{R}^p$. The other coordinates x^j being of class C^r on U, we get the relations

$$(13.10) \qquad y^j = f^j \left(y^1, \ldots, y^p\right) \text{ for all } y \in U \quad (p+1 \leq j \leq q) \,,$$

where $f^j \in C^r(U')$. Any $y \in Y$ sufficiently near a so that $(y^1, \ldots, y^p) \in U'$ and satisfying relations (10) is then in U since there is a one (and only one) $x \in U$ whose first p coordinates are given in U', its other $q - p$ coordinates satisfying (10). In other words,

$$(13.11) \quad y \in U \Longleftrightarrow (y^1, \ldots, y^p) \in U' \ \& \ y^j = f^j \left(y^1, \ldots, y^p\right) \,.$$

If $(f^{p+1}, \ldots, f^q)$ is considered to be a map f from U' to $\mathbb{R}^{q-p}$, this is equivalent to saying that *in the neighbourhood of a, the submanifold X of $\mathbb{R}^q$ is the graph of a map from $\mathbb{R}^p$ to $\mathbb{R}^{q-p}$*. Simply using coordinate changes given by canonical permutations, precursors were right to believe that, locally, any "good" curve or surface is the graph of a "good" function. As seen in section (i), this is in particular the case if X is defined globally by the "implicit equations" $F^k(y^1, \ldots, y^q) = 0$ provided the rank of the map $y \mapsto (F^k(y))$ is constant in the neighbourhood of X or else X is the image of an open subset of $\mathbb{R}^p$ under an open immersion.

Having said this, the elementary classical definition of a tangent vector u to a submanifold X of Y at x, for example to a sphere, says that there is a curve $t \mapsto \mu(t)$ in X such that $\mu(0) = x$, $\mu'(0) = u$, where, $\mu'(0) = \lim[\mu(t) - \mu(0)]/t$ here. The set $T_x(X)$ of these vectors is a vector subspace[65] of Y, not to be confused with the abstract vector space $X'(x)$. Indeed, in the neighbourhood of x, X may be supposed to the graph of a map f from $\mathbb{R}^p$ to $\mathbb{R}^{q-p}$. As $\mathbb{R}^q$ is identified to $\mathbb{R}^p \times \mathbb{R}^{q-p}$, $x = (a, b)$ with $a \in \mathbb{R}^p$ and $b = f(a) \in \mathbb{R}^{q-p}$. Similarly, $\mu(t) = (\mu_1(t), \mu_2(t))$ with $\mu_1(t) \in \mathbb{R}^p$ and $\mu_2(t) = f[\mu_1(t)]$ since $\mu(t) \in X$. μ_1 can be arbitrarily chosen and, setting $k = \mu_1'(0) = \lim [\mu_1(t) - \mu_1(0)] / t$, it follows that

$$(13.12) \qquad\qquad u = \mu'(0) = (k, f'(a)k) \,.$$

We, therefore, conclude that $T_x(X)$ is the image of $\mathbb{R}^p$ under the linear map $k \mapsto (k, f'(a)k)$, which gives the result. We thus recover the fact that, for $q = 2$, $p = 1$, the slope of the tangent to the graph of the function f at the point $(a, f(a))$ is the number $f'(a)$ the linear map $f'(a)$ of the general case can be identified to in dimension 1. Observe also that, contrary to what

---

[65] In classical geometry, the set $T_x(X)$ thus defined is not the "tangent plane" to X at x; depending on the point of view taken, it is a set either of points of Y or of vectors with initial point x. $T_x(X)$ is the vector subspace of Y which can be deduced from the traditional tangent plane by using the translation mapping the point x to 0.

M. Mandelbrot seems to think, we do not eliminate "geometry" from our pre-occupations: our definitions and even our notation directly generalize those of the 17th century.

But as seen in (ii), the curve μ drawn in X makes it also possible to define some $h \in X'(x)$ independently from the embedding of X into a Cartesian space: under every local charts (U, φ) of X at x, μ becomes a curve $t \mapsto \varphi[\mu(t)]$ in $\mathbb{R}^p$ and, cf. (12.8),

$$h(\varphi) = \frac{d}{dt}\varphi\,[\mu(t)] \quad \text{for } t = 0.$$

If, as above, the submanifold X is defined in the neighbourhood of $x = (a, b)$ by an equation $y = f(x)$ and if $\mu(t) = (\mu_1(t), \mu_2(t))$, the map $(x, y) \mapsto x$ from X to $\mathbb{R}^p$ can be chosen as (U, φ) since it defines the manifold structure on X. Then $\varphi[\mu(t)] = \mu_1(t)$ and so $h(\varphi) = \mu_1'(0)$, where this is the usual derivative. Hence, by (12),

(13.13) $$\mu'(0) = (h(\varphi), f'(a)h(\varphi))\,,$$

where $f'(a)$ is the tangent linear map to f at a in the usual sense of n° 2, (i). As the map $h \mapsto h(\varphi)$ from $X'(x)$ to $\mathbb{R}^p$ is linear and bijective, and as formula (12) likewise defines a linear bijection $k \mapsto h$ from $\mathbb{R}^p$ onto $T_x(X)$, we thereby obtain by composition an isomorphism $h \mapsto u$ from the "abstract" vector space $X'(x)$ onto the "concrete" vector subspace $T_x(X)$ of $\mathbb{R}^q$. For any curve μ drawn in X and such that $\mu(0) = x$, this isomorphism maps the "abstract" tangent vector $\mu'(0) \in X'(x)$ onto the vector

$$\lim\,[\mu(t) - \mu(0)]\,/t$$

also written $\mu'(0)$ by everyone. This also proves that the isomorphism of vector spaces $X'(x) \longrightarrow T_x(X)$ defined thereby is absolute, i.e. depends only on the embedding of X into $\mathbb{R}^q$ and not on the choice of the chart.

And it is not understandable why this assimilation is a false trail, in the words of Dieudonné: it in particular suggests that the vector spaces $X'(x)$ and $X'(y)$ tangent to X at two different points x and y could, like the subspaces $T_x(X)$ and $T_y(X)$ of $\mathbb{R}^q$, have common elements; this is not at all the case: an element h of $X'(x)$ is a *couple* consisting of a point $x \in X$ and of a family of vectors $h(\varphi)$ of $\mathbb{R}^p$ depending on a local chart at x; two tangent vectors at x and y cannot be equal if $x \neq y$. In fact, the set of tangent vectors to an n-dimensional manifold is a $2n$-dimensional manifold [n° 12(v)].

(v) *Riemann spaces.* The definition of tangent spaces makes it possible to define Riemann spaces. For this, in each $X'(x)$ take an Euclidean scalar product $(h|k)$ compelled to depend on x in a reasonable manner: suppose that the functions

$$g_{ij}(\xi) = (a_i(\xi)|a_j(\xi))$$

are at least C^2 on each chart (U, φ). If $h = h^i a_i(\xi)$ and $k = k^i a_i(\xi)$ are two tangent vectors to X at x, then

$$(h|k) = g_{ij}(\xi) h^i k^j$$

and in particular, for the metaphysical vector $dx = a_i(\xi) d\xi^i$,

$$ds^2 = (dx|dx) = g_{ij}(\xi) d\xi^i d\xi^j ,$$

i.e. the square of the length of dx. The length of a curve $\mu : [a, b] \mapsto X$ can then be defined by the formula

$$m(\mu) = \int_a^b \left(\mu'(t) | \mu'(t) \right)^{1/2} dt$$

and it is possible to look for the geodesics, i.e. the curves of minimal length connecting two given points of X, if they exist. It is also possible to define the covariant derivative of a tensor field, etc. All this has given rise to an extensive literature whose latest avatar seems to be Serge Lang's book, *Riemannian Geometry* (Springer, 1999).

If X is a submanifold of an Euclidean space E (i.e. a Cartesian space equipped with a Hilbert scalar product) and if $h, k \in X'(x)$ are defined by the curves μ and ν as mentioned in (iii) of n° 12, it is natural to set

$$(h|k) = (\mu'(0)|\nu'(0)) ,$$

where $\mu'(0)$ and $\nu'(0)$ are defined as limits. Equivalently, note that $X'(x)$ can be canonically identified to a vector subspace of $E'(x)$, hence to E, which gives the scalar product sought in $X'(x)$. The explicit computation of the ds^2 of X is particularly simple when X is defined by a parametric representation $x = \sigma(t)$ in the neighbourhood of point a, where t varies in an open subset of $\mathbb{R}^d$ and where σ is an open immersion. Then $X'(x)$ is the image of $\mathbb{R}^d$ under $\sigma'(t)$ if $x = \sigma(t)$, so that if, in Leibniz style, we set $dx = \sigma'(t) dt$, then

$$ds^2 = (\sigma'(t) dt | \sigma'(t) dt) .$$

For example, for the unit sphere in $\mathbb{R}^3$ in spherical coordinates

$$x = \cos \varphi \cos \psi , \quad y = \sin \varphi \cos \psi , \quad z = \sin \psi ,$$

we differentiate x, y and z, we simply calculate $dx^2 + dy^2 + dz^2$ and we find that $ds^2 = \cos^2 \psi d\varphi^2 + d\psi^2$.

It can be shown that for any connected Riemann space X, there is a diffeomorphism from X onto a submanifold Y of some space $\mathbb{R}^n$ which transforms the given ds^2 in X into that of Y (John Nash, 1956). Dieudonné (vol. 4, XX.15) proves a far weaker result (É. Cartan) as it is purely local, but it is already difficult.

14 – Vector Fields and Differential Operators

Since we have defined tensors at a point of a manifold X of class C^r in n° 12, (i), we can now define tensor *fields* of type (p,q) on X or, more generally (?), on an open subset of X. They associate to each point $x \in X$ a tensor of type (p,q) at the point x considered, either as a multilinear function of p vectors and q covectors at x, or as a system of numbers $T_{ij...}^{kh...}$ depending on x and on a local chart at x, and compelled to transform by the Italian formulas under chart change. T will be said to be of class C^s if in every local chart, its components are C^s functions on the corresponding open Cartesian subspace; we need to assume that $s \leq r-1$ because if changes of charts are of class C^r, their first partial derivatives are only of class C^{r-1}; thus there is no hope of the Italian formulas transforming C^r functions into C^r functions.

In practice, the most important tensor fields are the *vector fields* and the differential forms that will be defined below. A vector field L is of type $(0,1)$, so that it is obtained by associating to each $x \in X$ a tangent vector $L(x) \in X'(x)$ to X at x. Hence, if (U, φ) is a chart, then for each $x \in U$ there is a vector

$$L(x)(\varphi) = L^i(\xi)e_i \quad \text{where} \xi = \varphi(x),$$

i.e. a vector field (in the sense used by physicists) on the open subset $\varphi(U)$ of $\mathbb{R}^d$, $d = \dim(X)$ with the obvious formulas for chart change. L will also be said to be of class C^s if so are the functions L^i .

If f is function of class $C^s (1 \leq s \leq r)$ on an open set $W \subset X$, then set

$$Lf(x) = df\,[x; L(x)]$$

for all $x \in X$; many authors prefer to denote the function Lf by $D_L f$. If (U, φ) is a chart, with $U \subset W$, and if F is the function which expresses f in $\varphi(U)$, so that $f = F \circ \varphi$, then, by the multivariate chain rule,

$$Lf(x) = dF\,[\varphi(x); \varphi'(x)L(x)] \; ;$$

but by (12.13), we know that for $h \in X'(x)$,

$$\varphi'(x)h = h(\varphi) = h^i(\varphi)e_i \,.$$

Setting $\xi = \varphi(x)$, it follows that

(14.1) $Lf(x) = dF\,[\xi; L^i(\xi)e_i] = L^i(\xi)dF\,(\xi; e_i) = L^i(\xi)D_iF(\xi)\,,$

where the D_iF are the usual partial derivatives of F and the $L^i(x)$ the components of $L(x)$ in the chart considered. Obviously,

$$L(fg) = Lf.g + f.Lg$$

for all f and g.

Relation (14.1) suggests a generalization to manifolds of (linear) classical differential operators. For simplicity's sake, on a manifold X of class C^∞, a *differential operator of order* $\leq p$ and class C^∞ defines, for every open subset U, a linear map $L : C^\infty(U) \longrightarrow C^\infty(U)$ satisfying the following two conditions: (i) it is compatible with the restriction maps $C^\infty(U) \longrightarrow C^\infty(V)$ for $V \subset U$, so that for $f \in C^\infty(U)$, the value of the function $Lf \in C^\infty(U)$ at $x \in U$ only depends on the behaviour of f in the neighbourhood of x, (ii) in each chart (U, φ) of X, there is a relation of the form

$$(14.2) \qquad Lf(x) = \sum_{k=0}^{p} \sum_{i_1, \ldots, i_k} L^{i_1 \cdots i_k}(\xi) D_{i_1} \cdots D_{i_k} F(\xi),$$

where the functions $L^{i_1 \cdots i_k}$ only depend on the chart considered and where ξ, F and the D_i are defined as in (1).

Obvious operations can be defined on these operators: sum $L+M$, product LM, product fL by a function. If L and M are of order p and q respectively, their product $LM : f \mapsto L(Mf)$ is generally of order $p + q$, but their *Jacobi bracket*

$$(14.3) \qquad [L, M] = LM - ML$$

is of order $\leq p + q - 1$. It suffices to check this in $\mathbb{R}^n$; we can then suppose that

$$L = \varphi D_{i_1} \ldots D_{i_p} = \varphi D_{(i)}, \quad M = \psi D_{j_1} \ldots D_{j_q} = \gamma D_{(j)}$$

with given functions φ and ψ, and it all amounts to checking that, if we calculate

$$LMf - MLf = \varphi D_{(i)} \left[\psi D_{(j)} F \right] - \psi D_{(j)} \left[\varphi D_{(i)} F \right],$$

then the terms $\varphi \psi D_{(i)} D_{(j)} f$ and $\psi \varphi D_{(j)} D_{(i)} f$ cancel each other.

In particular, if L and M are defined by vector fields, the same holds for $[L, M]$; using Einstein's convention, it can be immediately seen that , in any chart,

$$(14.4) \qquad L = L^i D_i, \ M = M^i D_i \implies [L, M] = N^i D_i$$

with

$$(14.5) \qquad N^i = L^j . D_j \left(M^i \right) - M^j . D_j \left(L^i \right).$$

Hence, if D_L denotes the differential operator $f \mapsto Lf$, the vector field $[L, M]$ satisfies

$$D_{[L,M]} = D_L D_M - D_M D_L.$$

Exercise 1. Verify by a direct calculation that the vector field defined by (5) does not depend on the chart used.

Exercise 2. Let $L_1, \ldots, L_n$ be C^∞ vector fields on an open subset U of X; suppose that, for all $x \in U$, the $L_i(x)$ form a basis for $X'(x)$. Show that all differential operators on U can be written uniquely as a finite sum

$$\sum_{p_i \geq 0} a_{p_1 \ldots p_n}(x) L_1^{p_1} \ldots L_n^{p_n}.$$

Note that as $[L, M] = N$ involves the derivatives of the components of L and M, the value $N(x) \in X'(x)$ of N at x does not only depend on the vectors $L(x)$ and $M(x)$. The bracket $[h, k]$ of two tangent vectors is not well-defined.

The existence of vector fields satisfying global conditions poses problems related to the topology of manifolds: does there exist vector fields in X such that $L(x) \neq 0$ for all x? If X has dimension n, are there n vector fields L_i such that the $L_i(x)$ form a basis for $X'(x)$ for all $x \in X$? The answers are already in the negative for the 2-dimensional sphere.

15 – Vector Fields and Differential Equations

Vector fields serve to generalize the theory of first order differential equations to manifolds; we can start by searching for *integral curves* or *trajectories* $t \mapsto \gamma(t)$ from a vector field L onto a p-dimensional manifold X; they are defined by the condition

$$(15.1) \qquad\qquad \gamma'(t) = L[\gamma(t)],$$

where the left hand side is defined as in n° 12, (ii). In a local chart (U, φ) such that $\varphi(U) = \mathbb{R}^p$, this equivalent to looking for a function $x(t) = \varphi[\gamma(t)]$ satisfying

$$(15.2) \qquad\qquad Dx(t) = L[x(t)],$$

where $D = d/dt$ and where the functions $x(t)$ and $L(x)$ have values in $\mathbb{R}^p$. When L is C^1, the following results hold:

(a) *for all $t_0 \in \mathbb{R}$ and $x_0 \in X$, there exists a solution of (1) defined in the neighbourhood of t_0 and such that $\gamma(t_0) = x_0$;*

(b) *two such solutions coincide on the interval on which they are simultaneously defined.*

Statement (a) will follow from the analogous statement for equation (2). Similarly for (b), because if two solutions of (1) defined on the same interval I and equal at some point $t \in I$ are known to be also equal on a neighbourhood of t, then the set of $t \in I$ where they are equal is both open and close (continuity) in I, and so are equal to I.

Statement (b) shows that the solutions of (1) with a given value at a given point t_0 are the restrictions to their intervals of definition of a unique solution, defined on the union I of these intervals; I is the largest interval on which (1) has a solution, which is, therefore, said to be *maximal*. I is clearly open because of (a), but in general $I \neq \mathbb{R}$: if X is an open subset of $\mathbb{R}^2$ and L is a vector field whose vectors can de deduced from each other by parallelism, the trajectories are open line segments contained in X having as endpoints border points of X.

If, for all $x \in X$, $\gamma_x(t)$ denotes the maximum trajectory such that $\gamma_x(0) = x$ and I_x is its interval of definition, we are led to introduce the set $\Omega \subset \mathbb{R} \times X$ of (t, x) such that $t \in I_x$ and to set

$$(15.3) \qquad\qquad \gamma(t, x) = \gamma_x(t)$$

for $(t, x) \in \Omega$. This give a map $\gamma : \Omega \longrightarrow X$, the *global flow* of the vector field L. The following additional result then holds :

(c) *If L is of class C^k, then Ω is open in $\mathbb{R} \times X$ and γ is of class C^k in Ω.*

Similarly to (a) and (b), this statement is of a local nature: it amounts to showing that the solution of (2) satisfying $x(0) = \xi$ is defined and a C^k function of (t, ξ) on the product of an interval centered at 0 and a ball with given centre.

There are analogous statements for more general differential equations: instead of a function $L(x)$ of the only variable $x \in \mathbb{R}^p$, L can be supposed to depend on t, x and a parameter z varying in a Cartesian space. The purpose is then to find a function $x(t)$ satisfying

$$(15.4) \qquad\qquad x'(t) = L\,[t, x(t), z]\,, \quad x(0) = \xi$$

and to show that if L is C^k, then $x(t)$ is a C^k function of t, ξ and z.

Everything on the subject and even more can be found in Dieudonné, vol. 1, X.4 à X.8, but as this reference is not very easy to read, here we will substitute a *low-powered* proof to this type of *high-powered* one (expression used by Spivak for Serge Lang's *Analysis II*) in the manner of the inventors of the successive approximation method, for example Émile Picard. We have already used it in a particular case related to Bessel's equation in Chap. VI, n° 10.

(i) *Reduction to an integral equation.* If L is assumed to be continuous and if we consider $x(0) = \xi$, (4) amounts to solving

$$(15.5) \qquad\qquad x(t) = \xi + \int_0^t L\,[u, x(u), z]\,du\,,$$

where this is an oriented integral. Setting

$$(15.6) \qquad x(t) = y(t) + \xi\,, \quad L\,(t, y + \xi, z) = L\,(t, y, \xi, z)\,,$$

we are led to solve

$$y(t) = \int_0^t L\,[u, y(u), \xi, z]\,du\,.$$

As ξ and z are now parameters varying in Cartesian spaces, we might as well consider the parameter to be the couple (ξ, z) and remove ξ from the notation. Hence, we will suppose that the new function $L(t, y, z)$ is defined and C^k on a compact subset $|t| \leq a$, $\|y\| \leq b$, $\|z\| \leq c$ of $\mathbb{R} \times \mathbb{R}^p \times \mathbb{R}^q$.

To solve

$$(15.7) \qquad\qquad y(t) = \int_0^t L[u, y(u), z]\, du\,,$$

define the functions $y_n(t, z)$ by

$$(15.8) \qquad y_0(t, z) = 0\,, \quad y_{n+1}(t, z) = \int_0^t L[u, y_n(u, z), z]\, du$$

and hope they converge uniformly on every compact interval, in which case their limit is the solution of the problem.

(ii) *Existence of solutions.* Temporarily omitting the parameter z from the expressions for y_n, we get

$$(15.9) \quad y_{n+1}(t) - y_n(t) = \int_0^t \{L[u, y_n(u), z] - L[u, y_{n-1}(u), z]\}\, du\,.$$

An upper bound for the integral can be found by using the mean value formula: if $D_2 L(t, y, z)$ denotes the derivative of the map $y \mapsto L(t, y, z)$ and if

$$\|D_2 L(t, y, z)\| \leq M' \text{ for } |t| \leq a, \|y\| \leq b, \|z\| \leq c,$$

then

$$(15.10) \quad \|L[u, y_n(u), z] - L[u, y_{n-1}(u), z]\| \leq M'\|y_n(u) - y_{n-1}(u)\|$$

for $|u| \leq a$ and $\|z\| \leq c$ as long as the $y_n(u)$ remain in the ball $\|y\| \leq b$.

But let $M = \sup \|L(t, y, z)\|$ for $|t| \leq a$, $\|y\| \leq b$, $\|z\| \leq c$; if $\|y_n(t)\| \leq b$, (8) shows that $\|y_{n+1}(t)\| \leq M|t| \leq b$ if $|t| \leq b/M$. Setting

$$a' = \inf(a, b/M)\,,$$

we see that if the relation

$$(15.11) \qquad\qquad |t| \leq a', |z| \leq |c| \Longrightarrow \|y_n(t)\| \leq b$$

is true for some n, then it is true for $n + 1$. As $y_0(t) = 0$, (11) holds for all n, which makes its possible to use (10) within the limits indicated for t and z.

(8) primarily shows that $\|y_1(t)\| \leq M|t|$. Integrating from 0 to t in all integrals in u of this n°, we then get

$$\|y_2(t) - y_1(t)\| \leq M' \int \|y_1(u)\| du \leq MM't^2/2! \,,$$

$$\|y_3(t) - y_2(t)\| \leq M' \int \|y_2(u) - y_1(u)\| du \leq MM'^2 t^3/3!$$

and so on until

(15.12) $\|y_{n+1}(t) - y_n(t)\| \leq MM'^{n-1}t^n/n!$ for $|t| \leq a'$.

The series $\sum[y_{n+1}(t) - y_n(t)]$, dominated by an exponential series, thus converges normally in $|t| \leq a'$ to a solution $y(t) = y(t, z)$ of (7) defined for $|t| \leq a'$ and $|z| \leq c$ and with values in $|y| \leq b$, proving statement (a). Note that this result only assumes L to be continuous and to have a continuous derivative $D_2L(t, x, z)$.

A very particular and important case is that of a *linear differential equation*, i.e. for which L is of the form

$$L(t, y, z) = A(t, z)y + b(t, z)$$

with functions $A(t, z)$ and $b(t, z)$ defined and continuous for $|t| \leq a$, $|z| \leq c$. As $L(t, y, z)$ is defined without any restrictions on y, successive approximations $y_n(t, z)$ are defined and continuous; hence there are no other restrictions on the domain of existence of solutions apart from those imposed on the coefficients $A(t, z)$ and $b(t, z)$.

(iii) *Uniqueness of the solution.* A similar calculation to (9) and (10) shows that if y' and y'' (these are not derivatives) are solutions of (7), then setting

$$k(r) = \sup_{|t| \leq r} \|y'(t) - y''(t)\| \,,$$

gives

$$\|y'(t) - y''(t)\| \leq \int \|L[u, y'(u), z) - L[u, y''(u), z]\| du$$

$$\leq M'k(r)|t|$$

for $|t| \leq r$ as long as $y'(u)$ and $y''(u)$ remain in the ball $\|y\| \leq b$, which is the case for sufficiently small r since (7) implies $y(0) = 0$. Taking the sup of the left hand side for $|t| \leq r$, $k(r) \leq M'k(r)r$ follows, and so $k(r) = 0$ if $r \leq 1/M'$. This proves statement (b).

(iv) *Dependence on initial conditions.* It is now a matter of showing that, for t and z in the neighbourhood of 0, like L, the solution $y(t, z)$ of (7) is of class C^k as a function of (t, z). Suppose first that $k = 1$; it all amounts to showing that the functions (8) are C^1 and that their first derivatives converge uniformly (Chap. III, n° 22, theorem 23).

Formula (8) shows that, if L and $y_n(t, z)$ are C^k, so are $y_{n+1}(t, z)$, proving the first point. As $Dy_{n+1} = y_n$, where $D = d/dt$, the $Dy_n(t, z)$ converge uniformly on $|t| \leq a'$, $|z| \leq c$, and so the continuity of $Dy(t, z)$ follows. The case of derivatives with respect to the parameter z cannot be dealt with so easily. In what follows, D_2 (resp. D_3) will denote the operator transforming a function $f(t, y, z)$ into the derivative of the map y (resp. z) $\mapsto f(t, y, z)$; they correspond to the partial differentials of n° 2, (iii). We, therefore, need to prove the convergence of the functions

$$Y_n(t, z) = D_3 Y_n(t, z) : \mathbb{R}^q \longrightarrow \mathbb{R}^p.$$

Applying D_3 to the function $z \mapsto L[u, y_n(u, z), z]$ under the $\int$ sign in (8), first

$$(15.13) \qquad Y_{n+1}(t, z) = \int \{ D_2 L[u, y_n(u, z), z] . Y_n(u, z) +$$

$$+ D_3 L[u, y_n(u, z), z] \} \, du =$$

$$= \int U_n(u, z) Y_n(u, z) + V_n(u, z)] du,$$

where

$$U_n(u, z) = D_2 L[u, y_n(u, z), z] : \mathbb{R}^p \longrightarrow \mathbb{R}^p,$$
$$V_n(u, z) = D_3 L[u, y_n(u, z), z] : \mathbb{R}^q \longrightarrow \mathbb{R}^p.$$

However, $D_2 L$ and $D_3 L$ are uniformly continuous on the compact subset on which they are defined and $y_n(u, z)$ converges uniformly to $y(u, z)$. Hence

$$(15.14') \qquad D_2 L[u, y(u, z), z] = U(u, z) = \lim U_n(u, z)$$
$$(15.14'') \qquad D_3 L[u, y(u, z), z] = V(u, z) = \lim V_n(u, z)$$

uniformly on $|u| \leq a'$, $|z| \leq c$. If the left hand side of (13) converges uniformly to a limit $Y(t, z)$, it will, therefore, satisfy

$$(15.15) \qquad Y(t, z) = \int_0^t [U(u, z) . Y(u, z) + V(u, z)] \, du.$$

However, (15) is an integral equation of type (8), where $L(t, y, z)$ is replaced by a linear (affine) function $M(t, Y, z) = U(t, z)Y + V(t, z)$ in Y. As seen at the end of section (ii) of the proof, (15) has a unique solution, defined for $|t| \leq a$, $|z| \leq c$ and so

$$(15.16) \qquad D_n(t, z) = \|Y_n(t, z) - Y(t, z)\|$$

needs to be shown to converge uniformly to 0 on the compact subset $|t| \leq a'$, $|z| \leq c$.

However, (13) and (15) show that

$$D_{n+1}(t,z) \leq \int \|U_n(u,z)Y_n(u,z) - U(u,z)Y(u,z)\| \, du +$$

$$+ \int \|V_n(u,z) - V(u,z)\| \, du.$$

$$\leq \int \|U_n(u,z) - U(u,z)\| \cdot \|Y(u,z)\| \, du +$$

$$+ \int \|U_n(u,z)\| \cdot D_n(u,z) du +$$

$$+ \int \|V_n(u,z) - V(u,z)\| \, du$$

with extended *unoriented* integrals over the interval $I(t)$ with endpoints 0 and t; as $I(s) \subset I(t)$ for $s \in I(t)$, the right hand side is even an upper bound for $D_{n+1}(s,z)$ for all $s \in I(t)$. If k is a constant upper bound for $Y(u,z)$ and the $U_n(u,z)$ – they converge uniformly – for $|u| \leq a'$ and $|z| \leq c$, then

$$D_{n+1}(s,z) \leq k \int \|U_n(u,z) - U(u,z)\| \, du + k \int D_n(u,z) du +$$

$$+ \int \|V_n(u,z) - V(u,z)\| \, du ,$$

where integration is over $I(t)$, and not only over $I(s)$.

Let $r > 0$. There exists $N = N(r)$ such that

$$\|U_n(u,z) - U(u,z)\| \leq r \ \& \ \|V_n(u,z) - V(u,z)\| \leq r$$

for $n \geq N$, $|u| \leq a'$, $|z| \leq c$. The previous relation then shows that, for $n \geq N$,

$$(15.17) \qquad D_{n+1}(s,z) \leq (k+1)|t| + k \int_{I(t)} D_n(u,z) du$$

for $s \in I(t)$. Set $A = k + 1$ and

$$\Delta_n(t) = \sup_{\substack{u \in I(t) \\ |z| \leq c}} D_n(u,z).$$

The function being integrated on the right hand side of (17) is $\leq \Delta_n(u)$. Taking the sup of the left hand side for $s \in I(t)$ and $|z| \leq c$, it can be deduced that

$$(15.18) \qquad \Delta_{n+1}(t) \leq rA|t| + A \int_{I(t)} \Delta_n(u) du$$

for $n \geq N$, and so, iterating,

$$\Delta_{n+2}(t) \leq rA|t| + A \int_{I(t)} du \left\{ rA|u| + A \int_{I(u)} \Delta_n(v)dv \right\}$$

$$= r\left(A|t| + A^2|t|^2/2!\right) + A^2 \int_{I(t)} du_1 \int_{I(u_1)} \Delta_n(u_2)du_2 \,,$$

and so on. As $\Delta_n(u) \leq \Delta_n(t)$ for $u \in I(t)$, it follows that

$$\int_{I(t)} du_1 \int_{I(u_1)} du_2 \ldots \int_{I(u_{p-1})} \Delta_n(u_p)du_p \leq$$

$$\leq \Delta_n(t) \int_{I(t)} du_1 \int_{I(u_1)} du_2 \ldots \int_{I(u_{p-1})} du_p = \Delta_n(t)\left(|t|\right)^p /p! \,.$$

Hence iterating (18) gives

$$\Delta_{n+p}(t) \leq r\left(A|t| + \ldots + A^p|t|^p/p!\right) + \Delta_n(t)A^p|t|^p/p!$$
$$\leq r\left[\exp\left(Aa'\right) - 1\right] + \Delta_n(t)\left(Aa'\right)^p /p!$$

for $n \geq N(r)$, $p \geq 1$ and $|t| \leq a'$. (The reader will recognize these to be arguments given by Liouville already outlined in Chap. VII, n° 18). On the right hand side, the first term is arbitrarily small if r is conveniently chosen, and the second one tends to 0 as p increases. Hence $\lim \Delta_n(t) = 0$ for all t and in particular for $t = \pm a'$. So $D_3 y_n(t, z)$ is uniformly convergent. Hence, the function $y(t, z)$ is indeed C^1.

It remains to show that it is C^k if L is C^k. For $k = 2$, the right hand side of the differential equation

$$Dy(t, z) = L\left[t, y(t, z), z\right] \,, \quad y(0, z) = 0$$

is C^1 like L and y, so that $D^2 y(t, z)$ and $D_3 D_2 y(t, z)$ exist and are continuous. On the other hand, $D_3 y(t, z) = Y(t, z)$ satisfies (16), i.e.

(15.19) $$DY(t, z) = U(t, z)Y(t, z) + V(t, z) \,, \quad Y(0, z) = 0 \,.$$

As $U(t, z)$ and $V(t, z)$ are C^1 by (15) if L is C^2 and if y is C^1, solution $Y(t, z)$ of (19) is C^1. The function $y(t, z)$ is, therefore, C^2. And so on, which finishes the proof of statements (a), (b) and (c). I give up turning them into a theorem whose statement would take half a page.

(v) *Matrix exponential.* As seen above, the derivative $Y(t, z) = D_3 y(t, z)$ satisfies a differential equation (19) whose right hand side is an affine linear function of Y. Equations of the form

$$x'(t) = A(t)x(t) + b(t) \,, \quad x(0) = \xi \,,$$

where $x(t), b(t) \in \mathbb{R}^n$ and $A(t) \in M_n(\mathbb{R})$, are dealt with by using the general method, but there is an additional result in this case: the integrals are defined on every interval on which $A(t)$ and $b(t)$ are defined and continuous.

A particularly simple case is that of an equation

$$x'(t) = Ax(t), \quad x'(0) = \xi$$

with a constant matrix A. The successive approximation method leads to functions

$$x_1(t) = \xi + \int A\xi.du = (1 + At)\xi,$$

$$x_2(t) = \xi + \int A(1 + Au)\xi.du = \xi + At\xi + A^2 t^{[2]}\xi,$$

etc., whence the solution

(15.20) $$x(t) = \exp(tA)\xi,$$

where for any matrix or linear operator A, we set

(15.21) $$\exp(A) = \sum A^{[n]} = \sum A^n/n!;$$

the series converges since $\|A^n\| \le \|A\|^n$.

The binomial formula shows that, like in dimension one,

(15.22) $$\exp(A + B) = \exp(A)\exp(B) \quad \text{if} \quad AB = BA.$$

As $\exp(0) = 1$, the operator $\exp(A)$ is, therefore, always invertible, with

$$\exp(A)^{-1} = \exp(-A).$$

On the other hand, if A is replaced by UAU^{-1}, where the matrix U is invertible, each term A^n of the series is replaced by UA^nU^{-1}, and so

(15.23) $$\exp(UAU^{-1}) = U\exp(A)U^{-1}.$$

(22) also shows that the map $t \mapsto \exp(tA) = X(t)$ is a continuous homomorphism, and is even C^∞, from the additive group $\mathbb{R}$ to the multiplicative group $GL_n(\mathbb{R})$; it is called a *one-parameter subgroup* of $GL_n(\mathbb{R})$. It is the only one.

Indeed, if $X(t)$ is supposed to be differentiable at $t = 0$, formula $X(t+h) = X(t)X(h)$ shows that, like in dimension one, $X(t)$ is differentiable everywhere and that $X'(t) = X'(0)X(t) = AX(t)$, where $A = X'(0)$. Although $X(t)$ now has values in $M_n(\mathbb{R})$ rather than in $\mathbb{R}^n$, $X(t) = \exp(tA)$ again holds since the two sides satisfy the same differential equation with the same initial condition $X(0) = 1$.

If the only assumption made is that the function $X(t)$ is continuous, it is "regularized" by choosing a function φ on $\mathbb{R}$ in the Schwartz space $\mathcal{D}$ and by considering the integral

$$\int \varphi(t - u)X(u)du = \int \varphi(v)X(t - v)dv = X(t)\int \varphi(v)X(-v)dv.$$

The first integral, extended to $\mathbb{R}$ and in fact to a compact subset, is a C^∞ function of t like φ, and the third one can be made to approach $X(0) = 1$ by using a Dirac sequence φ_n (Chap. V, § 8, n° 27). So $X(t)$ is indeed C^∞. This remark has already been made in the classical case [Chap. VII, § 1, n° 2, (iii)].

Finally note the useful formula

$$(15.24) \qquad \det[\exp(X)] = \exp[\mathrm{Tr}(X)] ,$$

where, for any square matrix X, $\mathrm{Tr}(X)$ denotes the sum of the diagonal elements of X. Setting $D(t) = \det[\exp(tX)]$ indeed defines an obviously continuous homomorphism from $\mathbb{R}$ to $\mathbb{R}$, and so $D(t) = \exp(ct)$ for some $c = D'(0)$ to be computed. As the map $\det : M_n(\mathbb{R}) \longrightarrow \mathbb{R}$ is much more than differentiable, the multivariate chain rule shows that $D'(0) = \det'(1)\exp'(0)X$, where $\det'(1)$ is the derivative at $X = 1$ of the map $X \mapsto \det X$ and $\exp'(0) = 1$ that of the exp map at the origin. Hence $c = \det'(1)X = T(X)$ with a real valued *linear* function of X. By (23),

$$(15.25) \qquad T(UXU^{-1}) = T(X) \quad \text{for all } U \in GL_n(\mathbb{R}),$$

and so $T(UX) = T(XU)$. As it is not hard to check that every $Y \in M_n(\mathbb{R})$ is the sum of invertible matrices ($Y + \lambda 1$ is invertible provided $-\lambda$ is not an eigenvalue of Y), it can be deduced that

$$(15.26) \qquad T(XY - YX) = 0$$

for all matrices X and Y. Setting $T(X) = a_i^j X_j^i$, where the X_j^i are the coefficients of X, and by writing (26) explicitly for a matrix Y all of whose entries, apart from one, are non-zero, it immediately follows that $a_i^j = 0$ for $i \neq j$, so that $T(X)$ is a linear combination of the diagonal entries of X. Its value remains invariant if its entries undergo an arbitrary permutation, for this amounts to replacing X by UXU^{-1}, where the matrix U permutes the vectors of the canonical basis for $\mathbb{R}^n$. Hence $T(X)$ is proportional to the trace of X. It remains to check that $T(X) = \mathrm{Tr}(X)$ for *a* matrix X with non-zero trace; we leave it to the reader to choose this matrix in such a way as to minimize calculations.

This proof of (24) is slightly longer than the classical proof, but it teaches the reader, if he does not already know it, that, up to a constant factor, relation (26) characterizes the function $X \mapsto \mathrm{Tr}(X)$.

Exercise 1. Show that, for every linear functional $X \mapsto f(X)$ on $M_n(\mathbb{R})$, where $\mathbb{R}$ denotes an arbitrary commutative field, there is a unique matrix A such that $f(X) = \mathrm{Tr}(AX)$.

Exercise 2. Let E be an n-dimensional vector space over $\mathbb{R}$, $M(E)$ the set of linear maps $E \longrightarrow E$, (a_i) a basis for E and $f(x_1, \dots, x_n)$ the unique n-linear alternating form equals 1 on the basis vectors, i.e. is the determinant of the x_i with respect to this basis. Hence if $u \in M(E)$, then

$$\det(u) = f\,[u(a_1)\ldots, u(a_n)] = f\,(u_1,\ldots,u_n)\,,$$

where $u_i = u(a_i)$. Show that the derivative

$$\det{}'(u) : h \longmapsto \frac{d}{dt}\det(u+th)\quad\text{for } t=0\,,$$

which takes $M(E)$ linearly onto $\mathbb{R}$, is given by

$$h \longmapsto f\,(u_1+h_1,u_2,\ldots,u_n)+\ldots+f\,(u_1,\ldots,u_{n-1},u_n+h_n)\,,$$

where $h_i = h(a_i)$. Deduce that

$$(15.27)\qquad\qquad\det{}'(u)h = \det(u)\mathrm{Tr}(h)\,.$$

Exercise 3. Show that the group $SL_n(\mathbb{R})$ of $X \in M_n(\mathbb{R})$ such that $\det(X)=1$ is a closed submanifold of $M_n(\mathbb{R})$.

The § where we will discuss Lie groups in vol. IV will provide an opportunity to apply the results of this n° in a particularly important case.

16 – Differential Forms on a Manifold

The general definition of tensor fields makes the notion of a *differential form of degree p on a manifold* X obvious: such a form associates to each point x of the open subset of X a p-linear alternating form $\omega(x; h_1,\ldots, h_p)$ on $X'(x)$; hence, in every local chart (U,φ) at x, for $p=3$ for example,

$$(16.1)\qquad \omega(x; h, k, l) = a_{ijk}(\xi)h^i(\varphi)k^j(\varphi)l^k(\varphi)$$

with *antisymmetric coefficients* depending on the chosen chart. The right hand side of (2) is a differential form $\omega(\varphi)$ on $\varphi(U)$ and if (U,φ) is replaced by (V,ψ), under the chart change diffeomorphism $\varphi(x) \mapsto \psi(x)$, $\omega(\psi)$ is transformed into $\omega(\varphi)$ as inverse images ; the $a_{ijk}(\xi)$ are the components of a tensor. Conversely, considering in every chart a form $\omega(\varphi)$ which, under chart change, is transformed as above, defines a form ω on X.

The definition of the exterior product of two forms generalizes in an obvious way. This makes it possible to write

$$(16.2)\qquad \omega = \sum_{i<j<k} a_{ijk}d\xi^i \wedge d\xi^j \wedge d\xi^k = \frac{1}{3!}a_{ijk}d\xi^i \wedge d\xi^j \wedge d\xi^k$$

as in $\mathbb{R}^n$.

The notion of the inverse image of a form under a map $f:X \longrightarrow Y$ also easily generalizes: if ω is a form of degree 3 for example in Y, the form $\varpi = \omega \circ f$ on X is defined by

$$(16.3)\qquad \varpi(x; h_1, h_2, h_3) = \omega\,[f(x)\,; f'(x)h_1, f'(x)h_2, f'(x)h_3]\,.$$

This is the only formula likely to still be well-defined here.

To define the exterior derivative $d\omega$ of a given form ω on X, the forms $\omega(\varphi)$ expressing ω in the charts of X may be used. As in Cartesian space, the operation "inverse image" transforms an exterior derivative into an exterior derivative, the $d\omega(\varphi)$ clearly define a form on X, namely the derivative $d\omega$ we were looking for. If for example

$$(16.4) \qquad \omega(\varphi) = a_{jk}d\xi^j \wedge d\xi^k$$

in (U, φ), then

$$(16.5) \qquad d\omega(\varphi) = da_{jk} \wedge d\xi^j \wedge d\xi^k =$$
$$= \frac{1}{3}\left(D_i a_{jk} + D_j a_{ki} + D_k a_{ij}\right) d\xi^i \wedge d\xi^j \wedge d\xi^k .$$

To define $d\omega$ as we did in a Cartesian space, the covariant derivative ω' of ω should first be defined, and more generally that of a tensor field T on X. But the definition

$$(16.6) \qquad T'(x; h, k, u) = \frac{d}{ds}T(x + sh; k, u) \quad \text{for } s = 0$$

of T' is not well-defined in a manifold, and defining T by differentiating its components in each local chart would involve second derivatives. Formula (6) interpreted as we have done will, therefore, not lead to the components of a new tensor. The solution of the problem can be found in the theory of "connectedness", which will not be presented here. Let us only observe that when we investigate the manner in which the partial derivatives of the coefficients of a differential form are transformed under chart change, the second derivatives, which for an arbitrary tensor field occur in the formulas, disappear: a miracle of the antisymmetric character of the coefficients, as the reader can check with some patience.

17 – Integration of Differentiable Forms

All that has been done at the start of § 2 on "curvilinear" integrals over open subsets of a Cartesian space generalizes immediately to differential forms of degree 1 on a manifold X. If ω is such a form and $\gamma : I \longrightarrow X$ is a path of class C^1 in X, for simplicity's sake, the integral of ω along γ is obtained by replacing ω with its inverse image $\omega \circ \gamma$ under γ and by integrating the result over $[0, 1]$. The extended integral of a form of degree 2 over a 2-dimensional path $\sigma : I \times I = K \longrightarrow X$ is defined likewise. Then for ω of degree 1,

$$\iint_\sigma d\omega = \int_{\partial\sigma} \omega ,$$

as in (9.20). The invariance of the integrals of ω under homotopy now follow if ω is *closed*. Notice that ω is an exact differential if and only if its integral along every closed path is zero. Etc.

A much more serious problem consists in defining the extended integral of a form of maximum degree on X without taking a "parametric representation" of X for granted for otherwise we would only need to integrate over a cube.

(i) *Orientable manifolds.* Let X be an n-dimensional manifold and ω a differential form of degree n on X and of class C^0. Whatever be the final definition of the integral of ω, if X is not compact, we will clearly come across convergence problems at infinity unrelated to the problem at hand. This is already obvious if $X = \mathbb{R}$. It is, therefore, prudent to suppose that ω has compact support[66] in order to get rid of them, as we did at the start of Chapter V.

The simplest case is obtained by supposing the existence of a chart for all of X, in other words, that X is diffeomorphic to an open subset of a Cartesian space. For this, let us choose a diffeomorphism φ from X onto $\varphi(X) \subset \mathbb{R}^n$; it transforms ω into a form

$$\omega(\varphi) = a(\xi)d\xi^1 \wedge \ldots \wedge d\xi^n,$$

on $\varphi(X)$ and we wish to set

(17.1) $$\int_X \omega = \int_{\varphi(X)} a(\xi)d\xi^i \ldots d\xi^n.$$

This a an ordinary multiple integral over the open set $\varphi(X)$ of a function which vanishes outside a compact subset of it.

If ψ is another diffeomorphism from X onto an open subset of $\mathbb{R}^n$, then there is a form

$$\omega(\psi) = b(\eta)d\eta^1 \wedge \ldots \wedge d\eta^n$$

on $\psi(X)$. As $\omega(\varphi)$ is the inverse image of $\omega(\psi)$ under the chart change diffeomorphism $\theta : \varphi(X) \longrightarrow \psi(X)$

$$a(\xi) = b\left[\theta(\xi)\right] J_\theta(\xi).$$

Since the change of variable formula for multiple integrals showing that

$$\int_{\psi(X)} b(\eta)d\eta^1 \ldots d\eta^n = \int_{\varphi(X)} b\left[\theta(\xi)\right] |J_\theta(\xi)|d\xi^1 \ldots d\xi^n,$$

we are led to conclude that the two values proposed for the integral of ω are equal only if $J_\theta(\xi) > 0$ everywhere. This is not a good sign in every sense of the word since we want a result independent of the coordinate system used.

---

[66] Recall that it is the largest *closed* set Supp$(\ldots)$ outside which the function (or differential form, or…) is zero.

The same difficulty arises in the case of a general manifold. A simple case is that of a form ω whose compact support is contained in the domain of a chart (U, φ); it would be natural to set

$$\int_X \omega = \int_U \omega,$$

where the right hand side is defined by formula (1) applied to U. If the chart (U, φ) is replaced by the chart (V, ψ) such that $\mathrm{Supp}(\omega) \subset V$, then, applying (1) to U or V, it clearly suffices to integrate over $U \cap V$; the arguments used above then show that, here too, both definitions of the integral coincide only if the Jacobian of change of coordinates $\varphi(x) \mapsto \psi(x)$ is everywhere > 0 in $U \cap V$.

To overcome this difficulty, we are led to allow only changes of local charts whose Jacobian is everywhere > 0, more precisely, instead of all possible charts, to use only an atlas all of whose changes of charts have positive Jacobian. The existence of such an atlas – recall that the charts in an atlas must cover the manifold – is not obvious and may be false for some manifolds. When such an atlas exists, the manifold is said to be *orientable*.

The relation with the classical notion of orientable surface in $\mathbb{R}^3$ recalled in n° 9, (iv) is easy to see. Indeed, let X be a 2-dimensional submanifold of $\mathbb{R}^3$ and (U, φ) a local chart for X; if f is the inverse map of φ, in the open subset U, the surface X is given by the parametric representation

$$x = f^1(s, t), \quad y = f^2(s, t), \quad z = f^3(s, t),$$

where (s, t) vary in the open subset $\varphi(U)$ of the plane. Then the derivatives $D_1 f$ and $D_2 f$ of f with respect to s and t at each point $x \in U$ are two non-proportional tangent vectors to X at x (in the classical sense); their classical vector product is a normal vector to X at x, which is a continuous function of the point x and thus coherently orients the normals to X at points of U. Similarly, if (V, ψ) is another chart and if $g = \psi^{-1}$, the product $D_1 g \wedge D_2 g$ leads to an orientation of normals in V. In $U \cap V$, the $D_j g$ are linear combinations of the $D_i f$ whose coefficients are the entries of the Jacobian matrix of change of coordinate $\theta : \varphi(x) \longrightarrow \psi(x)$. As the vector product of two vectors is an alternating bilinear form of these,

$$D_1 g(\eta) \wedge D_2 g(\eta) = J_\theta(\xi) D_1 f(\xi) \wedge D_2 f(\xi),$$

which shows that, in $U \cap V$, the orientations of normals to X defined by the charts (U, φ) and (V, ψ) are identical only if $J_\theta > 0$ everywhere. So if X can be covered by charts such that the chart change formulas have positive Jacobians, then the normals at all points of X can be coherently oriented, which is the classical definition of an orientable surface.

Conversely, suppose that this condition holds; among all charts of X, only consider those leading to the chosen coherent orientation of normals;[67] the previous arguments then show that the relation $J_\theta > 0$ holds for any two of these charts, and the general definition of orientability given above is recovered.

In the general case, which we now return to, consider two atlases (U_i, φ_i) and (V_p, ψ_p) with positive Jacobians. For $x \in X$, choose indices i and p such that $x \in U_i \cap V_p$; let $\varepsilon(x) = \pm 1$ be the sign at the point x of the Jacobian of the chart change $\varphi_i(x) \mapsto \psi_p(x)$. It only depends on x, since if $x \in U_j \cap V_q$ for another couple of indices, the Jacobians of the chart changes $(U_i, \varphi_i) \longrightarrow (U_j, \varphi_j)$ and $(V_p, \psi_p) \longrightarrow (V_q, \psi_q)$ are positive by assumption. The result is then an immediate consequence of the multiplication formula for Jacobians. Having said this, observe that as the Jacobian of every chart change $(U_i, \varphi_i) \longrightarrow (V_p, \psi_p)$ is a continuous function on the open subset $U_i \cap V_p$, its sign remains constant in the neighbourhood of every $x \in U_i \cap V_p$; hence $\varepsilon(x)$ is a *continuous* function on X. The two atlases considered will be said to define the same orientation (resp. opposite orientations) if $\varepsilon(x) = +1$ (resp. -1) for all $x \in X$. If X is *connected*, there is clearly no other possibility; in other words, there are at most two ways of orienting a connected manifold.

Whether X is connected or not, an equivalence relation can be defined on the set of all charts for X by setting two charts to be equivalent if and only if the Jacobian of chart change is everywhere positive. This makes it possible to divide the charts into equivalence classes, each of these classes being an atlas of X with the following "maximality" property: given an atlas, if the Jacobians of change of coordinates between a chart for X and those in the atlas are all positive, then the chart belongs to it. Orienting a manifold then consists in choosing one of these classes; the charts which belong to it will then be said to be compatible with the orientation of X.

The simplest manifolds being Cartesian spaces, the orientation problem already arises for them and, in the same way, for their open subsets; in this case, there is a purely algebraic definition of orientation.

.Indeed, let (a_i) be a basis for the Cartesian space E; this immediately gives global chart for E (or for an open subset of E) by associating to each $x \in E$ its coordinates with respect to this basis; it could be called the *linear chart* for E associated to the chosen basis. This chart being by itself an atlas of E, the signs of all the Jacobians of its changes of charts are easy to compute... Hence, E is orientable, and if this chart is used to orient E, then the choice of a basis for E defines an orientation of E. Now let (b_i) be another basis for E; there is another linear chart for E associated to it and the latter can be obtained from the former by formulas known by everyone. Hence the orientations defined by the two bases considered are identical or

---

[67] If a chart (U, φ) does not satisfy this condition, it suffices to compose it with the diffeomorphism $(s, t) \longrightarrow (t, s)$ to obtain in the same open subset of X a chart compatible with the orientation of the normals.

opposite according to whether the determinant of the change of basis matrix is positive or negative. From this point of view, two bases can be considered to be equivalent if the determinant of the matrix taking one to the other is > 0. The set of all bases for E is thereby divided into two equivalence classes, one which plays a distinguished role; orienting E then amounts to choosing one of the these classes. The bases belonging to it are said to be *direct*, by analogy with physicists' "direct trihedrons". There is a distinguished class of bases in the spaces $\mathbb{R}^n$: that of the canonical basis. Hence it is possible to choose a *canonical orientation* in $\mathbb{R}^n$ – but it is impossible to do so in any other Cartesian space.

The case of a general manifold X can be presented in the same manner. Let (U, φ) be a chart for X. As seen in (i) of n° 13, at each point x of U, it gives rise to a basis $(a_i(\xi))$ for $X'(x)$ such that

$$h = h^i(\xi) a_i(\xi) \quad \text{for all } h \in X'(x).$$

If (V, ψ) is another chart, then $h = h^\alpha(\eta) b_\alpha(\eta)$ also holds in the basis corresponding to this chart. But as the formula $h^i(\xi) = \rho^i_\alpha(\eta) h^\alpha(\eta)$ whose coefficients are the $d\xi^i/d\eta^\alpha$ transforms the $h^i(\xi)$ to $h^\alpha(\eta)$, the determinant of the matrix taking the first basis to the second one is clearly the Jacobian of the change of basis at x (or of its inverse, which does not change the sign). Hence, if in every local chart (U, φ), the basis $(a_i(\xi))$ is used to orient the tangent space $X'(x)$ at each $x \in U$ as done above to orient a Cartesian space, two such charts are seen to define the same orientation in $U \cap V$ if and only if they orient $X'(x)$ in the same manner for all $x \in X$.

The conclusion is clear : *orienting a manifold X amounts to orienting each tangent space $X'(x)$* so that X can be covered by charts (U, φ) satisfying the following condition: for all $x \in U$, the orientation of $X'(x)$ is defined by the basis $(a_i(\xi))$ for $X'(x)$ associated to (U, φ).

(ii) *Integration of differential forms.* These arguments show that in order to define the integral of a differential form ω of maximum degree on a manifold X, X needs to be assumed to be *oriented* and ω to have compact support. For lack of better, the arguments of (1) then show that the integral of ω can given an absolute meaning in the particular case when the support of ω is contained in an open subset U of X; apply formula (1) after having chosen a chart (U, φ) *compatible with the orientation of X* ; the result is the same for all open Cartesian subsets U such that $\text{Supp}(\omega) \subset U$ and for all possible diffeomorphisms φ.

However, the method supposes that the compact subset $K = \text{Supp}(\omega)$ is contained in an open Cartesian set. In the general case, even that of a sphere in $\mathbb{R}^3$, only the existence of a covering of K by a finite number of such open subsets U_i can be guaranteed using BL. The method then consists in constructing forms ω_i with compact support satisfying

$$(17.2) \qquad \text{Supp}(\omega_i) \subset U_i \quad \& \quad \omega = \sum \omega_i$$

and in setting by definition,

$$(17.3) \qquad \int_X \omega = \sum \int_{U_i} \omega_i$$

since, in any reasonable interpretation of the integral of ω_i, it suffices to integrate over the open subset U_i outside which ω_i vanishes. But to give an "absolute" meaning to the left hand side of (3), the right hand side still needs to be shown to remain invariant if the U_i are replaced by the open Cartesian subsets V_p covering K and the ω_i by the ω_p satisfying (2) for the new covering, which is not at all obvious. *Partitions of unity* need to be used for this, a technique that can prove useful elsewhere.

Lemma 1. *Let A and B be two disjoint closed subsets in a metric space X. There is a function f defined and continuous on X satisfying $f(x) = 1$ on A, $f(x) = 2$ on B and $1 \le f(x) \le 2$ everywhere.*

The lemma is trivial if A is empty (take $f = 2$ everywhere) or if B is empty (take $f = 1$ everywhere). Otherwise, choose a distance function $d(x,y)$ defining the topology on X and set

$$f(x) = \inf\left[d(x,A), 2d(x,B)\right] / \inf\left[d(x,A), d(x,B)\right]$$

for $x \in X - (A \cup B)$, an open set on which f is continuous since so are the functions $d(x,A)$ and $d(x,B)$ that do not vanish there. In the neighbourhood of every point of A, $d(x,A) < d(x,B)$ and so $f(x) = 1$; in the neighbourhood of every point of B, $2d(x,B) < d(x,A)$ and so $f(x) = 2$. Setting $f(x) = 1$ on A and $f(x) = 2$ on B, we find the function sought[68] in all of the space X.

Lemma 2. *Let U be an open set and A a closed one contained in U. There is an open set V such that $A \subset V \subset \bar{V} \subset U$. If X is locally compact[69] and A is compact, $\bar{V}$ may be assumed to be compact.*

Here too, the first statement is trivial if $U = X$ (take $V = X$) or if A is empty (take $V = \{x\}$ with $x \in U$). Otherwise, set $B = X - U$, choose a function f by lemma 1 and take $V = \{f(x) < 3/2\}$, an open set containing A trivially and whose closure, contained in the open set $\{f(x) \le 3/2\}$, does not meet $B = X - U$; so $\bar{V} \subset U$.

---

[68] A more general result (Urysohn's theorem): if f is a real, bounded continuous function defined on a closed set $F \subset X$, there is a continuous extension of f to X. Assuming $f(F) \subset [1,2]$, the case it reduces to, formula

$$f(x) = d(x,F)^{-1} . \inf\left[f(u)d(x,u)\right] \text{ for } x \in X - F,$$

where inf relates to the $u \in F$, provides a solution. Dieudonné, IV, 5. The lemma correspond to the case $F = A \cup B$.

[69] i.e. such that every $x \in X$ has a compact neighbourhood V. Then any neighbourhood of x contains a compact neighbourhood, for example $V \cap B$, where B is a closed ball centered at x contained in W.

If X is locally compact and A is compact, for every $x \in A$ choose an open neighbourhood $V(x)$ of x, with compact closure $\overline{V(x)} \subset U$; By BL, A can be covered by finitely many $V(x_i)$; their union V is the answer to the question since $\bar{V} = \bigcup V(x_i)$ is compact and contained in U.

Lemma 3. *Let $U_0, \ldots, U_n$ be open sets and X their union. There exist open sets $V_0, \ldots, V_n$ whose union is X and such that $\bar{V}_i \subset U_i$ for all i.*

The V_p are constructed by induction on p by requiring them to also satisfy

$$V_0 \cup \ldots \cup V_p \cup U_{p+1} \cup \ldots \cup U_n = X .$$

As V_0 only needs to satisfy

$$X - (U_1 \cup \ldots \cup U_n) \subset V_0 \subset \bar{V}_0 \subset U_1 ,$$

it is obtained by applying lemma 2 to $A = X - (U_1 \cup \ldots \cup U_n)$ and $U = U_0$. Now if $V_0, \ldots, V_{p-1}$ are constructed so that

$$X = U_p \cup V_0 \cup \ldots \cup V_{p-1} \cup U_{p+1} \cup \ldots \cup U_n ,$$

the V_p are obtained by arguing in the same way about this new covering.

Lemma 4. *Let X be a locally compact metric space, A a compact subset of X and $(U_i)_{1 \le i \le n}$ a finite open covering of A. Then, there exist positive valued continuous functions f_i on X, with compact support and such that*

$$(17.4) \qquad \mathrm{Supp}(f_i) \subset U_i , \quad \sum f_i(x) = 1 \quad \text{on } A .$$

Set $U_0 = X - A$. Lemma 3 gives open sets $V_i (0 \le i \le n)$ covering X and such that $V_i \subset \overline{V}_i \subset U_i$. For $0 \le i \le n$, lemma 1 proves the existence of continuous functions g_i on X, with values in $[0, 1]$, and such that

$$(17.5) \qquad g_i(x) = 1 \text{ if } x \in \overline{V}_i , \quad g_i(x) = 0 \text{ if } x \in X - U_i .$$

Consider the function $g = \sum g_i$. It has ≥ 0 values and is even > 0 on the V_i, hence on their union X. The functions $h_i = g_i / g (0 \le i \le n)$ are, therefore, defined and continuous on X and satisfy $\sum h_i(x) = 1$ for all $x \in X$; but as $h_0 = g_0 / g$ vanishes on $X - U_0 = A$, the functions $h_i (1 \le i \le n)$ have sum 1 on A, each vanishing outside the corresponding open set U_i. The h_i still need to be transformed into functions with compact support. But as A is compact, lemma 2 shows the existence of an open set W with compact closure such that

$$A \subset W \subset \bar{W} \subset U_1 \cup \ldots \cup \mathring{U}_n$$

and lemma 1 that of a function p equal to 1 on A and vanishing outside W. Multiplying the h_i by p gives functions f_i vanishing outside the U_i and whose supports, contained in W, are compact; their sum on A is obviously again equal to 1, qed .

Thanks to Lemma 4, it is possible to give an unambiguous definition of the integral over an oriented manifold X of a differential form ω of class C^0, maximum degree $n = \dim(X)$ and with compact support. For this, in conformity with lemma 4, choose open Cartesian sets U_i covering the support of ω, as well as functions f_i and apply formula (3) choosing $\omega_i = f_i \omega$, a form whose support is a compact subset of U_i. If the U_i and the f_i are replaced by the V_p and the g_p satisfying the same condition, then obviously $\omega_i = \sum g_p \omega_i$. As the support of ω_i is a compact subset of U_i, the integral of ω_i over U_i is the sum of the integrals of $g_p \omega_i$ since, as seen at the start of (ii), the situation in an open Cartesian set is similar to that in an open subset of $\mathbb{R}^n$. But the support of $g_p \omega_i$ being contained in the open Cartesian space $U_i \cap V_p$, its integral over U_i, defined as in (i), is equal to its integral over $U_i \cap V_p$. Hence finally,

$$\sum \int_{U_i} f_i \omega = \sum_{i,p} \int_{U_i \cap V_p} g_p f_i \omega .$$

Now, permuting the roles played by the U_i and the V_p, similarly

$$\sum \int_{V_p} g_p \omega = \sum_{p,i} \int_{V_p \cap U_i} f_i g_p \omega$$

follows. So the two partitions of unity used to compute the integral of ω indeed lead to the same result.

At the same time, if ω and ϖ are forms with compact support, then

$$(17.6) \qquad \int \omega + \int \varpi = \int \omega + \varpi ;$$

the union of the supports of ω and ϖ being compact, the same partition of unity can be used to compute the three integrals in question; this reduces to the trivial case of an open subset of a Cartesian space.

Finally note that if ω is a form of degree $p \leq n$ and if Y is a p-dimensional submanifold of X, by inverse image, the immersion $Y \longrightarrow X$ leads to a form of maximum degree on Y. If the support of ω is compact and if Y is closed, then $\int_Y \omega$ can be defined.

18 – Stokes' Formula

It states that if ω is a differential form of degree $n - 1$ on an n-dimensional oriented manifold X and if Ω is an open set with compact closure whose border $\partial \Omega$ is an $n - 1$-dimensional submanifold of X in the sense defined at the end of n° 12, then

$$(18.1) \qquad \int_{\partial \Omega} \omega = \int_{\Omega} d\omega ,$$

obviously provided the edge of Ω is oriented properly and a further condition is imposed; its intuitive meaning is that in the neighbourhood of any of its border points, the open set Ω is located "on one side" of it. Corollary: *if X is an n-dimensional compact manifold, then*

$$\int_X d\omega = 0$$

for any form ω of degree $n - 1$ on X, since ∂X is empty.

Before moving on to the proof, we make some remarks about what happens in the neighbourhood of a point $x_0 \in \partial\Omega$.

Since, by assumption, $\partial\Omega$ is an $n - 1$-dimensional submanifold of X, as seen at the end of n° 12, there is a cubic chart (U, φ) at x_0[70] such that $\varphi(x_0) = 0$ and $\varphi(U \cap \partial\Omega)$ is the subset of K^n defined by relation $\xi^1 = 0$. Let U_0, U_+ and U_- be the subsets of U defined by the conditions $\xi^1 = 0$, $\xi^1 > 0$ and $\xi^1 < 0$, respectively. So $U_0 = U \cap \partial\Omega$. Since Ω does not meet its border, $U \cap \Omega \subset U_+ \cup U_-$ and, for the same reason,

$$U_+ \cap \Omega = U_+ \cap (\Omega \cup \partial\Omega).$$

The intersections of Ω with U_+ and U_- are, therefore, both closed and open in these connected open sets, and so only two cases are possible: (a) $U \cap \Omega = U_+$ or U_-, (b) $U \cap \Omega = U_+ \cup U_-$.

As will be seen, Stokes' formula supposes that case (a) holds everywhere. However, if one of these two cases holds at $x_0 \in \partial\Omega$, then is also holds at $x \in \partial\Omega$ sufficiently near x_0. The set S of $x \in \partial\Omega$ where case (b) holds is, therefore, an open subset of $\partial\Omega$. Similarly for the set S' of points where case (a) holds. This gives a partition of $\partial\Omega$ into two open sets, i.e into two closed ones. Hence, like $\partial\Omega$, S and S' are $n - 1$-dimensional compact submanifolds of X. Besides, $S \cup \Omega$ is clearly a connected open subset of X if Ω is connected. Replacing X by $S \cup \Omega$, the following question arises: *In an n-dimensional connected manifold, can the complement Ω of an $n - 1$-dimensional compact submanifold S be connected?* If the answer was always no, $S = \varnothing$ would hold and the "good" case (a) would also hold. If $X = \mathbb{R}^2$, then S is a smooth curve without multiple points. Hence, it may be assumed to be the finite union of pairwise disjoint simple closed curves, namely its connected components; the answer to the question is, therefore, no by Jordan's theorem, which we alluded to without proof at the end of Chap. IV, § 4; the same result holds for a sphere. But if we remove from the surface a two-dimensional torus, a circle whose plane contains the rotation axis of the torus, or is orthogonal to it, then case (b) holds: the complement of such a circle is connected and located on "both sides" of it. To have $S = \varnothing$, two circles would need to be removed from the torus; the open complement then has two connected

---

[70] i.e. such that $\varphi(U)$ is the cube $K^n : |\xi^i| < 1$ of $\mathbb{R}^n$.

components, and case (a) holds for each of them. Algebraic topology (duality theory) has long resolved and generalized this problem: the relation between the topology of a submanifold and that of its complement.

In what follows, case (a) will be assumed to always hold. $\partial\Omega$ is then said to be the *boundary* of Ω, a more restrictive expression than "border", and that Ω is an *open set with boundary* in X. It may then be assumed that $\xi^1 > 0$ on $U \cap \Omega$, if need be by replacing the function ξ^1 by $-\xi^1$ and by carrying out an odd permutation on the other coordinates, operations that leave the orientation of the chart considered and its cubic character unchanged. Then $U \cap \Omega = U_+$.

Having said that, the compact set $\Omega \cup \partial\Omega$ can be covered by a finite number of cubic charts (U, φ) compatible with the orientation of X and such that $U \cap \Omega = U$ or U_+. Let G be the union of these charts; it is open in X. By lemma 4 above, in G, ω can be decomposed into a sum of differential forms whose supports are compact sets contained in the charts considered. By the additive formula (15), it, therefore, suffices to prove Stokes' formula for these forms. In other words, the support of ω can be assumed to be compact and contained in one of these cubic charts (U, φ).

Let us first consider the case $U \cap \Omega = U$. As the support of ω does not meet $\partial\Omega$, the left hand side of Stokes' formula is zero. To show that the same is true for the right hand side, consider the open cube $\varphi(U) = \{|\xi^i| < 1\}$ of $\mathbb{R}^n$. Thus ω is replaced by a form

$$(18.2) \qquad \sum p_i(\xi) d\xi^1 \wedge \ldots \wedge \widehat{d\xi^i} \wedge \ldots \wedge d\xi^n ,$$

where the accent indicates that $d\xi^i$ must be omitted, and $d\omega$ by

$$(18.3) \qquad \sum (-1)^{i-1} D_i p_i(\xi) d\xi^1 \wedge \ldots \wedge d\xi^n .$$

To find the Lebesgue-Fubini integral of the ith term of (3), we can first integrate with respect to ξ^i, which gives the variation over $]-1, +1[$, up to sign, of the function

$$(18.4) \qquad t \longmapsto p_i\left(\xi^1, \ldots, \xi^{i-1}, t, \xi^{i+1}, \ldots, \xi^n\right) ;$$

as the support of ω is *compact* in the *open* set $\varphi(U)$, this function vanishes if t is sufficiently near 1 or -1. The result follows.

The $U \cap \Omega = U_+$ remains to be proved. Once again, there is a form (2) on $\varphi(U)$, but now we integrate its exterior derivative (3) over the open set $\xi^1 > 0$, so that integration with respect to the variable ξ^i must be extended to the interval $]-1, +1[$ if $i \neq 1$ and to the interval $]0, 1[$ if $i = 1$. If $i \neq 1$, the result is zero as in case (a). If, however, $i = 1$, the result is the variation of (4) over $]0, 1[$; the support of $d\omega$ being compact in $\varphi(U)$, function (4) vanishes for t in the neighbourhood of 1 as in case (a), but not in the neighbourhood of $\xi^1 = 0$; so, applying the FT,

(18.5) $$\int_\Omega d\omega = -\int_{K^{n-1}} p_1\left(0,\xi^2,\dots,\xi^n\right) d\xi^2 \dots d\xi^n .$$

This is the extended ordinary multiple integral over the cube K^{n-1} of $\mathbb{R}^{n-1}$.

As for the extended integral of ω over $\partial\Omega$, i.e. over $U\cap\partial\Omega$, it is computed by using the cubic chart (U_0,φ_0) of the manifold $\partial\Omega$, where $U_0 = U\cap\partial\Omega$ as above and where

$$\varphi_0(x) = \left(0,\varphi^2(x),\dots,\varphi^n(x)\right) .$$

The image of U_0 under this chart is precisely the cube K^{n-1} appearing in (5), and φ_0 transforms ω into the form that can be deduced from (2) by replacing the ξ^1 by 0, which cancels all the terms of (2) for which $i \neq 1$. Hence

(18.6) $$\int_{\partial\Omega} \omega = \varepsilon \int_{K^{n-1}} p_1\left(0,\xi^2,\dots,\xi^n\right) d\xi^2 \dots d\xi^n ,$$

with a sign ε depending on the orientation of $\partial\Omega$, which has not yet been defined. Hence to obtain Stokes' formula in this case, it is necessary to ensure that $\varepsilon = -1$, in other words to orient $\partial\Omega$ so that the chart (U_0,φ_0) is *incompatible* with this orientation.

The result can be stated differently. Let f be the inverse map of φ. Consider the point $x_0 = f(0) \in U\cap\partial\Omega$ and set $a_i = f'(0)e_i$, where (e_i) is the canonical basis for $\mathbb{R}^n$. This gives a basis for $X'(x_0)$ defining its orientation since the chart (U,φ) is compatible with it; besides, the vectors $a_2,\dots,a_n$ form a basis for the tangent subspace to $Y = \partial\Omega$ at x_0 and define the orientation of $\partial\Omega$ opposite to the one appropriate for Stokes' theorem. Having said this, consider the curve

$$\gamma : t \longmapsto f(-t,0,\dots,0)$$

in X passing through x_0 for $t = 0$; The assumptions made about the chart (U,φ) imply that for sufficiently small $|t|$, $\gamma(t) \in \Omega$ for $t < 0$ and $\gamma(t) \notin \Omega$ for $t > 0$. The curve γ is, therefore, the trajectory of a moving object *coming out of* Ω by crossing the boundary of Ω at x_0 at time $t = 0$; its velocity vector at $t = 0$ is $-a_1$. The basis $(-a_1,a_2,\dots,a_n)$ being incompatible with the orientation of Ω, it can be made compatible by an odd permutation of $a_2,\dots,a_n$ which transforms these vectors into a basis for $Y'(x_0)$ compatible with the orientation of $\partial\Omega$. The rule to be applied can, therefore, be stated as follows: let $h_1 \in X'(x)$ be the velocity at x of a moving object *coming out of* Ω at the point x. A basis $(h_2,\dots,h_n)$ for the tangent space to $\partial\Omega$ at x defines the orientation of $\partial\Omega$ if and only if the basis $(h_1,\dots,h_n)$ for $X'(x)$ is compatible with the orientation of X.

Consider the simplest case: X is a 2-dimensional submanifold in $\mathbb{R}^3$, i.e. what physicists mean by a "surface", $\partial\Omega$ being a curve drawn in X and limiting an open subset Ω of X. Choose a basis (h_1,h_2) at a point $x \in \partial\Omega$ for the traditional tangent plane $T_x(X)$ defining the orientation of X and for which h_2 is tangent to the curve limiting Ω. If the vector h_1 "comes out of" Ω, $\partial\Omega$ must be oriented like h_2. But if (h_1, h_2) defines the orientation of X, this means that the orientation of its normal is that of the vector $h_1 \wedge h_2$. In other words, the vectors h_1, h_2 and the unit vector of the oriented normal at x, written in this order, must form a "direct" trihedron. We thus recover the classical rule described in n° 9, (iv): a passerby following the boundary of Ω in the direction prescribed by Stokes' formula and remaining oriented like the normal to the surface, when looking in front of him or her, must leave Ω on his or her left, which until further notice is the same for both genders.

In particular, this rule applies to an open subset Ω of $\mathbb{C}$ limited by one or several closed simple curves $\gamma_0, \ldots, \gamma_p$, as can be encountered in the theory of holomorphic functions; if Ω is assumed to be interior to γ_0 exterior[71] to $\gamma_1, \ldots, \gamma_p$ and if Ω is given the usual orientation, then γ_0 must be "positively" oriented (counterclockwise) and the other γ_k negatively.

Suppose now that $X = \mathbb{R}^3$, which is another classical case, so that Ω is an open bounded set whose border is a 2-dimensional compact submanifold of X, for example a sphere, a torus, etc. It is natural to orient Ω like the canonical basis for X. It then remains to orient $\partial\Omega$ in such a way that Stokes' (for that matter, Ostrogradsky's) formula holds. The general result shows that if we choose a basis (h_1, h_2, h_3) at $x \in \partial\Omega$, oriented like the canonical basis and such that (i) h_1 is the velocity vector at x of a moving object departing from x and come out of Ω, (ii) (h_2, h_3) form a basis for the tangent plane to $\partial\Omega$ at x, then it needs to be oriented like the basis (h_2, h_3). This amounts to saying that the normal vector $h_2 \wedge h_3$ to $\partial\Omega$ at x must *come out of* Ω.

---

[71] In the sense given to these terms in Chapter VIII: $\mathrm{Ind}_\gamma(z) = +1$ or -1 according to whether z is interior or exterior to γ.

X – The Riemann Surface of an Algebraic Function

1 –Riemann Surfaces

Let X be a 2-dimensional C^0 manifold in the sense of Chap. IX, n° 11, (ii). If (U, φ) is a chart for X, φ may be considered a homeomorphism from the open set U onto an open subset of $\mathbb{C}$: if $\xi_1(x), \xi_2(x)$ are the coordinates of $\varphi(x) \in \mathbb{R}^2$ for $x \in U$, it suffices to agree that $\varphi(x) = \xi_1(x) + i\xi_2(x)$.

Let (U, φ) and (V, ψ) be two charts for S. The change of charts $\varphi(U \cap V) \longrightarrow \psi(U \cap V)$ is a priori not C^0, so that if, by miracle, a function defined on $U \cap V$ can be expressed holomorphically in (U, φ), there is no reason this is also possible in (V, ψ). For this, the change of charts would need to transform holomorphic functions into holomorphic functions, in other words would need to be a *conformal representation* of $\varphi(U \cap V)$ on $\psi(U \cap V)$.

This brings us to the notion of a Riemann surface (or of a complex analytic manifold with complex dimension 1): It is a 2-dimensional *connected*[1] manifold X of class C^0 with an atlas (U_i, φ_i) all of whose changes of charts

$$\varphi_i(U_i \cap U_j) \longrightarrow \varphi_j(U_i \cap U_j)$$

are holomorphic, in which case these are conformal representations (permute i and j). This leads to the more general definition of *a holomorphic chart* (U, φ) of X by the condition that, for all i, coordinate changes $\varphi_i(x) \mapsto \varphi(x)$ and $\varphi(x) \mapsto \varphi_i(x)$ be holomorphic on the open sets on which they are defined. When (U, φ) is a holomorphic local chart at $a \in U$ such that $\varphi(a) = 0$, the function φ is said to be a *local uniformizer* at a; for this case, some authors adopt the notation q_a instead of φ. We will also occasionally do so despite the fact that it could give the wrong idea that q_a is determined by a. Holomorphic functions being C^∞, a Riemann surface is first of all a C^∞ manifold.

The most obvious example, apart from that of an open subset of $\mathbb{C}$, is the Riemann sphere $\hat{\mathbb{C}} = \mathbb{C} \cup \{\infty\}$: it has an atlas with two charts (U, φ) and (V, ψ), where $U = \mathbb{C}$, $\varphi(z) = z$, $V = \hat{\mathbb{C}} - \{0\}$, $\psi(z) = 1/z$.

A more useful case as it controls the theory of elliptic functions consists in choosing a lattice L in $\mathbb{C}$, i.e. a discrete subgroup generated by two non-proportional numbers ω_1 and ω_2 (Chap. II, n° 23) and in considering the

---

[1] Not assuming this would lead to ridiculous complications. To start with, theorem 1 below would become false.

quotient space $X = \mathbb{C}/L$ of classes mod L. Equip it with the obvious topology: $U \subset X$ is open if and only if its inverse image is open in $\mathbb{C}$ or, what amounts to the same in these circumstances, if it is the image of an open subset of $\mathbb{C}$; X becomes a compact space homeomorphic to the torus $\mathbb{T}^2$. To define a complex structure on X, first note that if $D \subset \mathbb{C}$ is a *sufficiently small* open disc centered at a, its images under the translations $z \mapsto z + \omega$, with $\omega \in L$, are pairwise disjoint, so that $p: \mathbb{C} \longrightarrow \mathbb{C}/L$ maps D homeomorphically onto an open subset U of S. If φ denotes the inverse map $U \longrightarrow D$ to p, the couple (U, φ) is a chart of X, and to find the analytic structure sought, it suffices to show that the charts thus defined satisfy the condition imposed above, which is obvious. The function $\varphi(x) - a$, where a is the centre of D, is then a local uniformizer at the point $p(a)$.

If X is a Riemann surface, for any open subset $O \subset S$, it is possible to define functions h, which will be said to be *holomorphic* (resp. *meromorphic*) on O, by requiring that they satisfy the following condition: for any holomorphic chart (U, φ), there exists a holomorphic (resp. meromorphic) function h_φ on the open subset $\varphi(U \cap O)$ of $\mathbb{C}$ such that $h(x) = h_\varphi[\varphi(x)]$ for all $x \in U \cap O$; checking it for the charts in an atlas would be enough. An equivalent definition: in the neighbourhood of every $a \in U \cap O$, the function h is the sum of a power series (resp. Laurent series with finitely many terms of degree < 0) in $\varphi(x) - \varphi(a)$:

$$(1.1) \qquad h(x) = \sum c_n \left[\varphi(x) - \varphi(a) \right]^n = \sum c_n q_a(x)^n .$$

Hence, like in $\mathbb{C}$, a meromorphic function on O is not defined everywhere, unless it is assigned the value ∞ at all points of a discrete subset of O. Denote by $\mathcal{H}(O)$ (resp. $\mathcal{M}(O)$) the set of holomorphic (resp. meromorphic) functions on O. Their defining property is clearly of a local nature. Like on $\mathbb{C}$, the usual algebraic operations, including division, can be performed on meromorphic functions on X: if f is not the function 0, its zeros are isolated since X is connected, which removes all difficulties. To be correct, the result should also be defined at points, where, apparently this is not the case: if for example f and g have poles at $a \in X$ and if their polar parts in their Laurent series (1) cancel mutually, a value is assigned to $f + g$ at a, namely the sum of the constant terms of their Laurent series. Therefore, the set $\mathcal{M}(X)$ of meromorphic functions on X can also be considered a commutative field.

An essential difference between C^∞ theory and that of Riemann surfaces is the existence of holomorphic or meromorphic functions on a given open set O, while at the same time it being clear that if O is contained in the domain of a chart, then this is not obvious in the other cases. Showing – which we will not do here – that there are many meromorphic functions defined *globally* on a Riemann surface is the first major difficulty encountered as we progress in the theory.

This is not surprising. If, as above, we consider a lattice L in $\mathbb{C}$ and the Riemann surface $X = \mathbb{C}/L$, holomorphic or meromorphic functions on an open subset O of X are those which, composed with $p : \mathbb{C} \longrightarrow \mathbb{C}/L$, are holomorphic or meromorphic on the open subset $p^{-1}(O)$ of $\mathbb{C}$. They are invariant under translations $z \mapsto z + \omega$, $\omega \in L$. Any global existence proof of meromorphic functions on $\mathbb{C}/L$, therefore, shows, without the slightest calculation, the existence of *elliptic functions* , i.e. of meromorphic functions on $\mathbb{C}$ invariant under translations $z \mapsto z + \omega$. This result cannot be made trivial: either you prove the theorem holds for all Riemann surfaces, or else, like Weierstrass, you write series

$$\sum_{\omega \in L} 1/(z - \omega)^k , \quad k = 4, 6, 8, \ldots$$

answering the question (Chap. II, n° 23).

Let us now define the *order* $v_a(h)$ of a meromorphic function h at a point a. For this choose a local uniformizer q_a at a. This gives a Laurent series expansion $h(x) = \sum c_n q_a^n(x)$ in the neighbourhood of a. $v_a(h)$ is then the smallest integer for which $c_n \neq 0$. This definition does not depend on the choice of the chart.

Let ω be a complex-valued C^∞ differential form of degree 1 on X (or more generally on an open subset O of X: replace X by O). Let (U, φ) be a local chart of X, understood to be henceforth always holomorphic. Setting $\varphi(x) = \zeta$, (U, φ) transforms ω into a form ω_φ on $\varphi(U)$ which can be written

$$(1.2) \qquad \omega_\varphi = h_\varphi(\zeta)d\zeta + k_\varphi(\zeta)d\bar{\zeta} ,$$

with C^∞ functions h_φ and k_φ depending on the chart considered. ω will be said to be *holomorphic* if, for every chart (U, φ), $\omega_\varphi = h_\varphi(\zeta)d\zeta$ with a holomorphic function h_φ on $\varphi(U)$ or, equivalently, if ω is the inverse image under φ of a holomorphic differential form on $\varphi(U)$. If (V, ψ) is another chart, changes of charts

$$\theta : \varphi(U \cap V) \longrightarrow \psi(U \cap V), \quad \rho : \psi(U \cap V) \longrightarrow \varphi(U \cap V)$$

clearly (transitivity of inverse images) transform ω_φ into ω_ψ and conversely; hence

$$\omega_\varphi = h_\varphi(\zeta)d\zeta \Longrightarrow \omega_\psi = h_\varphi \left[\rho(\zeta) \right] d \left[\rho(\zeta) \right] = h_\varphi \left[\rho(\zeta) \right] \rho'(\zeta)d\zeta ,$$

so that the coefficient h_φ of ω in the local chart (U, φ) is transformed by

$$(1.3) \qquad h_\psi(\zeta) = h_\varphi \left[\rho(\zeta) \right] \rho'(\zeta) .$$

This is a very particular case of tensor calculus formulas. Like in $\mathbb{C}$, $d\omega = 0$.

More generally, meromorphic differential forms can be defined on X : these are holomorphic forms on $X - D$, where D is a discrete subset of X and such that, for all $a \in D$ $\omega_\varphi = h_\varphi(\zeta)d\zeta$ in a sufficiently small (hence in any) local chart (U, φ) at a, where h_φ is meromorphic on $\varphi(U)$ and has a unique pole at the point $\varphi(a)$. If for example h is a meromorphic function on X and if h is is represented by a meromorphic function $h_\varphi(\zeta)$ in (U, φ) then, by the multivariate chain rule, its differential dh is represented by $h'_\varphi(\zeta)d\zeta$; therefore, dh is a meromorphic differential form with the same poles as h. The product of a differential meromorphic form ω and of a meromorphic function is defined in an obvious way. For example, for any meromorphic function h, consider the form dh/h whose poles, like in $\mathbb{C}$, are clearly the poles and zeros of h.

For a differential form $\omega = h_\varphi(\zeta)d\zeta$, where h_φ is meromorphic on $\varphi(U)$, there is an expansion $h_\varphi(\zeta) = \sum c_n [\zeta - \varphi(a)]^n$ in the neighbourhood of all $a \in U$; by definition, the coefficient c_{-1} is the *residue* of ω at a, written $\operatorname{Res}(\omega, a)$. It also is independent of the choice of the chart [Chap. VIII, n° 5, (v): invariance of the residue under conformal representation]. For example, take $\omega = dh/h$, where h is meromorphic on X; if h is represented in the chart (U, φ) by a meromorphic function $h_\varphi(\zeta)$, then ω is clearly represented by the form $h'_\varphi(\zeta)d\zeta/h_\varphi(\zeta)$; ω is, therefore, a meromorphic differential form, and

$$(1.4) \qquad \operatorname{Res}_a(dh/h) = v_a(h)$$

as in $\mathbb{C}$.

We now prove a result generalizing what has been seen in Chap. 8, n° 5 about functions defined on the Riemann sphere:

Theorem 1. *Let X be a compact Riemann surface.*

(a) *Any function defined and holomorphic on X is a constant.*
(b) *For any meromorphic function f on X,*

$$(1.5) \qquad v(f) = \sum v_a(f) = 0.$$

(c) *For any meromorphic differential form ω on X,*

$$(1.6) \qquad \sum \operatorname{Res}_a(\omega) = 0.$$

(a) is obvious: a function h everywhere holomorphic reaches its maximum somewhere, and so is constant in the neighbourhood of its maximum. As X is by definition connected, classical arguments apply verbatim. (Corollary: the only *entire* elliptic functions on $\mathbb{C}$ are the constants).

To obtain (b), it suffices by (4), to prove (c). It will follow from Stokes' theorem. The latter can be applied since the Jacobians of holomorphic changes of charts are > 0, making it possible for X to be oriented by these charts.

Having said this, X being compact and the poles of ω being isolated, the latter are finite in number; let us denote them by $a_k (1 \le k \le n)$. For each k, choose a local chart (U_k, φ_k) such that $\varphi_k(a_k) = 0$ and, for sufficiently small $r > 0$, denote by $D_k(r)$ the set of $x \in U_k$ such that $|\varphi_k(x)| \le r$. These "discs" are closed in X and if r is sufficiently small, also pairwise disjoint; then ω has a unique pole at a_k in $D_k(r)$ and even in $D_k(r')$ for some $r' > r$.

Next consider the open subset G obtained by removing the discs $D_k(r)$ from X. Its border is the union of the "circles" limiting the "discs" $D_k(r)$. In the neighbourhood of a border point of $D_k(r) \subset D_k(r')$, the situation is similar to that of two concentric discs in $\mathbb{C}$. It is, therefore, clear that, on the one hand, the border of G is a (real) one-dimensional submanifold of of the union S of circles $|\varphi_k(x)| = r$, and that, on the other, in the neighbourhood of a border point of G, the open subset G is located on only one side of its boundary.

ω is holomorphic and a fortiori C^∞ on the manifold $X' = X - \{a_1, \ldots, a_n\}$,. G is open in X', with compact closure in X' since the latter being the complement in X of the open discs $|\varphi_k(x)| < r$, is closed in the compact set X and does not contain any a_k. Stokes' theorem can, therefore, be applied:

$$\int_{\partial G} \omega = \iint_G d\omega .$$

$d\omega = 0$ since ω is holomorphic, so that the sum of the integrals of ω along the "circles" limiting $D_k(r)$ is zero.

Using the chart (U_k, φ_k) which transforms ω into a form $\omega_k = h_k(\zeta)d\zeta$ in the disc $|\zeta| < r'$, it becomes clear that (U_k, φ_k) transforms the orientation of X, hence of X', into the standard orientation of $\mathbb{C}$; it also transforms the boundary of $D_k(r)$, oriented in conformity with Stokes' formula, into the circle $|\zeta| = r$ oriented counterclockwise. So the diffeomorphism φ_k transforms the extended integral ω over the boundary of $D_k(r)$ into the extended integral of ω_k over the circumference $|\zeta| = r$, equal to $2\pi i \operatorname{Res}_0(h_k)$. But the residue of h_k at $0 = \varphi_k(a_k)$ is, by definition, the residue of ω at a_k. The theorem follows.

Statement (b) has important corollaries. First, if f is a meromorphic function on X, for any $c \in \mathbb{C}$, the functions f and $f - c$ clearly have the same poles with same multiplicities. In conclusion, as $\sum v_a(f)$ is the difference between the number of zeros and the number of poles of f (counted with their multiplicities), for a compact surface, *the number of solutions of $f(x) = c$ is independent of c and equal to to the number of poles of f.*

On the other hand, if ω is a meromorphic form on X and if (U, φ) is a holomorphic local chart at a, with $\varphi(a) = 0$, the image $\omega_\varphi = h_\varphi(\zeta)d\zeta$ of ω in $\varphi(U)$ is meromorphic on $\varphi(U)$. The *order* $v_a(\omega)$ at a is then defined by the relation

$$v_a(\omega) = v_0(h_\varphi) .$$

Since the coefficient $\rho'(\zeta)$ in (3) is holomorphic and $\neq 0$ at a, the definition does not depend on the choice of the chart (U, φ). The $a \in X$, where $v_a(\omega) \neq 0$ form a discrete and hence a finite set if X is compact. Obviously, $v_a(h\omega) = v_a(h) + v_a(\omega)$ for every meromorphic function h on S. Hence setting

$$v(\omega) = \sum v_a(\omega)$$

for any meromorphic form on X, $v(h\omega) = v(h) + v(\omega) = v(\omega)$ for all mero-morphic functions h. But if ω and ω' are two meromorphic forms, there is is a meromorphic function h such that $\omega' = h\omega$. This is obvious in any local chart (h is the ratio between the coefficients of ω and ω'), and the functions h obtained in the local charts can be "glued" to give a globally defined func-tion because of the transformation formula (3) which is identical for ω and ω'. The conclusion is that *the integer $v(\omega)$ is the same for all meromorphic differentials on S*. Set

$$v(\omega) = 2g - 2,$$

where g is the *genus* of the *compact* Riemann surface X.

We prove that g is an integer ≥ 0 and, what is far less obvious, that two compact Riemann surfaces are homeomorphic[2] if and only if they have the same genus. The Riemann sphere has genus 0. Indeed, the differential form $\omega = dz$ has a double pole at infinity since it must be computed by using the local uniformizer $\zeta = 1/z$, and as, at all $a \in \mathbb{C}$, $\omega = 1.d(z - a)$ with $1 \neq 0$, the pole at infinity is the only contribution to the calculation of $v(\omega)$; hence $v(\omega) = -2$ and $g = 0$. Conversely, any compact Riemann surface of genus 0 is isomorphic (and not only homeomorphic) to the Riemann sphere. For $g = 1$, we get the quotients $\mathbb{C}/L$ of the theory of elliptic functions; this classical case will be studied in volume IV. For $g \geq 1$, X is homeomorphic to a sphere with g handles.

2 – Algebraic Functions

Riemann imagined his surfaces in order to study algebraic functions of one variable and in particular, to make them uniform, though in his work they were far less clearly defined than here. To understand this, what is meant by an *algebraic function* $\zeta = \mathcal{F}(z)$ of a complex variable z should be first understood. The foremost characteristic of an algebraic "function" is that it is not a function: like $\mathcal{L}og\, z$, which is not algebraic, or like $z^{1/3}$ or

$$\left[\left(z^2 + 1\right)^{1/2} - \left(z^3 - 2z + 1\right)^{1/3}\right]^{1/4}$$

that are so, it can take several values (an infinite number in the first case, 3 in the second and 24 in the third) for a given value of z; the notation $\mathcal{F}(z)$ can, therefore, only represent one *set* of complex numbers, the only notation making sense being $\zeta \in \mathcal{F}(z)$ and not $\zeta = \mathcal{F}(z)$. By definition, the elements of $\mathcal{F}(z)$ are the roots (possibly including ∞ as will be seen later) of an equation

(2.1) $P(z, \zeta) = 0,$

---

[2] But not isomorphic as complex manifolds (an isomorphism being a holomor-phic homeomorphism whose inverse is also holomorphic). The classification of Riemann surfaces, up to isomorphism, is far more complicated.

where

(2.2) $$P(X,Y) = \sum a_{pq}X^pY^q = P_0(X)Y^n + \ldots + P_n(X) =$$
$$= Q_0(Y)X^m + \ldots + Q_m(Y)$$

is a given polynomial[3] in two variables or "indeterminates", with complex coefficients, for example $Y^3 - X$ in the case of $z^{1/3}$ and an equation of degree 24 in the third example.[4] If P_0 is not identically zero, in which case $\mathcal{F}$ will be said to be of degree n (it would perhaps be better to say $1/n$ to avoid confusion with the degree of a polynomial), equation (1) has at most n roots and, if $P_0(z) \neq 0$, then for given z, the number of possibly multiple roots is exactly n.

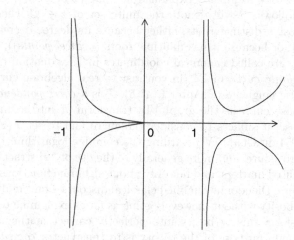

Fig. 2.1.

Equation (1) has multiple roots for the values of z for which the equations $P(z, \zeta) = 0$ and $D_2 P(z, \zeta) = 0$ have common roots at ζ, where D_2 is the

---

[3] In what follows, assuming that P is *irreducible*, i.e cannot be non-trivially written as the product $P = QR$ of two other polynomials, will prove useful since if that were the case, equations $Q = 0$ and $R = 0$ would need to be considered separately; as will be seen, this is also a necessary condition for the Riemann surface that will be constructed to be connected. Any polynomial P is a product of irreducible factors: consider a factor Q of minimum total degree and apply an induction argument on the total degree of P. The *total degree* is the largest integer d such that $a_{pq} \neq 0$ for a couple (p, q) such that $p + q = d$.

[4] To compute it, construct a polynomial in Y having as roots the six differences $u - v$, where $u = \varepsilon(X^2 + 1)^{1/2}$ with $\varepsilon \in \{1, -1\}$ and $v = \omega(X^3 - 2X + 1)^{1/3}$, where ω is a cubic root of unity. The result (it can a priori be shown) is then seen to be a polynomial $p(X, Y)$ – the "irrationals" disappear since the coefficients of p are the elementary symmetric functions of the six differences $u - v$. The equation sought is then $p(X, Y^4) = 0$.

differential operator with respect to ζ. The classical result from algebra[5] – that the set of these values of z is finite – follows. For the equation $\zeta^2(z^2 - 1) - z^5 = 0$, whose graph in $\mathbb{R}^2$ has a cusp point at the origin, that is the case if $z = 0, 1$ or $+1$; for z in the neighbourhood of 0, the equation obviously has two roots in the neighbourhood of 0, and though they can be distinguished by their sign in the real domain, this is impossible in the complex one. Indeed, following by continuity one of the roots along a circle centered at 0, we end up with the opposite root since the number $z^5/(z^2 - 1)$ with ζ as "the" square root describes a curve around the origin. In the particularly simple case when $P(X, Y) = Y^2 - X$, which corresponds to $\zeta = z^{1/2}$, the curve does not have any singularity at the origin, but its tangent at 0 is vertical and the conclusion is the same: the "function" $z^{1/2}$ is not well-defined in the neighbourhood of 0 as already explained in Chapter IV.

In what follows, S will denote the finite set of $z \in \mathbb{C}$ where equation (1) has at most n distinct roots, either because its degree decreases (points canceling P_0), or because it has multiple roots (*critical points*).

If z and ζ are called canonical coordinates in $\mathbb{C}^2$, equation (1) defines a complex *algebraic curve* in $\mathbb{C}^2$ (in contrast to real algebraic curves in $\mathbb{R}^2$). In set theoretic language (Chapter I, n° 5), $\mathcal{F}$ is a *correspondence*[6] between $\mathbb{C}$ and $\mathbb{C}$ whose curve is the graph. To transform $\mathcal{F}$ into a function F in the strict sense, it suffices to consider the curve and to set $F(z, \zeta) = \zeta$, as was done in § 4 du Chap. IV regarding the complex logarithm. Ignorance of this simple procedure, and more generally of the *abc* of "abstract" set theory has long confused matters, and not only about algebraic functions. Il explains why somewhere, Dieudonné qualified classical discourses on "multiform functions" as verbosity without, however, going as far as explaining to his readers how to transform this verbosity into perfectly correct mathematical arguments. The main purpose of the theory is to construct a *compact* Riemann surface on which z, the "function" $\zeta = \mathcal{F}(z)$ and more generally any rational expression in z and ζ become genuine meromorphic functions. The graph of $\mathcal{F}$ is only a first approximation in the construction, which, as we will see, is considerably more difficult.

The first step consists in considering the open set $B = \mathbb{C} - S$ defined by the condition

$$z \in B \iff \operatorname{Card} \mathcal{F}(z) = n$$

---

[5] If $P(Y)$ and $Q(Y)$ are polynomials of degrees p and q with coefficients in an integral ring, for example $\mathbb{C}[X]$, and if y is a common root of P and Q, successive multiplication of $P(y)$ by $1, y, \ldots, y^{q-1}$ and of $Q(y)$ by $1, y, \ldots, y^{p-1}$ give $p + q$ linear equations homogeneous in $y^n (0 \le n \le p + q - 1)$; as this system admits a non-zero solution (since $1 \ne 0$), its determinant, a polynomial in coefficients of P and Q, must be zero. Do the calculations for $p = q = 2$.

[6] Besides this term was used in algebraic geometry long before the invention or propagation of set theory.

and in showing that the subset X of the graph of $\mathcal{F}$ located over B admits a natural complex analytic structure (in other words, is a non-compact Riemann surface) for which $p : (z, \zeta) \mapsto z$ and $F : (z, \zeta) \mapsto \zeta$ are holomorphic. The compact Riemann surface sought will then be obtained by adjoining a finite number of points to X, an operation analogous to the construction of the Riemann sphere $\hat{\mathbb{C}}$ from $\mathbb{C}$.

Let us first prove the following result:

Lemme 1 (Continuity of the roots of an algebraic equation). *Let a be a point of $\mathbb{C}$ and $\alpha \in \mathbb{C}$ a multiple root of order p of $P(a, \zeta) = 0$. There exist numbers $r > 0$ and $\rho > 0$ such that, for all z satisfying $0 < |z - a| < r$, the equation $P(z, \zeta) = 0$ has exactly p roots, all simple, in $0 < |\zeta - \alpha| < \rho$.*

As seen in Chap. VIII, n° 5, (viii) in a more general context, if the given $\rho > 0$ is sufficiently small, the number $\nu(z)$ of roots of $P(z, \zeta) = 0$ satisfying $|\zeta - \alpha| < \rho$ is a continuous function, with respect to the topology of compact convergence, of the function $P_z : \zeta \mapsto P(z, \zeta)$. But since P is a continuous function of the couple (z, ζ), hence uniformly continuous on every compact subset of $\mathbb{C} \times \mathbb{C}$, P_z is a continuous function of z with respect to the topology of compact convergence; $\nu(z)$ is, therefore, a continuous function of z, and so $\nu(z) = \nu(a)$ in the neighbourhood of a. Hence, for sufficiently small $|z-a| < r$, equation $P(z, \zeta) = 0$ has exactly p not necessarily distinct roots such that $|\zeta - \alpha| < \rho$. There being finitely many values of z for which the equation $P(z, \zeta) = 0$ has a multiple root, the only one of these values satisfying $|z-a| < r$ is a if r is sufficiently small. The p roots of $P(z, \zeta) = 0$ such that $|\zeta - \alpha| < \rho$ are, therefore, simple if $0 < |z - a| < r$, qed.

It will be seen later that $\mathcal{F}$ decompose into n uniform branches $f_k(z)$ in every *simply connected* open subset U of B, a *uniform branch* in U being, by definition, a genuine function f defined and (for the moment) *holomorphic* on U, such that $f(z) \in \mathcal{F}(z)$ for all $z \in U$; this is what had been proved in § 4 of Chap. IV for the pseudo-function $\mathcal{L}og\, z$ for $U \subset \mathbb{C}^*$.

A local result first needs to be proved:

Lemma 2. *Let $P(X, Y)$ be a polynomial with complex coefficients and $E \subset \mathbb{C}^2$ the set of simple points[7] of the curve $P(z, \zeta) = 0$. Then E is a submanifold of $\mathbb{C}^2 = \mathbb{R}^4$ and, for all $(a, \alpha) \in E$ where $D_2 P(a, \alpha) \neq 0$, there are open neighbourhoods V of a and W of α such that $E \cap (V \times W)$ is the graph of a function $\zeta = f(z)$ defined and holomorphic on V with values in W.*

Equivalently, there exists a unique holomorphic function f on V satisfying

$$(2.3) \qquad f(a) = \alpha \ \& \ P\,[z, f(z)] = 0 \quad \text{for all } z \in V.$$

f is called a local *uniform branch* at a of the algebraic function defined by P.

Since α is a simple root of $P(a, \zeta) = 0$, there are (lemma 1) neighbourhoods V and W of a and α such that, for all $z \in V$, the equation $P(z, \zeta) = 0$

---

[7] i.e. points where $D_1 P(z, \zeta)$ and $D_2 P(z, \zeta)$ are not both zero.

has exactly one root $\zeta \in W$. We need to show that this unique root ζ is a *holomorphic* function of z.

Regard the map

$$P : (z, \zeta) \longmapsto P(z, \zeta)$$

from $\mathbb{C}^2$ to $\mathbb{C}$ as a map from $\mathbb{R}^4$ to $\mathbb{R}^2$. At each point, it has a tangent linear map $P'(z, \zeta)$ which is $\mathbb{C}$-linear since P is holomorphic at z and ζ, namely (Chap. IX, formula (2.24))

(2.4) $(h, k) \longmapsto D_1 P(z, \zeta)h + D_2 P(z, \zeta)k$,

where h, k are vector variables in $\mathbb{C}$ and where the derivatives are taken to be in the complex sense; (4) shows that P' is surjective at every point of the open subset Ω of $\mathbb{C}^2$, where $D_1 P$ and $D_2 P$ are not both zero, so that $P : \Omega \longrightarrow \mathbb{R}^2$ is a *submersion*. Hence, every equation $P(z, \zeta) = c$ defines a closed submanifold of Ω having dimension $4 - 2 = 2$ [Chap. IX, n° 13, (ii)], hence a submanifold of $\mathbb{R}^4 = \mathbb{C}^2$; this is in particular the case of E. The tangent vector space to E at a point $(a, \alpha) \in E$ where $D_2 P \neq 0$ is the set of (h, k) such that

$$D_1 P(a, \alpha)h + D_2 P(a, \alpha)k = 0,$$

i.e. such that

$$k = -D_1 P(a, \alpha)h / D_2 P(a, \alpha);$$

hence the manifold E is the graph of a map $\zeta = f(z)$ in the neighbourhood of (a, α), where f is C^∞, with a tangent linear map given by

$$f'(a)h = -D_1 P(a, \alpha)h / D_2 P(a, \alpha)$$

[Chap. IX, n° 13, (iv)]. This formula shows that $f'(a)$ is $\mathbb{C}$-linear, and so f is holomorphic, qed.

Similarly, E could be shown to be the graph of a holomorphic function $z = g(\zeta)$ in the neighbourhood of a point (a, α) where $D_1 P \neq 0$.

Returning to the Riemann surface X, i.e. the graph of $\mathcal{F}$ over B, consider some $a \in B = p(X)$. Denoting by α_k the n simple roots of the equation $P(a, \zeta) = 0$, there are holomorphic functions $f_k(z)$, $1 \leq k \leq n$ in the neighbourhood of a, satisfying

(2.5) $P[z, f_k(z)] = 0, \quad f_k(a) = \alpha_k$.

If D is a sufficiently small disc centered at a, these n local uniform branches f_k at a are all defined on D and, being continuous, are pairwise distinct at all $z \in D$ since so are the $f_k(a) = \alpha_k$. Denoting by $D_k \subset X$ the image of D under $z \mapsto (z, f_k(z))$, the D_k are seen to be pairwise disjoint and to have as their union the set $p^{-1}(D)$ of points of X projecting onto D. The maps $p : D_k \longrightarrow D$ and $f_k : D \longrightarrow D_k$ being continuous and mutually inverse, p is a homeomorphism from D_k onto D, so that the D_k are connected by arcs like D. Finally, D_k is open in X since it is the set of $(z, \zeta) \in X$ satisfying $z \in D$ and $\zeta \in W_k$, where W_k is an open neighbourhood of α_k. As a result, *the D_k are the connected components of $p^{-1}(D)$ and p is a local homeomorphism from X onto B.*

On the other hand, for all k, the couple (D_k, p) is clearly a local chart of X; we thus obtain an atlas for X, a priori C^0. It is in fact *holomorphic*. Indeed, let (a, α), (b, β) be two points of X and f, g be the local uniform branches at a and b such that $f(a) = \alpha$, $g(b) = \beta$. They are defined on the discs U and V centered at a and b and map them homeomorphically onto open neighbourhoods $f(U)$ and $g(V)$ of (a, α) and (b, β). If $f(U) \cap g(V) \neq \varnothing$, f and g are equal at least at one point of $U \cap V$, hence on all of $U \cap V$ (uniqueness of uniform branches on a connected open set), and the change of charts taking the local chart $(f(U), p)$ to the local chart $(g(V), p)$ transforms the coordinate $p(z, \zeta) = z$ of a point of the first chart into its coordinate $p(z, \zeta) = z$ in the second. The change of coordinates is, therefore, the map $z \mapsto z$, which is as holomorphic as possible. The conclusion that follows from these arguments is that there a complex analytic structure on X turning X into a Riemann surface. It is not yet compact, but it is a start.

To go from here to the existence of global uniform branches on every simply connected open set contained in B and to complete X to obtain a *compact* Riemann surface $\hat{X}$ associated to P, it is helpful to develop some aspects of general topology that can also be useful elsewhere.

3 – Coverings of a Topological Space

As this theory requires quite a few explanations, I will break it down into several parts and confine myself to the essential minimum.[8] In particular, I will not mention the notion of a fundamental group of a space as it is not needed to construct Riemann surfaces of algebraic functions.

(i) *Definition of a covering.* The notion of a covering space of a topological space (separated, i.e. satisfying Hausdorff's axiom) generalizes the situation encountered at the end of the previous n°. Take two separated spaces X and B and a continuous and surjective map $p : X \longrightarrow B$; although this is not always necessary, we will suppose that the "base" B is *connected*[9] and *locally*

---

[8] This section follows quite closely Chapter XVI.28 in Dieudonné's *Eléments d'analyse*, which follows even more closely what N. Bourbaki has written on the subject when it was on its agenda in the 1950s. As at the time, many homotopy experts belonged to the group, to start with, Samuel Eilenberg and Jean-Pierre Serre, and other people who had seriously thought about the subject, it is unlikely that a anything better could be achieved.

[9] In all of this n° and in the rest of this §, "connected" will mean *arc-connected*

connected (i.e.there are arbitrarily small connected neighbourhoods of all points $z \in B$, which is for example the case with manifolds). To encourage the reader to compare the general theory to arguments already used for complex variables in the previous n° and in § 4 of Chap. IV, we will denote an arbitrary point of B by z and an arbitrary point of X by ζ.

In order for the triplet (X, B, p) to be a covering, we compel it to be *locally trivial*, more precisely:

(R) every point $z \in B$ has an open neighbourhood D whose inverse image $p^{-1}(D)$ is the union of a family $(D_i)_{i \in I}$ of pairwise disjoint open sets mapped homeomorphically onto D by p.

The most obvious consequence of (R) is that p transforms every neighbourhood of a point $\zeta \in X$ onto a neighbourhood of $p(\zeta)$. In particular, p transforms every open subset of X onto an open subset of B. Moreover, for all $z \in B$, the *fiber* $p^{-1}(\{z\})$ of z in X is a *discrete* subset of X since its intersections with the open subsets D_i reduce to a point.

Example 1. Choose $B = \mathbb{C}^*$, $X = \mathbb{C}$ and p to be the map

$$\mathbf{e} : \zeta \longmapsto \exp(2\pi i \zeta)$$

from X onto B. As $\mathbf{e}'(\zeta) \neq 0$ everywhere, p is a local homeomorphism (Chap. VIII, n° 5, theorem 7 applied locally). If $a \in B$ is the image of some $\alpha \in X$ and if $D \subset B$ is a sufficiently small disc centered at a, then there is neighbourhood D' of α homeomorphically mapped onto D by p. The periodicity of the exponential function shows that

$$p^{-1}(D) = \bigcup_{\mathbb{Z}} D' + n = \bigcup D_n \,,$$

and if D' (i.e. D) is sufficiently small. the translates D_n of D' are pairwise disjoint and homeomorphically mapped onto D by p. Hence $\mathbb{C}$ can be regarded as a covering space, moreover simply connected, of $\mathbb{C}^*$. If we choose a lattice L of periods like in the theory of elliptic functions, then $\mathbb{C}$ becomes a simply connected covering space of the torus $\mathbb{C}/L$. Finally, the map $t \mapsto \mathbf{e}(t)$ transforms $\mathbb{R}$ into a covering space of $\mathbb{T} = \mathbb{R}/\mathbb{Z}$.

Example 2. For a given integer $k > 0$, consider the quotient $P/k\mathbb{Z}$ of P by the group of horizontal translations $\zeta \mapsto \zeta + nk$, where $n \in \mathbb{Z}$, with the obvious topology. The function $\mathbf{e}(\zeta/k)$ is invariant under these translations, and so defines a continuous map $p : P/k\mathbb{Z} \longrightarrow D^*$. It is more or less obvious that we thus obtain a " k-sheeted " covering space of D^*, as it used to be called earlier, i.e. the *canonical covering of order* k of D^*. It can also be constructed, up to isomorphism, by using the map $z \mapsto z^k$ from D^* onto D^* ; axiom (R) holds by lemma 1 of n° 2 applied to the polynomial $Y^k - X$ for $a = 0$; in fact, the canonical covering of order k of D^* is just the subset of the Riemann surface of the polynomial $Y^k - X$ located over D^*.

It may happen that condition (R) holds for $D = B$. If every $\zeta \in D_i$ is identified to the couple (z, i), where $z = p(\zeta)$, X is transformed into the

Cartesian product $B \times I$ equipped with the product topology induced by the topology of B and the discrete topology of I (any subset of I is an open set); conversely, the product $X = B \times I$, where I is a discrete space, is a covering space of B thanks to the map $p(z, i) = z$. Such a covering space is said to be (globally) *trivial* or a *decomposition*.

In the case of algebraic functions (see end of the previous n°), there is a "n-sheeted" covering space of B, but as shown by the graph x of the pseudo-function $\zeta = \mathcal{L}og\, z$ studied in Chapter IV, §4 in the general case, the number of D_is is not necessarily finite or even constant if B is not connected.[10] Here $B = \mathbb{C}^*$, X is the set of couples $(z, \zeta) \in \mathbb{C}^2$ such that $z = \exp(\zeta)$, and $p(z, \zeta) = z$. In any disc $D \subset \mathbb{C}^*$, the "multiform function" $\mathcal{L}og\, z$ decomposes into uniform branches $L_k(z)$ depending on $k \in \mathbb{Z}$ and the graphs $D_k \subset X$ of these L_k are the connected components of $p^{-1}(D)$. As an aside, note that this situation is just the same as that in example 1: associating the point $(e(\zeta), \zeta)$ of the graph of $\mathcal{L}og\, z$ to all $\zeta \in \mathbb{C}$, we get a homeomorphism from the first covering space onto the second one which commutes with the corresponding maps p. Two such covering spaces of a same space B are said to be *isomorphic.*.

In the general case, if condition (R) holds in a neighbourhood D of z, it obviously also holds in any neighbourhood $D' \subset D$. Hence if B is locally connected, D can be assumed to be connected; then so are the D_i. Since the D_i are pairwise disjoint, they are the connected components of the open subset $p^{-1}(D)$ of X.

In practice, all the spaces considered are metrizable. If the topology of B is defined by a distance $d(x, y)$, then, for all $z \in B$, there exist numbers $r > 0$ such that X is trivial over the ball $D(z, r)$. If $r(z)$ denotes the upper bound of these r, X is clearly trivial over $D(z, r)$ for all $r < r(z)$. If $d(z, z') < r < r(z)$, X is trivial over $D(z', r')$ for all $r' < r(z) - r$ since $D(z', r') \subset D(z, r)$. In conclusion, $r(z') > r(z) - d(z, z')$ and hence the function r is *lower semicontinuous* (Chap. V, n° 10). For any compact subset $K \subset B$,

$$(3.1) \qquad \inf_{z \in K} r(z) = r(K) > 0$$

because there exists $z \in K$, where $r(z)$ is minimum (same reference).

(ii) *Sections of a covering space.* Since $p : D_i \longrightarrow D$ is a homeomorphism, we can consider the inverse map $\varphi_i : D \longrightarrow D_i$; it satisfies $p \circ \varphi_i = id$; it is the analogue of a local uniform branch. More generally, if E is a subset of B, a *section of X over E* is any continuous map $\varphi : E \longrightarrow X$ such that $p[\varphi(z)] = z$ for all $z \in E$, in other words, the analogue of a uniform branch on E. Since p is a local homeomorphism, there are sections over all *sufficiently small* neighbourhoods of all $a \in B$, namely the φ_i; there is even a section

---

[10] Axiom (R) shows that the number of elements of $p^{-1}(\{z\})$ is a locally constant function of $z \in B$, and so is constant if B is connected. This number, whether finite or not, is generally call the *order* of the covering space (X, B, p).

taking a given value at a given point of D . If two sections φ and ψ defined on the same set $E \subset B$ are equal at $a \in E$, they are equal to the same φ_i at a: since D_i is open in X, the values of φ and ψ in the neighbourhood of a are in D_i, and so $\varphi(z) = \psi(z) = \varphi_i(z)$ for all $z \in E$ sufficiently near a.

The set of points of E where φ and ψ are equal is, therefore, open in E; it is also closed since φ and ψ are continuous. As a result, *two sections of X over a connected subset of X are identical if they are equal at some point.* For example, if $\mathcal{F}$ is an algebraic function and if there are two uniform branches f and g of $\mathcal{F}$ on a connected open subset $U \subset \mathbb{C} - S$ (using the notation of the previous n°), then the existence of some $a \in U$ where $f(a) = g(a)$ implies that $f = g$ is all of U. So there are at most n uniform branches on U if $n = d°(\mathcal{F})$ and, in fact, exactly n if U is simply connected (n° 4, theorem 6).

Let us now show that *if φ is a section of X over a connected subset E of B, then $\varphi(E)$ is a connected component of $p^{-1}(E)$.* As $(p^{-1}(E), E, p)$ is obviously a covering of E, it suffices to do so for $E = B$. Since $z \mapsto (z, \varphi(z))$ and $(z, \zeta) \mapsto z$ are continuous and mutually inverse, φ is a homeomorphism from B onto $Y = \varphi(B)$, which is, therefore, connected. By (R), Y is clearly open in X. On the other hand, if a sequence $\varphi(z_n) \in Y$ converges to a limit, the $z_n = p[\varphi(z_n)]$ converge to some $a \in B$, and so $\lim \varphi(z_n) = \varphi(a)$. As a result, Y is open and closed in X, and connected as the image of B, qed.

On the other hand, *if Y is a connected component of X, then $p(Y)$ is a connected component of B, and so $p(Y) = B$ if B is connected.* Firstly, Y being open in X and p being a local homeomorphism, $p(Y)$ is open in B. Let $a \in B$ be a closure point of $p(Y)$ and D a connected open neighbourhood of a in B satisfying (R). As D meets $p(Y)$, $p^{-1}(D) = \bigcup D_i$ meets Y. If some connected open D_i meets Y, then $Y \cup D_i$ is a connected open set containing Y. Hence $D_i \subset Y$ and $a \in p(D_i) \subset p(Y)$, so that $p(Y)$ is closed, qed.

Besides, Y is clearly a covering space of B: apply (R) by replacing X by Y.

Finally suppose that, for any connected component Y of X, the map $p : Y \longrightarrow B$ is injective. It is then bijective if B is connected, and as p is a local homeomorphism, it is a global homeomorphism from Y onto B. As a result, Y is the image of B under a global section φ of X and the covering (X, B, p) is trivial.

(iii) *Path-lifting.* Let $\gamma : I \longrightarrow B$ be a continuous path in B, where $I = [0, 1]$. A lifting of γ is a path $\mu : I \longrightarrow X$ such that $p \circ \mu = \gamma$ (§ 4 of Chap. IV for the case of $\mathcal{L}og\, z$). Such a lifting always exists, but there is better still:

Theorem 2. *Let (X, B, p) be a covering and $\gamma : I \longrightarrow B$ a path in B. There is a unique lifting μ from γ to X with a given initial point. If two paths γ_0 and γ_1 in B are fixed-endpoint homotopic and if μ_0 and μ_1 are liftings of γ_0 and γ_1 having the same initial point, then μ_0 and μ_1 have the same terminal point and are homotopic.*

We first prove the uniqueness of μ. In the neighbourhood of some $t_0 \in I$, a lifting $\mu(t)$ of γ, being a continuous function of t, takes its values in a neighbourhood of $\mu(t_0)$ mapped homeomorphically by $p : X \longrightarrow B$ onto a neighbourhood D of $\gamma(t_0)$ satisfying axiom (R) . Hence $\mu = \varphi \circ \gamma$ in D, where φ is the unique section of X over D mapping $\gamma(t_0)$ onto $\mu(t_0)$. It follows that the set of point where two liftings coincide is open and closed in I, proving uniqueness.

To prove the existence of μ, choose $\alpha \in X$ such that $p(\alpha) = \gamma(0) = a$ and, as in §4 du Chap. IV, consider all the couples (J, μ), where $J \subset I$ is an interval with initial point 0 and μ is a continuous map from J to X satisfying $p \circ \mu = \gamma$ on J, as well as $\mu(0) = \alpha$. The existence of such couples as well as their "coherence" is clear – Take J sufficiently small so that X is trivial over $\gamma(J)$: if (J', μ') and $(J", \mu")$ are two such couples, then, by the uniqueness of liftings, $\mu' = \mu"$ in $J' \cap J"$. The union of all these J gives a couple (J_0, μ_0) such that μ_0 cannot be extended beyond the terminal point b of J_0. But as X is trivial over the connected open neighbourhood D of $\gamma(b)$ in B, if $b' \in J_0$ is sufficiently near b for $\gamma(b') \in D$ to hold, then there is a section of X in D equal to $\mu_0(b')$ at $\gamma(b')$, which makes it possible to extend μ_0 beyond b if $b < 1$, and at b if $b = 1$. So $b = 1$.

Next, consider a homotopy $\sigma : I \times I = K \longrightarrow B$ between paths γ_0 and γ_1 of the statement, whose initial and terminal points and will be denoted by a and b; hence

$$(3.2) \qquad \sigma(s,0) = a, \quad \sigma(s,1) = b \quad \text{for all } s,$$
$$\sigma(0,t) = \gamma_0(t), \quad \sigma(1,t) = \gamma_1(t) \quad \text{for all } t.$$

Let α be the common initial point of μ_0 and μ_1. The problem consists in constructing a continuous map $\sigma' : K \longrightarrow X$ satisfying $p \circ \sigma' = \sigma$ and

$$(3.3) \qquad \sigma'(s,0) = \alpha \quad \text{for all } s,$$
$$\sigma'(0,t) = \mu_0(t), \quad \sigma'(1,t) = \mu_1(t) \quad \text{for all } t.$$

In fact there is no need to require $\sigma'(s,0)$ to be independent of s, because condition $p \circ \sigma' = \sigma$ shows that $s \mapsto \sigma'(s,0)$ is a continuous map from I to the *discrete* space $p^{-1}(a)$, and so is constant. Conditions (3) could even be replaced by the unique condition that $\sigma'(0,0) = \alpha$; indeed, relation $p \circ \sigma' = \sigma$ shows that (1) $\sigma'(s,0)$ takes its values in $p^{-1}(a)$, and so is constant, (2) $\sigma'(0,t)$ is a lifting of γ_0 having the same initial point as μ_0, and so is equal to μ_0, (3) $\sigma'(1,t)$ is a lifting of γ_1 having the same initial point as μ_1, and so is equal to μ_1.

Like in the previous case, the uniqueness of σ' is clear since $K = I \times I$ is connected.

As $\sigma(B)$ is compact and σ is uniformly continuous, (1) shows that there exists $r > 0$ such that X is trivial over $\sigma(D)$ for all discs D of radius r in K. If we draw a grid on K consisting of the lines $s = i/n$ and $t = j/n$, with $0 \le i, j \le n$, where n is sufficiently large, then X is trivial over the images

$\sigma(K_{ij}) = H_{ij}$ of the compact squares K_{ij} thereby obtained. If a section φ_{ij} of X over each H_{ij} is chosen, for the moment arbitrarily, setting $\sigma'_{ij} = \varphi_{ij} \circ \sigma$ in K_{ij} gives

$$p \circ \sigma'_{ij} = \sigma \quad \text{in} \ K_{ij}.$$

If some squares K_{ij} (at most four) have a common point x, taking n sufficiently large, they may be assumed to be all contained in the open disc centered at x and of radius r; as these K_{ij} are connected, and hence so are the H_{ij} and their union, if the sections φ_{ij} are taken to be equal at $\sigma(x)$, then they are the restrictions to these H_{ij} of the unique section over their union, whose value at x is the same as theirs.

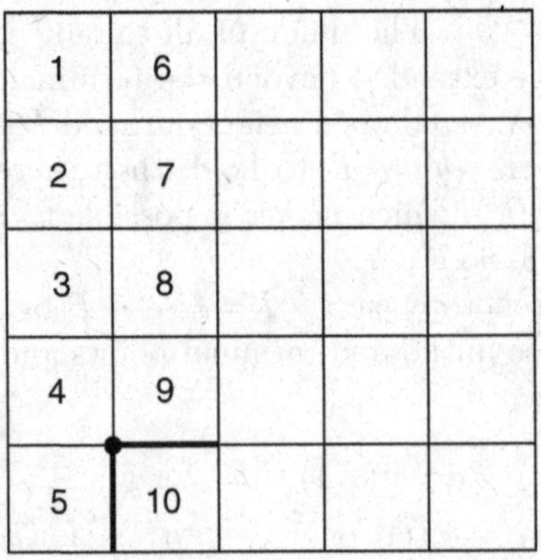

Fig. 3.2.

To choose the φ_{ij} so that the maps σ'_{ij} are the restrictions to K_{ij} of the map σ' sought, order the K_{ij} into a simple sequence $K_i (1 \leq i \leq n^2)$ as indicated in the above figure. In $H_1 = \sigma(K_1)$, choose the section φ_1 equal to $\alpha = \mu_0(0)$ at $\sigma(0,0)$ and define

$$\sigma'_1 = \varphi_1 \circ \sigma \quad \text{in} \ K_0.$$

As $t \mapsto \sigma'(0,t)$ is a lifting to the interval $[0, 1/n]$ of γ_0 whose initial point is the same as that of μ_0,

(3.4) $$\sigma'_1(0,t) = \mu_0(t) \quad \text{for} \ 0 \leq t \leq 1/n.$$

As $\sigma(s,0) = a$ for all s, similarly

(3.5) $$\sigma'_1(s,0) = \alpha \quad \text{for} \ 0 \leq s \leq 1/n.$$

Having done this, in $H_2 = \sigma(K_2)$, choose the section φ_2 equal to φ_1 over the image of the common side of the two squares K_1 and K_2, then the section φ_3

over H_3 equal to φ_2 over the image of the common side of the two squares K_2 and K_1, and so on, and define $\sigma_i' = \varphi_i \circ \sigma$ in K_i.

Suppose that the existence of a map σ' in $K_1 \cup \ldots \cup K_i$ coinciding with σ_j' in each $K_j, j \leq i$ has been proved. The definition of σ' will extend to $K_1 \cup \ldots \cup K_{i+1}$ if and only if $\sigma_{i+1}' = \sigma_j'$ in $K_{i+1} \cap K_j$, where $j \leq i$, when this intersection is non-empty. Let us do it for $i = 10$ and $j = 5$ (figure!). By construction, the partial liftings σ_{10}' and σ_9' are equal in $K_{10} \cap K_9$, but the induction hypothesis shows that σ_9' and σ_5' are equal in the set $K_9 \cap K_5$ reduced to a point and hence non-empty. As this point also belongs to K_{10}, σ_{10}' and σ_5' are equal at this point, and so in $K_{10} \cap K_5$ as expected.

This argument shows that there is continuous map $\sigma' : I \times I \longrightarrow X$ such that $p \circ \sigma' = \sigma$. As it satisfies $\sigma'(0,0) = \alpha$, this proves the theorem.

An immediate corollary is that *all covering maps over I are trivial*, as they have global sections: take $B = I$ and for σ take the identity map from I to B. The same result holds for I^n, where $n \in \mathbb{N}$.

More importantly:

Corollary 1. *Let (X, B, p) be a covering and suppose that X is connected and simply connected.*[11] *Let γ_0 and γ_1 be two paths in B with the same endpoints and let μ_0 and μ_1 be liftings to X of γ_0 and γ_1 having the same initial point. γ_0 and γ_1 are homotopic if and only if μ_0 and μ_1 have the same terminal point.*

If the condition holds, the two liftings are homotopic, since X is simply connected, and thus that is also the case of the given paths in B. The converse, which makes no assumptions on X, is the second statement of theorem 2.

In particular, *a closed path γ in B is homotopic to a point if and only if some (and hence all) lifting of γ to X is closed*, of course provided that X is simply connected.

To state the next corollary which solves an essential problem in Cauchy theory, take $B = \mathbb{C}^*$ and consider the path

$$\mathbf{u} : t \longmapsto r. \exp(2\pi i t)$$

in B, i.e. the circle centered at 0 and of radius $r > 0$ traveled once counter-clockwise; let $n\mathbf{u}$ be the path $t \mapsto r. \exp(2\pi i n t)$, i.e. the circle centered at 0 traveled n times counterclockwise if $n \geq 0$, or $-n$ times clockwise if $n < 0$; obviously he choice of r has no impact on the homotopy class of $n\mathbf{u}$ in $\mathbb{C}^*$, the latter being all that matters to us in what follows. Hence $r = 1$ may be assumed.

Corollary 2. *For any closed path γ in $\mathbb{C}^*$, there is a unique integer n such that γ is homotopic to $n\mathbf{u} : t \mapsto \mathbf{e}(nt)$.*

---

[11] i.e. having the following property: two arbitrary paths with the same endpoints are always fixed end-point homotopic.

Indeed, the map $\mathbf{e} : \mathbb{C} \longrightarrow \mathbb{C}^*$ transforms $\mathbb{C}$ into a connected and simply connected covering space of $\mathbb{C}^*$ [(i), Example 1]. A lifting of γ is a path μ in $\mathbb{C}$ such that $\mathbf{e}[\mu(t)] = \gamma(t)$ for all t, in other words, is a *uniform branch of $\mathcal{L}og\, z$ along γ* in the sense of Chap. IV, §4, up to the factor $2\pi i$. As γ is closed, $\mu(1) = \mu(0) + n$ for some $n \in \mathbb{Z}$, so that μ is fixed-endpoint homotopic to the rectilinear path

$$t \longmapsto (1-t)\mu(0) + t\mu(1) = (1-t)\alpha + t\beta\,.$$

As a result, γ is fixed-endpoint homotopic to the path

$$t \longmapsto \mathbf{e}\,[(1-t)\alpha + t\beta] = \gamma(0)\mathbf{e}(nt)\,,$$

i.e. to $n\mathbf{u}$. The end of the proof can be left to the reader.

The corollary shows that γ is homotopic to a point in $\mathbb{C}^*$ if and only if the variation of $\mathcal{L}og\, z$ or, equivalently, of $\mathcal{A}rg\, z$ over γ is zero. If the variation of the argument is $2\pi n$, then γ is homotopic to a circle centered at 0 traveled n times.

In the previous statement, $\mathbb{C}^*$ could be replace by a pointed open disc centered at 0, for example $0 < |z| < 1$; it suffices to argue in the Poincare half-plane rather than in $\mathbb{C}$.

Exercise 2. Extract from the preceding arguments a direct elementary proof of corollary 2.

(iv) *Coverings of a simply connected space.* We are now ready to prove one the main results of the theory:

Theorem 3. *Every covering (X, B, p) of a simply connected and locally connected space is trivial.*

It all amounts to showing the existence of a global section with given value $\alpha \in X$ at a given point $a \in B$. To define it at an arbitrary $z \in B$, connect a to z by a path γ and consider *the* lifting μ of γ with initial point α. If γ is replaced by another path γ' connecting a to z, the terminal point $\mu(1)$ does not change because, B being simply connected, γ and γ' are fixed-endpoint homotopic, and hence so are their liftings. A map $f : B \longrightarrow X$ can, therefore, be defined without any ambiguity by setting $f(z) = \mu(1)$. So $p[f(z)] = z$ for all $z \in B$. For good reasons, this is very much like the construction of a primitive of a holomorphic function on a simply connected open set.

To show that f is a section, it suffices to prove that it is *continuous* at every $z \in B$. But let D be a connected open neighbourhood of z over which X is trivial. To calculate $f(z')$ for $z' \in D$, choose once for all a path γ connecting a to z. Let it be followed by a path γ' connecting z to z' in D. To lift the new path to X, lift γ onto a path μ with initial point α and terminal point $f(z)$ by definition, then lift γ' onto a path with initial point $f(z)$; its terminal point is $f(z')$. But $f(z)$ belongs to one of the connected components D_i of $p^{-1}(D)$, hence so does the lifting of γ' since its image is a connected subset of $p^{-1}(D)$. As a result, $f(z')$ is *the* point of D_i projecting onto z', qed.

In the next statement, B will be said to be *locally simply connected* if all neighbourhoods of $z \in B$ contain a simply connected open neighbourhood. A trivial but fundamental example: any C^0 manifold since, in a Cartesian space, a ball is simply connected and even contractible.

Theorem 4. *Every connected and locally simply connected space B has a connected and locally simply connected covering (X, B, p). It is unique up to isomorphism. If (Y, B, q) is a connected covering of X and if $\alpha \in X$ and $\beta \in Y$ are such that $q(\beta) = p(\alpha)$, then there is a unique continuous map f from X onto Y such that*

$$p = q \circ f, \quad f(\alpha) = \beta,$$

and (X, Y, f) is a covering of Y.

The proof consists of several stages. We will indicate their outline and leave to the reader the task of completing them.

(a) Choose a point $a \in B$ and consider the set of all paths $\gamma : I \longrightarrow B$ with initial point a. Two homotopic (understood to be fixed-endpoint homotopic) paths are considered to be equivalent. Let X be the set of classes of these paths. Define $p : X \longrightarrow B$ by associating to each path its terminal point. α will denote the class of the constant path $t \mapsto a$.

(b) Let $D \subset B$ be a simply connected open subset, z a point of D and $\zeta \in X$ the class of a path γ connecting a to z. Denote by $D(\zeta)$ the set of classes ζ' of paths $\gamma' = \gamma.\delta$ obtained by adjoining to γ a path δ with initial point z in D; ζ' does not depend on the terminal point of δ since D is simply connected. From this, the map p from $D(\zeta)$ to D is deduced to be bijective.

The set $D(\zeta)$ remains invariant if z is replaced by some $z' \in D$ and γ by $\gamma' = \gamma.\delta$, where δ connects z to z' in D; because if we connect z and z' to a $z'' \in D$ by the paths ϵ and ϵ' in D, the paths $\gamma.\varepsilon$ and $\gamma.\delta.\varepsilon'$ connecting a to z'' are homotopic. Hence

$$(3.6) \qquad \zeta' \in D(\zeta) \Longrightarrow D(\zeta) = D(\zeta').$$

To simplify the language, we say that a set of the form $D(\zeta)$ is, so to speak, a "*disc*" in X. Let $D(\zeta)$ and $D'(\zeta')$ be two "discs" in X, γ and γ' paths with initial point a and terminal points z and z' in B with homotopy class ζ and ζ', ζ'' a point of $D(\zeta) \cap D'(\zeta')$ corresponding to a path γ'' in B with terminal point $z'' \in D \cap D'$, and $D'' \subset D \cap D'$ a simply connected neighbourhood D'' of z''. By definition, there exist paths δ and δ' with initial points z and z' in D and D' and terminal point z'' such that γ'' is homotopic to both $\gamma.\delta$ and $\gamma'.\delta'$. Then

$$(3.7) \qquad D''(\zeta'') \subset D(\zeta) \cap D'(\zeta').$$

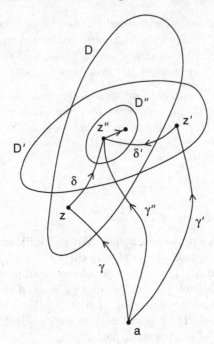

Fig. 3.3

The above figure shows the constructions needed to get the result.

(c) For given D and $z \in D$, the $D(\zeta)$ with $p(\zeta) = z$, whose union is $p^{-1}(D)$, are *pairwise disjoint*. Indeed, if ζ and ζ' are the classes of the two paths γ and γ' with terminal point z and if $\zeta'' \in D(\zeta) \cap D(\zeta')$, there is a path δ in D with initial point z and terminal point $p(\zeta'')$ such that $\gamma.\delta$ and $\gamma'.\delta$ are homotopic; then so are γ and γ' as well (exercise!). Hence $\zeta = \zeta'$.

(d) Let us say that a set $U \subset X$ is *open* if, for all $\zeta \in U$, $D(\zeta) \subset U$ for sufficiently small D. Verifying axioms (unions and intersections) is easy – use (7) for intersections –, and (6) shows that any "disc" is open in X. As the open subsets D are arbitrarily small, every open subsets of X is the union of the $D(\zeta)$ contained in it.

To check that the space X is separated, it suffices to show that relation $D(\zeta) \cap D'(\zeta') = \varnothing$ holds if $\zeta \neq \zeta'$. This is clear if $p(\zeta) \neq p(\zeta')$: choose D and D' such that $D \cap D' = \varnothing$. If $z = z'$, this follows from (c).

(e) The map $p : D(\zeta) \longrightarrow D$ is a homeomorphism. Continuity follows from the fact that, if $z' \in D' \subset D$, then the set

$$p^{-1}(D') = \bigcup_{p(\zeta')=z'} D'(\zeta')$$

is open in X. Since, on the other hand, p maps every "disc" $D(\zeta)$ onto some open set D, p maps every open subset of X, and in particular of $D(\zeta)$ onto some open subset of B. As $p : D(\zeta) \longrightarrow D$ is bijective, it is a homeomorphism.

The fact that (X, B, p) is a covering of B is now clear.

(f) X is connected. Indeed, denoting by α the class of the constant path $t \mapsto a$, if γ connects a to $z \in B$, the classes of the paths

$$\gamma_s : t \longmapsto \gamma(st), \qquad s, t \in I,$$

define a path in X connecting α to the class ζ of γ.

To show that X is simply connected, observe that if there are two paths μ and μ' in X with initial point α and the same terminal point ζ, then these are the liftings of two paths γ and γ' with initial point a and terminal point $p(\zeta)$; however, the endpoints of μ and μ' in X are, by definition, the homotopy classes of γ and γ'; γ are γ' are, therefore, homotopic, hence so are μ and μ' as well (theorem 2). Taking into account the connectedness of X, we move on from here to paths having an arbitrary initial point.

(g) Let (Y, B, q) be a connected covering of B and let $\alpha \in X$, $\beta \in Y$ be such that $q(\beta) = p(\alpha)$. For all $\zeta \in X$, there exists a path μ connecting α to ζ. It is unique up to homotopy since X is simply connected. If γ is the projection from μ onto B, then there is a unique lifting ν of γ to Y having β as initial point. As γ is unique up to homotopy, the terminal point $f(\zeta)$ of ν only depends on ζ. This gives the map f sought. Its uniqueness is due to the fact that this is obviously the only way it could possibly be defined. That (X, Y, f) is a covering is more or less obvious.

If Y is simply connected, oppositely, there is a map $g : Y \longrightarrow X$ such that $p \circ g = q$, $g(\beta) = \alpha$. Clearly, f and g are mutually inverse. This gives the isomorphism of the simply connected coverings considered, qed.

If B is a connected and locally arc-connected space, the connected and simply connected covering (X, B, p) of theorem 4 is the *universal covering* of B; by part (g) of the proof, it "dominates" all the others. Example 1 above shows that due to the exponential map, $\mathbb{C}$ is a universal covering space of $\mathbb{C}^*$, while, due to the map $t \mapsto \mathbf{e}(t)$, $\mathbb{R}$ the universal covering space of $\mathbb{T}$.

Exercise 3. Let $G \subset \mathbb{C}$ be a domain and $(\tilde{G}, G, p)$ its universal covering. Show that there is a complex analytic structure on $\tilde{G}$ such that p is a holomorphic submersion. Let f be a holomorphic function on G and ω_f the inverse image under p of the differential form $f(z)dz$. Show that there is holomorphic function F on $\tilde{G}$ such that $dF = \omega_f$. What about the case $G = \mathbb{C}^*$?

(v) *Coverings of a pointed disc.* Let D^* be a pointed disc, i.e. either an open disc in $\mathbb{C}$ with its centre removed, or else the exterior of a closed disc in $\mathbb{C}$ (pointed disc "centered at ∞"). Using a conformal representation, we need only consider the disc $0 < |z| < 1$. Like all coverings of a Riemann surface, a covering (Y, D^*, q) of D^* has a natural Riemann surface structure: the analytic structure of the discs' projections onto D^* is transferred to the discs and the compatibility clause is immediate. The following result will play an essential role in the construction of the compact Riemann surface of an algebraic function:

Theorem 5. *Let (Y, D^*, q) be a connected covering of order $k < +\infty$ of the pointed disc $D^* : 0 < |z| < 1$. Then there is a holomorphic function φ in Y such that (i) $\zeta \mapsto \varphi(\zeta)$ is a conformal representation of Y on D^*, (ii) $\varphi(\zeta)^k = q(\zeta)$.*

This means that (Y, D^*, q) is isomorphic to the covering of D^* obtained by taking $Y = D^*$ and $q(z) = z^k$ [(i), Example 2], or to the covering obtained by constructing the Riemann surface of the algebraic function $z^{1/k}$ using the method of n° 2, or finally that the algebraic function $z^{1/k}$, which obviously does not have uniform branches on the pointed disc, becomes uniform on Y: a kth root of $z = q(\zeta)$ can be associated to each $\zeta \in Y$ so that it only depends holomorphically on ζ (but obviously not on z).

(a) Choose an arbitrary point a of D^* and some $\beta \in Y$ such that $q(\beta) = a$ and associate to every *closed* path γ with initial point a in D^* the terminal point of its lifting ν with initial point β in Y. As the terminal point of ν only depends on the homotopy class of γ, by theorem 2 and its corollary 2, only the paths $\gamma_n : t \mapsto ae(nt)$ need to be considered; let ν_n be the lifting of γ_n. The path $\gamma_m.\gamma_{n-1}$ consisting in γ_m followed by the opposite of γ_n is clearly the path γ_{m-n}, up to parameterisation. If ν_m and ν_n have the same terminal point, the lifting of γ_{m-n} is obviously the *closed* path $\nu_m.\nu_{n-1}$. If, conversely, the lifting ν_{m-n} of γ_{m-n} is closed, the path $\nu_{m-n}.\nu_n$ with initial point β is a lifting of $\gamma_{m-n}.\gamma_n$, a path, identical to γ_m, up to parameterisation. Then $\nu_m = \nu_{m-n}.\nu_n$ up to parametrization, so that the paths ν_m and ν_n have the same terminal point.

This leads to two conclusions. First, if ν_m and ν_n are closed, so is ν_{m-n}; the set of n such that γ_n is lifted onto a *closed* path is, therefore, the subgroup $p\mathbb{Z}$ of $\mathbb{Z}$. Secondly, the terminal point of ν_n only depends on the class of $n \bmod p$. As any point of Y projecting onto a is the terminal point of such a lifting, we conclude that there are an many classes $\bmod\, p$ as points of Y over a, i.e. that $p = k$.

(b) Having made this point, let us consider the universal covering $(P, D^*, \mathbf{e})$ of D^*, where P is the half-plane $\mathrm{Im}(\zeta) > 0$ and $\mathbf{e}$ the map $\zeta \mapsto \mathbf{e}(\zeta)$ from P onto D^*. If we choose some $\alpha \in P$ such that $\mathbf{e}(\alpha) = a$, then theorem 4 shows the existence and uniqueness of a continuous map $f : P \longrightarrow Y$ such that $f(\alpha) = \beta$ and $q[f(\zeta)] = \mathbf{e}(\zeta)$ for all $\zeta \in P$. Let ζ' and ζ'' be two points such that $f(\zeta') = f(\zeta'')$. So $\zeta'' = \zeta' + n$ for some $n \in \mathbb{Z}$. Setting $\mathbf{e}(\zeta') = z'$, $\mathbf{e}(\zeta'') = z''$, $f(\zeta') = f(\zeta'') = \eta \in Y$, the map f transforms every path μ connecting ζ' to ζ'' into a closed path ν with initial point η in Y, which is a lifting of the path γ, the image of μ under $\mathbf{e}$. Choosing as μ the rectilinear path $[\zeta', \zeta''] = [\zeta', \zeta' + n]$, clearly $\gamma(t) = z'\mathbf{e}(nt)$. Since γ can be lifted to a closed path ν in Y, $n = 0 \bmod k$ by part (a) of the proof, which can be applied to all points of D^* and in particular to z'. Conversely, if n is a multiple of

k, f transforms $[\zeta', \zeta' + n]$ into a necessarily closed lifting of $t \mapsto z'\mathbf{e}(nt)$. So $f(\zeta') = f(\zeta'')$.

It follows that f is a homeomorphism of the quotient $P/k\mathbb{Z}$ onto Y, which is obviously compatible with the complex analytic structures of both spaces and with the projections $q : Y \longrightarrow D^*$ and $p_k : P/k\mathbb{Z} \longrightarrow D^*$, so that the given covering (Y, D^*, q) is isomorphic, including from a complex analytic point of view, to the canonical covering $(P/k\mathbb{Z}, D^*, p_k)$ of example 2 of this n°. Hence it remains to construct the function φ of the theorem for this particular covering. However, the map $z \mapsto \mathbf{e}(z/k)$ from P onto D^* is invariant under $k\mathbb{Z}$; so there is one and only one map $\varphi : P/k\mathbb{Z} \longrightarrow D^*$ which, for all $z \in P$, transforms the class of $z \bmod k\mathbb{Z}$ into $\mathbf{e}(z/k)$. The reader is left to check that φ is holomorphic and satisfies conditions (i) and (ii) of the theorem.

It applies to all disc centered at $a \in \hat{\mathbb{C}}$; replace relation $\varphi(\eta)^k = q(\eta)$ by

$$\varphi(\eta)^k = q(\eta) - a \;\; \text{if} \;\; a \neq \infty, \quad \varphi(\eta)^k = 1/q(\eta) \;\; \text{if} \;\; a = \infty.$$

It applies also to coverings of $\mathbb{C}^*$: replace P by $\mathbb{C}$ in the preceding arguments.

4 – The Riemann Surface of an Algebraic Function

(i) *Global uniform branches.* Once again, consider an irreducible equation

(4.1) $$P(z, \zeta) = P_0(z)\zeta^n + \ldots + P_n(z) = 0$$

of degree n in ζ and, as in n° 2, remove from $\mathbb{C}$ the finite set of the values of z where (1) does not have n distinct roots. This gives an open subset B. As shown, the subset $X \subset \mathbb{C}^2$ of the curve $P = 0$ which projects onto B is both a Riemann surface and a covering of order n of B. If $(a, \alpha) \in X$ and if D is a sufficiently small open disc centered at a, there is a unique holomorphic function $f(z)$ in D satisfying $P[z, f(z)] = 0$ and $f(a) = \alpha$, and the couple $(f(D), p)$ is a local holomorphic chart of X in the neighbourhood of (a, α), defined in the "disc" $D(\alpha) = f(D)$. As $(z, \zeta) \in f(D)$ means that $\zeta = f(z)$, the functions $p : (z, \zeta) \mapsto z$ and $F : (z, \zeta) \mapsto \zeta$ are holomorphic in this chart, and hence in X in the sense of n° 1. More generally, if f and g are polynomial, the functions $(z, \zeta) \mapsto f(z, \zeta)$ and $(z, \zeta) \mapsto g(z, \zeta)$ are holomorphic on X, so that $(z, \zeta) \mapsto f(z, \zeta)/g(z, \zeta)$ is meromorphic if g does not identically vanish on X (i.e. is not a multiple of the polynomial P).

A first immediate consequence of these results concerns *global uniform branches* of an algebraic function. Their existence is governed by the following result, where we keep the above notation:

Th 6. *Let U be a simply connected open set contained in B. There are n holomorphic functions f_k in U whose values at each $z \in U$ are the n roots of the equation $P(z, \zeta) = 0$.*

As X is a covering space of B, $p^{-1}(U)$ is a covering space of U, hence is trivial (theorem 3). So it has n sections $z \mapsto (z, f_k(z))$, with functions f_k defined on U and locally, hence globally, holomorphic, qed.

Our task is now going to be to complete X to obtain a compact Riemann surface $\hat{X}$ for which the map $p : X \longrightarrow B = \mathbb{C} - S$ can be extended to a map $\hat{X} \longrightarrow \hat{\mathbb{C}}$ which will be holomorphic without however satisfying axiom (R) of coverings in the neighbourhoods of the points $z \in \hat{\mathbb{C}} - B$. For this, finitely many points need to be adjoined to X "over" the $z \in \hat{\mathbb{C}} - B$ and the holomorphic charts in the neighbourhood of these points need to be defined.[12]

This supposes that the structure of X over a neighbourhood of any $a \in \hat{\mathbb{C}} - B$ is known; theorem 5 provides the answer. It will nonetheless be somewhat long, but there is no quick method, especially if we want to explain everything.

(ii) *Definition of a Riemann surface* $\hat{X}$. Let us return to the Riemann surface (X, B, p) of the equation $P(z, \zeta) = 0$ over the open subset B of $\mathbb{C}$. For any given $a \in \hat{\mathbb{C}} - B$, choose an open disc $D(a)$ centered at a not containing any other point of $\hat{\mathbb{C}} - B$ apart from a, and let Y be a connected component of $p^{-1}(D^*(a))$; it is a connected covering space of order $k \leq n$ of $D^*(a)$. Let φ_Y be a holomorphic function on Y satisfying the properties of theorem 5 with respect to $D^*(a)$; since

$$(4.2) \qquad \varphi_Y(z, \zeta)^k = z - a \text{ resp. } 1/z$$

for all $\eta = (z, \zeta) \in Y$, $\varphi_Y(\eta)$ tends to 0 as $p(\eta) = z$ tends to a. An "ideal point" (so Forster says) then needs to be added to Y as is done to go from $\mathbb{C}$ to $\hat{\mathbb{C}}$. Denote this point by η_Y, set $\hat{Y} = Y \cup \{\eta_Y\}$ and assume that $\varphi_Y(\eta_Y) = 0$. This gives a bijection from $\hat{Y}$ onto the open disc centered at 0 in $\mathbb{C}$; thus the holomorphic structure of this disc can be transferred to $\hat{Y}$. The connected component $Y = \hat{Y} - |\eta_Y|$ becomes an open subset of $\hat{Y}$ and the holomorphic functions on an open subset U of $\hat{Y}$ are those that can be expressed holomorphically by using φ_Y: equip $\hat{Y}$ with the holomorphic structure for which $(\hat{Y}, \varphi_Y)$ is a *holomorphic chart* of $\hat{Y}$ and φ_Y is a local uniformizer at η_Y.

We show that the holomorphic functions on every open subset U of Y, that we are already acquainted with from the end of n° 2 (an open subset of Y is also open in X) are just the holomorphic functions of φ_Y. First of all, U is the union of "discs" of X. The notion of holomorphy being a local one, we can confine ourselves to the case of a disc U, hence suppose that $U = D'(\beta)$ for some $(b, \beta) \in Y$, where $D' \subset D^*(a) \subset B$ is a sufficiently small open disc centered at $b \neq a$ over which the covering X, and hence Y, is trivial. $D'(\beta)$ is the image of D' under $z \mapsto (z, f(z))$, where f is *the*

[12] The following developments, and even some of the preceding ones, are strongly influenced by Chap. I of Otto Forster's *Lectures on Riemann Surfaces* (Springer, 1981). Let us also mention Hershel M. Farkas and Irwin Kra, *Riemann Surfaces* (Springer, 1980).

uniform branch in D' of the algebraic function $P(z, \zeta) = 0$ which takes the value β at b. If U is regarded as an open subset of X, the end of n° 2 tells us that the holomorphic functions of (z, ζ) on U are the holomorphic functions of z on D'. So it amount to showing that the holomorphic functions of $\varphi_Y(z, \zeta)$ on $D'(\beta)$ are identical to the holomorphic functions of z. However, $\varphi_Y(z, \zeta) = \varphi_Y(z, f(z))$ in $D'(\beta)$. As φ_Y is holomorphic on Y in the sense of n° 2 (theorem 5) and as $z \mapsto (z, f(z))$ is a holomorphic map of z from D' to X, the left hand side is a holomorphic function of z on D'. Oppositely, relation $\varphi_Y(z, f(z))^k = z - a$ resp. $1/z$ shows that z is a holomorphic function of $\varphi_Y(z, f(z))$ on $D'(\beta)$, qed.

The complete Riemann surface $\hat{X}$ sought is then obtained as follows: for each $a \in \hat{\mathbb{C}} - B$, the points η_Y corresponding to the connected components of X over $D^*(a)$ are adjoined to X. *These η_Y are considered to be pairwise distinct.* If the covering space Y of $D^*(a)$ is of order k, $(\hat{X}, \hat{\mathbb{C}}, p)$, which is a covering only over B and perhaps over a neighbourhood of ∞, is said to be a *branched covering* of $\hat{\mathbb{C}}$ at the *branch point η_Y of order k*. This is not a property of the Riemann surface $\hat{X}$, which is as smooth as possible a manifold; it is a property of the map $p : \hat{X} \longrightarrow \hat{\mathbb{C}}$.

For example, if we take the algebraic equation $\zeta^2 - z = 0$, then X is the set of $(z, \zeta) \in \mathbb{C}^2$ such that $\zeta^2 = z$, $z \neq 0$. Over a disc D which does not contain 0, the surface has two disjoint connected components corresponding to the two uniform branches on D of the pseudo-function $z^{1/2}$. This is no longer the case over a disc D centered at 0, since, when z circles once around the point 0, the determination chosen at the start for $z^{1/2}$ becomes the opposite determination at termination; this means that two points of the surface projecting onto z can be connected by a curve, as in the case of the logarithm (Chap. IV, §4). The surface X is, therefore, connected over $D^* = D - \{0\}$; this is also the case over D since as z tends to 0, the two possible values of $z^{1/2}$ also tend to 0, and only one point, namely $(0, 0)$, remains over the origin. There is a "branch point" in this case only because we are trying to express ζ by using z in the neighbourhood of 0; if we tried to express z as a function of ζ, all would become normal again. But, even in the real domain, a general algebraic curve can have singularities (multiple points, cusp points, etc.) otherwise more complicated than a vertical tangent.

In the general case, $\hat{X}$ being the union of X and of the $\hat{Y}$, we define the topology of $\hat{X}$ by declaring $U \subset \hat{X}$ to be open if $U \cap X$ is open in X and if, for all Y such that $\eta_Y \in U$, the set $U \cap \hat{Y}$ is open in $\hat{Y}$. The Hausdorff axiom and the fact that X and the $\hat{Y}$ are open in $\hat{X}$ are immediately verified. Finally, the complex analytic structure of $\hat{X}$ is obtained by adjoining the charts $(\hat{Y}, \varphi_Y)$ to the charts already available in X. As shown, these are compatible with the complex analytic structure of the open subsets Y, hence of X, defined in n° 2; they are also compatible with each other as they are pairwise disjoint. This provides $\hat{X}$ with the structure of a Riemann surface, which coincides with that of n° 2 in the open set X.

The map p from $\hat{X}$ onto $\hat{\mathbb{C}}$ is obtained by setting $p(\eta_Y)$, for each open set $\hat{Y}$, to be the point $a \in \hat{\mathbb{C}}$ from which Y was obtained. If $D(a)$ is the disc centered at a chosen to get the connected components Y over a neighbourhood of a, the inverse image $p^{-1}(D(a))$ is the union of open sets $\hat{Y}$, and so is open in $\hat{X}$, obviously so is also $p^{-1}(D')$ for any disc $D' \subset D(a)$. As a result, p is continuous and, even better, maps every open subset of $\hat{X}$ onto an open subset of $\hat{\mathbb{C}}$. But $\hat{X}$ is not, strictly speaking, a covering of $\hat{\mathbb{C}}$.

Let us show that $\hat{X}$ is *compact*. For each $a \in \hat{\mathbb{C}} - B$, consider the open sub-set $p^{-1}(D(a))$ of $\hat{X}$. It is the union of open subsets of type $\hat{Y}$. Let $D_a \subset D(a)$ be a *closed* disc, hence compact in $\hat{\mathbb{C}}$, centered at a; for all $Y \subset p^{-1}(D(a))$, the chart (Y, φ_Y) of $\hat{X}$ transforms $p^{-1}(D_a) \cap \hat{Y}$ homeomorphically into a closed, hence compact, disc centered at 0. As a result, $p^{-1}(D_a)$ is the finite union of compact sets, and so is compact. Similarly, for all $a \in B$, there is a closed disc D_a centered at a such that $p^{-1}(D_a)$ is compact. Since the interiors of the D_a cover the compact space $\hat{\mathbb{C}}$, $\hat{\mathbb{C}}$ can be covered by finitely many discs D_a. Thus $\hat{X}$ is the union of finitely many compact sets, qed.

(iii) *The algebraic function $\mathcal{F}(z)$ as a meromorphic function on $\hat{X}$.* Let us now show that the functions $(z, \zeta) \mapsto z$ and $F : (z, \zeta) \mapsto \zeta$, defined and holomorphic on X have meromorphic extensions on $\hat{X}$. It is enough to prove this on each open set $\hat{Y}$, using the the chart $(\hat{Y}, \varphi_Y)$ to verify it. If $\hat{Y}$ corresponds to a point $a \in \hat{\mathbb{C}} - B$ and if Y is of order k, then $\varphi_Y(z, \zeta)^k = z - a$ or $1/z$, whence the result related to $(z, \zeta) \mapsto z$; in particular, $(z, \zeta) \mapsto z$ has a pole of order k at η_Y if this point projects onto the point ∞ of $\hat{\mathbb{C}}$. As an aside, note that the point ∞ having been excluded during the construction of X, $\hat{X}$ may very well be a genuine covering space of a neighbourhood of ∞, in other words that $k = 1$ for each η_Y projecting onto ∞; the Riemann surface of $\zeta(\zeta - 1)z = 1$ has two points over ∞, the function ζ taking values 0 and 1 at these points.

The case of the algebraic function $F : (z, \zeta) \mapsto \zeta$ is less obvious. We will suppose that η_Y projects onto a point $a \neq \infty$, the other case being similarly dealt with. As $(\hat{Y}, \varphi_Y)$ is a holomorphic chart of $\hat{Y}$ and as F is holomorphic on $Y = \hat{Y} - \eta_Y$, there is a Laurent series expansion

$$(4.3) \qquad F(z, \zeta) = \zeta = \sum_{\mathbb{Z}} c_n q^n = h(q), \quad q = \varphi_y(z, \zeta)$$

in Y, and it all amounts to showing that $c_n = 0$ for sufficiently large $n < 0$. Since $P(z, \zeta) = 0$ and $q^k = z - a$,

$$P_0\left(q^k + a\right) h(q)^n + \ldots + P_n\left(q^k\right) = 0$$

and so

$$(4.4) \qquad h(q)^n + s_1(q)h(q)^{n-1} + \ldots + s_n(q) = 0 \text{ for } q \neq 0.$$

The rational functions

$$s_i(q) = P_i \left(q^k + a\right) / P_0 \left(q^k + a\right)$$

are meromorphic at the origin; these are elementary symmetric functions of the roots of (4), up to sign. To show that $h(q)$ has at most one pole at $q = 0$, it suffices to show that there is an upper bound of the form $h(q) = O(q^{-N})$. Now, the coefficients $s_i(q)$ are of this form. Therefore, the following result remains to be proved. It probably dates back to time immemorial and for example can be found in Dieudonné's exercise in *Calcul infinitésimal*, Chap. III:

Lemma 1. *Let $\zeta_i (1 \leq i \leq n)$ be the roots of an equation*

(4.5)
$$\zeta^n + c_1 \zeta^{n-1} + \ldots + c_n = 0$$

with complex coefficients. Then

(4.6)
$$\sup |\zeta_i| \leq \max \left(1, |c_1| + \ldots + |c_n|\right).$$

Let M be the left hand side of (6). Each root of (5) satisfies

$$|\zeta_i|^n \leq |c_1| \cdot |\zeta_i|^{n-1} + \ldots + |c_n| \leq |c_1| \cdot M^{n-1} + \ldots + |c_n|,$$

whence

$$M^n \leq |c_1| \cdot M^{n-1} + \ldots + |c_n|.$$

If $M \leq 1$, there is nothing to prove. If $M \geq 1$, we then write

$$M \leq |c_1| + \ldots + |c_n| / M^{n-1} \leq |c_1| + \ldots + |c_n|,$$

qed.

Note that if $P_0(a) \neq 0$, the coefficients $s_i(q)$ of (5) are holomorphic at $q = 0$, and so are bounded for sufficiently small q, hence so is $h(q)$. As a result, the function $F(z, \zeta) = \zeta$ *is holomorphic at the point* $\eta_Y \in \hat{X}$ *if* $P_0(a) \neq 0$, in other words if the equation $P(a, Y) = 0$ is effectively of degree n at the point a. Since $P(z, \zeta) = 0$ in Y, the value of ζ at the point η_Y is obtained by passing to the limit as (z, ζ) converges to η_Y, hence is a root of $P(a, \zeta) = 0$.

On the contrary, if $R_0(0) = P_0(a) = 0$, then $P_0(X + a) = X^m Q_0(X)$ with $Q_0(0) \neq 0$, the coefficients

$$s_i(q) = P_i \left(q^k + a\right) / q^{mk} Q_0 \left(q^k\right)$$

can have poles of order $\leq mk$ at the origin and so are $O(q^{-mk})$. Thus so is $h(q)$ as well. The exact order of ζ at the point η_Y can theoretically be calculated by the Newton polygon method,[13] but the latter has fortunately long disappeared from presentations on Riemann surfaces.

In all cases, the situation over the point a is then easily elucidated. Let us return to the disc $D(a)$ centered at a used above and let $Y_1, \ldots, Y_m$ be the various connected components of X over $D^*(a)$. Each Y_i is a connected covering space of order k_i of $D^*(a)$, and $k_1 + \ldots + k_m = n$ since there are n points of X over each point of B. If $q_i(z, \zeta)$ denotes the function φ_Y corresponding to $Y = Y_i$, so that (Y_i, q_i) is a chart of $\hat{X}$, then

$$(4.7) \qquad q_i(z, \zeta)^{k_i} = z - a \text{ if } a \neq \infty, \ = 1/z \text{ if } a = \infty;$$

$(z, \zeta) \mapsto \zeta$ is a meromorphic function $h_i(q_i)$ on Y_i, possibly with a pole at $q_i = 0$, hence a meromorphic Laurent series in the local uniformizer q_i. The points k_i of Y_i located over a given $z \in D^*(a)$ correspond to the k_i values of q_i satisfying (7); they can be deduced from each other by taking the product of q_i and of the k_i-th roots of unity. For $z \in D^*(a)$, equation $P(z, \zeta) = 0$ has exactly k_i distinct roots such that $(z, \zeta) \in Y_i$; they correspond to the distinct k_i points of Y_i located over z; these are the numbers obtained by replacing q_i, in the series h_i, by the k_i k_i-th roots of $z - a$ or of $1/z$. As $(z, \zeta) \in Y_i$ tends to the point $\eta_Y = \eta_i$ adjoined to $Y = Y_i$, ζ tends either to the finite limit $\zeta_i = h_i(0)$ satisfying $P(a, \zeta_i) = 0$, or to infinity. If $P_0(a) \neq 0$ and $a \neq \infty$, we have seen that the h_i are holomorphic at the origin and we deduce that, for all $r > 0$, there exists $\rho > 0$ such that, for $|z - a| < \rho$, the equation $P(z, \zeta) = 0$ has at least k_i simple roots satisfying $|z - \zeta_i| < r$; as a result, by lemma 1 of n° 2, the order of multiplicity of the root ζ_i of $P(z, \zeta) = 0$ is at least k_i.

As $\sum k_i = n$, this result suggests that the connected components Y_i of X over $D^*(a)$ correspond bijectively to the various roots of the equation $P(a, \zeta) = 0$, each multiple root ζ_i of order k_i giving rise to a component Y_i of order k_i.

False: $\zeta_i = \zeta_j$ may hold for $i \neq j$, in which case the multiplicity of the root ζ_i is at least $k_i + k_j$.

Exercise 1. Consider the equation

$$(4.8) \qquad \zeta^2 - z\zeta - z^4 = 0.$$

It has double roots in ζ for $z = 0$, $i/2$ and $-i/2$. Therefore, the Riemann surface X constructed in n° 2 is the subset of the graph of relation (8) in $\mathbb{C}^2$ located over the open subset

$$B = \mathbb{C} - \{0, i/2, -i/2\}$$

---

[13] The shortest presentation, but not necessarily the most accessible, is that of Dieudonné in *Calcul infinitésimal*, Appendix to Chap. III. The fact that he only looks for real branches and limited expansions instead of Laurent series in q so as not to traumatize his novice readers at the outset does not change anything. Besides, the method applies to functions that are not necessarily algebraic.

of $\mathbb{C}$. Setting $\zeta = z\tau$, the equation becomes $\tau^2 - \tau = z^2$, which has two uniform branches in the disc D, i.e. the disc $|z| < 1/2$. Show that these are given by

$$(4.9) \qquad\qquad g_1(z) = 1 + z^2 + \ldots$$
$$(4.10) \qquad\qquad g_2(z) = -z^2 + z^4 + \ldots.$$

Show that $p^{-1}(D^*)$ is the union of the graphs Y_1 and Y_2 over D^* of the functions

$$f_1(z) = zg_1(z) = z + z^3 + \ldots \quad \text{and} \quad f_2(z) = zg_2(z) = -z^3 + z^5 + \ldots$$

and that they are the *two* connected components of $p^{-1}(D^*)$ despite the fact that the equation $P(0, \zeta) = 0$ has *a* root. Show that, like $(z, \zeta) \mapsto z$ and $(z, \zeta) \mapsto \zeta$, the meromorphic function $\tau : (z, \zeta) \mapsto \zeta/z$ on $\hat{X}$ takes values 1 and 0 at the two points η_1 and η_2 of $\hat{X}$ projecting onto $z = 0$. Can it be attributed a value greater than 0 by considering the graph of equation (8)?

When $P_0(a) = 0$, equation $P(a, \zeta) = 0$ has strictly less than n roots; to get n roots, replace ζ by $1/\zeta = \zeta'$, in other words replace the initial polynomial $P(X, Y)$ by

$$Q(X, Y) = Y^n P(X, 1/Y) = P_n(X)Y^n + \ldots + P_0(X).$$

For $X = a$, 0 is a root, which can be interpreted by saying that $P(a, \zeta) = 0$ admits the root ∞ with an order of multiplicity r equal to that of the root 0 of $Q(a, \zeta') = 0$; the latter is given by the relations

$$P_0(a) = \ldots = P_{r-1}(a) = 0, \quad P_r(a) \neq 0.$$

As shown by these relations, the degree of the equation $P(a, \zeta) = 0$ is $n - r$, it has n roots in all, namely $n - r$ finite roots and an infinite root of order r.

(iv) *Connectedness of $\hat{X}$.*

Theorem 7. *The Riemann surface of an* irreducible *algebraic equation is connected.*

It suffices to prove this for the open subset X of $\hat{X}$ because the point η_Y adjoined to X can obviously be connected to points of X by paths.

As seen in section (ii) of the previous n°, like X, any connected component X' of X is a covering space of B. Let k be its order, so that for any disc $D \subset B$, the open set $p^{-1}(D) \cap X'$ is the union of "discs" $D_i (1 \leq i \leq k)$ corresponding under $z \mapsto (z, f_i(z))$ to uniform branches on D of our algebraic "function" $z \mapsto \zeta$. The identity

$$\prod (T - f_i(z)) = T^k + c_1(z)T^{k-1} + \ldots + c_k(z),$$

where T is an indeterminate, shows that the $f_i(z)$ are the roots of the polynomial

$$(4.11) \qquad P'(T) = T^k + c_1(z)T^{k-1} + \ldots + c_k(z) = 0$$

with coefficients in the ring of holomorphic functions on D (nothing to do with the derivative of P). But for all $z \in D$, the $c_i(z)$ are the elementary symmetric functions of the $f_i(z)$, i.e. the roots of the equation $P(z, \zeta) = 0$ such that $(z, \zeta) \in X'$. It follows that the $c_i(z)$ are the same for all discs $D \subset B$ and hence are the restrictions to D of holomorphic functions on B. We show that these are *rational* and to do this, that they only have polar singularities at every $a \in \hat{\mathbb{C}} - B$.

As in (ii), consider the disc $D(a) = D$; over D^*, X is the union of open connected sets Y; hence the same holds for X'. If $Y \subset X'$ and if $q = \varphi_Y(z, \zeta)$ is the corresponding local uniformizer, ζ is a meromorphic function of q on Y with at most one pole at $q = 0$. For $z \in D^*$ the roots ζ_i of $P(z, \zeta) = 0$ such that $(z, \zeta_i) \in Y$ give an upper bound $\zeta_i = O(q^{-N})$ as q tends to 0. But if the order of the covering space Y of D^* is equal to r, then $q^r = z - a$ or $1/z$ as the case may be, whence $\zeta = O((z - a)^{-N})$ or $O(z^N)$ for another integer N. Hence elementary symmetric functions of $\zeta_i = f_i(z)$ such that $(z, \zeta_i) \in X'$ satisfy similar upper bounds properties. Thus, being holomorphic on B and at most of polynomial growth in the neighbourhood of $\hat{\mathbb{C}} - B$, the coefficients of (11) are indeed rational functions of z.

As a result, for each connected component X' of X, (11) is an algebraic equation with coefficients in the field K of *rational* functions of z. Denoting by X_j the connected components of X and by P_j the corresponding polynomials, it becomes clear that

$$(4.12) \quad \prod P_j(T) = \prod (T - \zeta) = T^n + s_1(z)T^{n-1} + \ldots + s_n(z) = P(z, T),$$

where the product is extended to all roots ζ of $P(z, \zeta) = 0$ and where, in consequence, the coefficients are the rational functions $s_i(z) = P_i(z)/P_0(z)$ already encountered in (4). Multiplied by $P_0(z)$, this identity between polynomials in T with coefficients in the field $K = \mathbb{C}(z)$ of one variable rational fractions shows that in the ring $K[T]$, the $P_j(T)$ divide $P(T)$. The coefficients of the P_j are not necessarily polynomials in z, but we know (exercise below) that if P is irreducible, i.e. does not have any non-trivial divisors with coefficients in $\mathbb{C}[z]$, then neither does it have any with coefficients in $\mathbb{C}(z)$. Hence there is a unique index j, and X is connected, qed.

Exercise 2. Let $f(Y) = \sum a_k Y^k$ be a polynomial with coefficients $a_k \in \mathbb{Z}$. The gcd $c(f)$ of these coefficients is said to be the *content* of f. (a) Let $f, g \in \mathbb{Z}[Y]$ and let p be a prime number dividing $c(fg)$, i.e. all the coefficients of fg. Show that p divides $c(f)$ or $c(g)$. [As p divides $a_0 b_0$, it divides a_0 or b_0; if p divides a_0 but not all the coefficients of f, let r be the largest integer such that p divides $a_0, \ldots, a_{r-1}$. By calculating the coefficients of $Y^r, Y^{r+1}, \ldots$ in fg, show that p divides b_0, then b_1, etc.] (b) f is said to be *primitive* if $c(f) = 1$. Show that if f and g are primitives, then so is fg ("Gauss' Lemma"). (c) Show that $c(fg) = c(f)c(g)$ for all $f, g \in \mathbb{Z}[X]$. (d) Show that the preceding arguments and results continue to hold if $\mathbb{Z}$ is replaced by the ring $A = k[X]$ of polynomials in one variable with coefficients in the field k [replace "prime" with "irreducible"] and more generally by an arbitrary principal ring A. (e) Let $P(X, Y)$ be an irreducible polynomial with coefficients in a field k. Show that as a polynomial in Y with coefficients in the field $K = k(X)$ of rational fractions, P is still irreducible. In other words: if P has a non-trivial divisor in $K[X] = k(X)[Y]$, then it also has one in $k[X, Y]$.

(v) *Meromorphic functions on $\hat{X}$.*

Theorem 8. *All meromorphic functions on $\hat{X}$ are rational functions of z and ζ.*

We will assume a result from the general theory of commutative field extensions though it is not hard to show. Otherwise the proof given will be complete.

Let φ be a meromorphic function on $\hat{X}$. It has finitely many poles projecting onto points $a_i \in \hat{\mathbb{C}}$. Let B_φ be the open set obtained by removing from B the points a_i belonging to it. If, for some $z \in B_\varphi$, the n (distinct) points of X over z are numbered by (z, ζ_k), then the expression

$$(4.13) \qquad \prod (T - \varphi(z, \zeta_k)) = T^n + c_1(z)T^n - 1 + \ldots + c_n(z)$$

is well-defined. Its coefficients, i.e. the elementary symmetric functions of $\varphi(z, \zeta_k)$ up to sign, are defined on all of B_φ. These are holomorphic functions on B_φ because, over a sufficient small disc $D \subset B_\varphi$ centered at a, the covering space X decomposes into "discs" $D(\alpha_k)$ on which φ is a holomorphic function of the local uniformizer q, and so of z, whence the result. As, on the other hand, the function φ is $O(q^{-N})$ in the neighbourhood of any point that does not project onto B_φ, where q is the local uniformizer at this point, the $c_i(z)$ have at most poles at the point of $\hat{C} - B_\varphi$, hence are *rational* functions of z.

If M denotes the field of meromorphic functions on $\hat{X}$ and if every rational function $f(z)$ on $\hat{\mathbb{C}}$ is identified with the function $(z, \zeta) \mapsto f(z)$ on $\hat{X}$, then it follows that *every $\varphi \in M$ is algebraic of degree $\leq n$ on $K = \mathbb{C}(z)$.*

On the other hand, M contains the field L of rational functions of z and ζ. As seen at the end of the previous section (iv), the polynomial $P(X, Y)$ is irreducible as a polynomial in Y with coefficients in K. The following very simple result then shows L has dimension n over K:

Lemma 2. *Let M be a commutative field, K a subfield of M and ζ an element of M satisfying an irreducible algebraic equation of degree n over K. Then the subfield L of M generated by K and ζ has dimension n over K and admits $1, \zeta, \ldots, \zeta^{n-1}$ as a basis over K.*

Since ζ^n is a linear combination of $1, \zeta, \ldots, \zeta^{n-1}$ with coefficients in K, the same is true for ζ^p for all $p > n$:

$$\zeta^{n+1} = \zeta \left(c_0 + \ldots + c_{n-1}\zeta^{n-1} \right) = c_0\zeta + \ldots + c_{n-1}\zeta^n =$$
$$= c_0\zeta + \ldots + c_{n-1}\zeta^{n-1} + c_{n-1} \left(c_0 + \ldots + c_{n-1}\zeta^{n-1} \right) ,$$

etc. (induction on p). As a result, the set A of polynomials in ζ with coefficients in K is a vector space of dimension $\leq n$ over K. It is a subring of M and, for all $a \in A$, the map $u : x \mapsto ax$ from A to A is linear over K; it is injective if $a \neq 0$ since we are in a field. As A is finite-dimensional, u is bijective. So there is some $x \in A$ such that $ax = 1$. As a result, A is a subfield and $L = A$. If L had dimension $< n$ over K, there would be a non-trivial linear relation between $1, \zeta, \ldots, \zeta^{n-1}$ with coefficients in K. In other words, ζ would satisfy an equation of degree $< n$ over K, a contradiction.

Hence, returning to meromorphic functions over $\hat{X}$,

$$K \subset L \subset M , \quad \dim_k(L) = n .$$

To prove that $M = L$, it, therefore, suffices to show that $\dim_K(M) \leq n$. However, $\varphi \in M$ is known to satisfy an algebraic (not necessarily irreducible) equation of degree n over K; so it suffices to prove (or to admit without proof, which is what we will do here) the next general result:

Lemma 3. *Let M be a commutative field of characteristic 0, K a subfield of M and n an integer ≥ 1. Suppose that all $x \in M$ satisfy an algebraic equation of degree $\leq n$ with coefficients in K. Then, the dimension of M as a vector space over K is $\leq n$.*

This lemma is itself based on the *primitive element theorem* due to Dedekind for the field of algebraic numbers and valid for all fields of characteristic 0: if all $x \in M$ are algebraic over K, then for any finitely number of elements $x_1, \ldots, x_p \in M$, there exists x such that the subfield $K[x_1, \ldots, x_p]$ generated by K and the x_i (these are obviously the polynomials in x_i with coefficients in K: apply lemma 2 p times) is equal to $K[x]$. If we admit this result, then with the assumptions of lemma 3, $\dim_K K[x_1, \ldots, x_p] \leq n$, which, for $p = n + 1$, shows that $n + 1$ elements of M can never be linearly independent over K, qed.

(vi) *The purely algebraic point of view.*[14] I will not go any further in this theory. Let us, however, say a few words about another method for associating a Riemann surface to any *algebraic function field of one variable* over $\mathbb{C}$. This is the name given to any field L containing the field $K = \mathbb{C}(X)$ of rational fractions in one variable and finite-dimensional over K. The primitive element theorem shows that L is necessarily isomorphic to the field of rational

---

[14] Serge Lang, *Introduction to Algebraic and Abelian Functions (2nd ed., Springer, 1982).*

fractions in z and ζ, where ζ is an algebraic function of z; hence, if the base field is $\mathbb{C}$, this is not a generalization. This point of view is different because we do not a priori choose some $\zeta \in L$ such that L is the set of polynomials in ζ with coefficients in $\mathbb{C}(z)$. That being the case, how can a Riemann surface $\hat{X}$ associated to L be constructed?

The basic idea is very simply and with some technical changes applies to far more general fields than $\mathbb{C}$. The $x \in L$ must be meromorphic functions on $\hat{X}$. If that is the case, a value $x(P) \in \hat{\mathbb{C}}$ can be assigned to x at each point $P \in \hat{X}$; it must satisfy the following conditions:

($P1$)   the set of x such that $x(P) \neq \infty$ is a subring $o(P)$ of L, and the map
       $P \mapsto x(P)$ is a homomorphism from $o(P)$ onto $\mathbb{C}$;
($P2$)   the relation $x(P) = \infty$ implies that $x^{-1}(P) = 0$;
($P3$) $x(P) = x$ for all $x \in \mathbb{C}$.

Any map $P : x \mapsto x(P)$ from L to $\hat{\mathbb{C}}$ satisfying the previous conditions is, by definition, a *place* of the field L. Having said that, the Riemann surface sought is the set of these places, a set on which a compact Riemann structure is then defined.

Some authors define the places de L by using the associated rings $o(P)$, which can be characterized directly: they must contain $\mathbb{C}$ and satisfy

$$x \notin o \implies x^{-1} \in o.$$

A subring of a field K with this property is a *valuation ring* of K. An example in the field $\mathbb{Q}$ of rational numbers is the set of fractions whose denominator does not contain a given prime p. An example in $\mathbb{C}(X)$ is the set of holomorphic rational fractions in some given $a \in \hat{\mathbb{C}}$.

Exercise 3. We thus obtain all the valuation rings of $\mathbb{Q}$, and of $\mathbb{C}(X)$ containing $\mathbb{C}$, which corresponds to the fact that the Riemann surface of $\mathbb{C}(X)$ is $\hat{\mathbb{C}}$.

Other authors prefer to define the places of a field L of algebraic function by using *valuations*. For them, a point of the Riemann surface is a map

$$v : L \longrightarrow \mathbb{Z}$$

satisfying the following conditions:

$$v(xy) = v(x) + v(y), \quad v(x + y) \geq \min(v(x), v(y))$$

and $v(x) = 0$ if $x \in \mathbb{C}$. This definition corresponds to the fact that at every point P of the Riemann surface of L, its order $v_P(x)$ at P can be associated to each point $x \in L$ since in the neighbourhood of P, the function x is a Laurent series in a local uniformizer. The reader will have no difficulty in determining all the valuations of $\mathbb{C}(X)$, or of $\mathbb{Q}$.

Ultimately, the points of a Riemann surface of a field L of algebraic functions are, interchangeably, the places, the valuations rings or the valuations of L. We go from a valuation ring o to a place by observing that the $x \in o$ that are not invertible in o form an ideal $\mathfrak{p}$ of o and that the quotient $o/\mathfrak{p}$ is isomorphic to $\mathbb{C}$; for $x \in o, x(P)$ is then the class of $x \bmod \mathfrak{p}$, and set $x(P) = \infty$ if $x \notin o$. On the other hand, it can be shown that the only ideals of o are the pairwise distinct powers $\mathfrak{p}^n$ of $\mathfrak{p}$; for $x \in o, v_P(x)$ is then the smallest n such that $x \in \mathfrak{p}^n$, and for $x \notin o$ set $v_P(x) = -v_P(x^{-1})$. In fact, $\mathfrak{p}$ is the set of the $x \in L$ that vanish at the point P of the Riemann surface, and $\mathfrak{p}^n$ the set of x having a zero of order $\geq n$ at P.

The relations with the theory of algebraic curves also need to be mentioned.

The algebraic point of view was invented by Richard Dedekind and Heinrich Weber[15] about thirty years after Riemann. His rather obscure and vague constructions, a fortiori the "verbosity" of his far less brilliant successors, must have annoyed the crystal clear minds of these two algebraists. As Dedekind had already provided a clear and in many respects final form of the theory of algebraic number fields,[16] he naturally tried to apply similar methods to algebraic functions fields of one variable by replacing $\mathbb{Q}$ by $\mathbb{C}(X)$. About thirty years later, Hermann Weyl, *Die Idee der Riemannschen Fläche*, introduced the first correct ideas about "abstract" 1-dimensional complex manifolds into this question and used Dirichlet's principle to prove a priori the existence of "many" meromorphic functions on Riemann surfaces. This result, which is obvious for surfaces associated to algebraic functions, requires even now a long proof in the general case. In the 1930s and 40s, in particular thanks to André Weil and Oscar Zarisky, what a n-dimensional algebraic variety over an arbitrary field is was beginning to be understood and a purely algebraic mechanism was being set up. It was an improvement on the doubtful geometric arguments of the Italian school (which nonetheless had discovered results and introduced very important ideas since 1870). Starting with the notion of a place and ending with that of a complex analytic manifold, Claude Chevalley published the first post-war modern presentation on algebraic functions in one variable. Reading it is still advisable. Serge Lang's book goes much further in about a hundred pages, which indicates how concise the proofs are. . .

---

[15] author of a *Lehrbuch der Algebra* where almost everything that was known in algebra around 1900 can be found.

[16] It has been substantially improved on and generalized, but without fundamentally changing its point of view other than by introducing the notion of valuation. Reading his main articles that can be found in his complete works, is still a most advisable exercise.

Towards the end of the 1950s, Alexandre Grothendieck, influenced by a seminar given by Chevalley on algebraic varieties and by Jean-Pierre Serre's introduction of sheaf theory into the question, entered the picture and made the theory so general and abstract that it can no longer be understood anymore, unless one reads his 30 NP,[17] written by Dieudonné whom nothing could ever stop, and those of his disciples; better enter a convent. Nevertheless, because of Grothendieck's theory of schemes which consists of algebraic geometry over a ring rather than a field, it has been possible to solve previously inaccessible classical problems in number modular functions or number theory.

---

[17] NP: abbreviation for "new page", a concept invented by the Bourbaki group when France, multiplied its currency by hundred and introduced "new francs".

Index

Table of Contents of Volume I

Table of Contents of Volume II